CONTENTS

CONSTITUTIVE LAWS FOR ENGINEERING MATERIALS
With Emphasis on Geologic Materials

CHANDRAKANT S. DESAI

Department of Civil Engineering and Engineering Mechanics
University of Arizona, Tucson, Arizona

HEMA J. SIRIWARDANE

Department of Civil Engineering
West Virginia University, Morgantown, West Virginia

Prentice-Hall, Inc., Englewood Cliffs, NJ 07632

Library of Congress Cataloging in Publication Data

Desai, C. S. (Chandrakant S.),
  Constitutive laws for engineering materials, with
emphasis on geologic materials.

  Includes bibliographical references and index.
  1. Materials—Mechanical properties—Mathematical
models. 2. Strains and stresses—Mathematical models.
I. Siriwardane, Hema J. II. Title.
TA417.6.D47   1984   620.1′123′0724   83-11016
ISBN 0-13-167940-6

Editorial / production supervision and
  interior design: Ellen Denning
Cover design: Edsal Enterprises
Manufacturing buyer: Anthony Caruso

© 1984 by Prentice-Hall, Inc., Englewood Cliffs, New Jersey 07632

Printed in the United States of America

10 9 8 7 6 5 4 3 2 1

ISBN 0-13-167940-6

PRENTICE-HALL INTERNATIONAL, INC., *London*
PRENTICE-HALL OF AUSTRALIA PTY. LIMITED, *Sydney*
EDITORA PRENTICE-HALL DO BRASIL, LTDA., *Rio de Janeiro*
PRENTICE-HALL CANADA INC., *Toronto*
PRENTICE-HALL OF INDIA PRIVATE LIMITED, *New Delhi*
PRENTICE-HALL OF JAPAN, INC., *Tokyo*
PRENTICE-HALL OF SOUTHEAST ASIA PTE. LTD., *Singapore*
WHITEHALL BOOKS LIMITED, *Wellington, New Zealand*

PREFACE

Constitutive or stress–strain laws or models of engineering materials play a significant role in providing reliable results from any solution procedure. Their importance has been enhanced significantly with the great increase in development and application of many modern computer-based techniques such as the finite element, finite difference, and boundary integral equation methods. It has been realized that the advances and sophistication in the solution techniques have far exceeded our knowledge of the behavior of materials defined by constitutive laws. As a consequence, very often, results from a numerical procedure that may have used less appropriate constitutive laws can be of limited or doubtful validity.

The foregoing realization has spurred active research and interest in the theoretical formulation of constitutive laws and determination of their parameters. The former involves use of the principles of mathematics and continuum mechanics, whereas the latter hinges on accurate identification and determination of parameters that define the constitutive models. The objective of this text is to present a simplified treatment of the theory and mathematics of various constitutive laws, together with the important aspects of the determination of constitutive parameters, their verification, and implementation. The latter aspects have not been covered adequately in previous treatments.

The simplest constitutive laws used in engineering are linear such as the Hooke's law. These laws are valid only for a very limited class of materials because most engineering systems are nonlinear and complex. The influence of nonlinear response becomes more prominent in the case of materials that are influenced by factors such as state of stress, residual or initial stress, volume changes under shear, stress history or stress paths, inherent and induced anisotropy, change in the physical state, and fluid in the pores.

Although the main aim of the text is to cover various recent models for complex (geologic) materials as influenced by the factors noted above, reviews and descriptions of conventional models are also included for the sake of completeness and to trace historical developments. Hence the text can be useful for the study of models for a wider range of engineering materials.

Tensor notation is used in the text; however, at many locations equivalent statements in matrix notation are presented. Since most readers of the text will be familiar with matrix notation, this approach should be useful.

A special feature of the book is the use of a number of laboratory test data for various (geologic) materials. Here the importance of laboratory test data is emphasized, and brief descriptions of various laboratory devices, including recently developed truly triaxial or three-dimensional devices, are included. Comprehensive series of test results for various materials are used to illustrate the determination of material parameters for constitutive models described in various chapters. Most of the laboratory test data are derived from recent results obtained by the authors and their associates using advanced test devices.

Another important feature is the discussion as to how to implement the models in solution (numerical) techniques. Here the results are often cast in incremental matrix notation for three-dimensional and various two-dimensional (plane strain, axisymmetric) idealizations. This will show the reader the use of the results in specific solution procedures. A number of applications are presented. Here numerical predictions with various constitutive models are compared with closed-form solutions and/or laboratory and field observations.

Chapter 1 presents an introduction to the text, including a simple representation of constitutive laws. Brief statements of the axioms of continuum mechanics are given in Chapter 2 simply to make the reader aware that the constitutive laws are derived such that the axioms and natural principles are satisfied.

Chapters 3 and 4 present reviews of strain and stress and definitions of their invariants as used in this text. Reasons for the study of advanced constitutive laws, their role in implementation in (numerical) solution schemes, the importance of laboratory testing with brief descriptions of test devices, and details of laboratory test results for an artificial soil obtained by using the truly triaxial device are included in Chapter 5. These test data are used in subsequent chapters to illustrate the determination of parameters for various models.

Higher-order elasticity (hyperelasticity) and hypoelasticity are the topics in Chapters 6 and 7, where detailed descriptions are given for models up to the second order. Determination of parameters for various models based on laboratory test results for the artificial soil from Chapter 5 as well as for other materials are presented by using a least-squares-fit program that is included in Appendix 3. Chapter 8 covers a number of quasilinear or piecewise linear models.

Chapters 9 to 11 cover details of the plasticity models. Introduction to the theory of plasticity together with classical models such as those of Tresca and von Mises are given in Chapter 9, while Chapter 10 covers other models such as the Mohr–Coulomb and Drucker–Prager. Critical state and cap models are discussed in Chapter 11.

A number of example problems are included in each of Chapters 3 to 11. These include the determination of parameters for various models for various materials, predictions of numerical methods using the models, and exercise problems for students together with partial solutions.

Chapter 12 gives a comprehensive review of recent developments, including kinematic hardening, kinematic and isotropic hardening, general anisotropic hardening, nonassociative models in plasticity and rate-type models involving a combination of hypoelasticity and critical state. Brief reviews of some recent developments, including viscoplastic and endochronic models and models for cyclic loading and for rocks, are presented at the end of this chapter.

Appendixes 1 and 2 give reviews of tensor notation, invariants of tensors, and the Caley–Hamilton theorem. Computer codes for least-squares fit and for computerized evaluation for parameters of the cap model are included in Appendixes 3 and 4, respectively.

It is expected that the reader has at least a bachelor's or master's degree in engineering with a background in the strength of materials and an introduction to the theories of elasticity and plasticity; current interest and activity in the area of constitutive laws can be helpful. A background in matrix algebra is required. A background in tensor analysis will be quite useful, although the elements of the notation used in this text can be understood by studying Appendix 1. Many present-day practitioners are involved in developing and using constitutive laws; hence the book can be useful to them. Significant research activity is under way at many academic institutions toward development and applications of constitutive laws. The book can be useful to such researchers.

With the growing need for advanced constitutive laws in research, teaching, and applications, new courses are expected to be developed on the subject. It may be noted that courses on the theoretical aspects of constitutive laws are already taught in many engineering (mechanics) curricula. This book would fill the need in bringing the theoretical concepts to implementation, thus providing a bridge between theory and application. The senior author has successfully taught such a course for the last several years.

Finally, this book endeavors to present a comprehensive treatment of various constitutive laws by considering the following five steps:

1. Mathematical formulation
2. Identification of material parameters
3. Determination of material parameters
4. Verification with respect to laboratory test data under various stress paths and physical conditions
5. Verification and evaluation with respect to boundary value problems

We wish to express deep appreciation to Dr. R. O. Davis, University of Canterbury, New Zealand, Dr. Y. F. Dafalias, University of California at Davis, and Dr. D. DaDeppo and Mr. M. O. Faruque, University of Arizona, for reading parts of the manuscript and for providing useful comments and suggestions. A number of graduate students have contributed directly or indirectly toward solutions of some of the examples and toward the laboratory test data used; it is difficult to list them all here. We express our special thanks to them. We appreciate the patience and encouragement of our wives, Patricia and Ranjani.

Complete appreciation of constitutive laws may not be possible without many concepts from physics and mathematics. Physics provides the fundamental basis and mathematics a concise way to express the physical phenomena. Most basic concepts governing physical phenomena can be essentially simple, and they can be understood through simple explanations. Our endeavor has been to work toward this goal.

C. S. DESAI
H. J. SIRIWARDANE

1

INTRODUCTION

Things that we can perceive, see, hear, or build can be explained by using certain principles and laws of nature: conservation of mass, energy, linear and angular momenta, the laws of electromagnetic flux, and the concept of thermodynamic irreversibility. These are among the fundamental principles on which the subject of *mechanics* is based. One of the first notable successes of mechanics was its application in the study of planetary motion of the solar system, which was indeed a mysterious problem in ancient times!

The subject of continuum mechanics is based on the foregoing governing principles, which are independent of the internal constitution of material. However, the response of a system or a medium subjected to (external) forces cannot be determined uniquely only with the governing field equations derived from the basic principles. The internal constitution of material plays an important role in the subject of continuum mechanics.

A number of concepts relating to constitutive laws are discussed in this chapter in a simplified manner. For additional reading on these concepts, the reader may consult many formative works (1–6). A schematic diagram showing the importance and place of constitutive laws in continuum mechanics is given in Fig. 1-1.

DEFINITION

A *constitutive law* or *model* represents a mathematical model that describes our ideas of the behavior of a material. In other words, a constitutive law

1

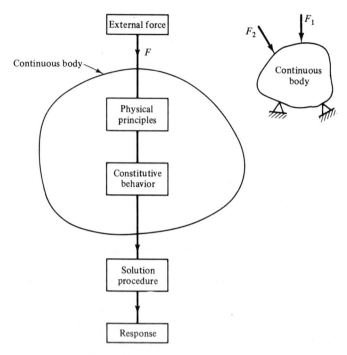

Figure 1-1 Place of constitutive laws in continuum mechanics.

simulates physical behavior that has been perceived mentally. The main advantage of establishing a mathematical model is to apply the ideas for solving (complex) events quantitatively. Therefore, the power of a constitutive model depends on the extent to which the physical phenomenon has been understood and simulated.

Study of the response of a substance or body under external excitation constitutes the major endeavor in engineering and sciences. The important ingredients involved in such a study are (a) external excitation, (b) the internal constitution of the medium, and (c) the response.

The external excitation can be any means by which the environment of the substance is changed. As an example, consider an application of a load on a certain medium. This load, which is an external excitation, tries to deform the object. Application of a differential fluid pressure (head) causing flow in a fluid (water) or in a porous medium is another example of an external excitation. Will two objects made of different materials but with identical geometry respond in a similar manner to the same external excitation? The answer to this question requires a knowledge of the internal constitution of the materials.

Generally, it has been observed that materials with the same geometry but with different internal constitution respond in different ways to the same external excitation. Therefore, in the study of response behavior in nature, the internal constitution of materials is of utmost importance. The primary goal of this book is to introduce the basic concepts of constitutive laws that describe the response of the body and to present a detailed description of widely used constitutive laws in engineering practice and their applications. In engineering applications, the response behavior can be studied at a macroscopic level without considering atomic and molecular structure. The subject of studying material behavior at the macroscopic level can be called *continuum mechanics* (1, 2).

A solution to a boundary value problem in continuum mechanics requires constitutive equations in addition to the governing field equations. The basic principles governing Newtonian mechanics are (a) conservation of mass, (b) conservation of momentum, (c) conservation of moment of momentum (or angular momentum), (d) conservation of energy, and (e) laws of thermodynamics; these principles are considered to be valid for all materials irrespective of their internal constitution (1–4). Therefore, a unique solution to a boundary value problem in continuum mechanics cannot be obtained only with the applications of governing field equations. Hence a unique determination of the response requires additional considerations that account for the nature of different materials. The equations that model the behavior of a material are called "constitutive equations" or "constitutive laws" or "constitutive models."

Consider the behavior of two bodies with the same geometry. As an example, consider two bricks, one made of rubber and the other of steel but with the same geometry. When an external excitation is applied to these two bodies, the rubber brick will deform more than the steel. The *load–deformation* behavior or in general the *cause-and-effect* relation is dependent on how the matter is constituted. The internal constitution of matter affects the deformation behavior of the body.

As another example, consider a pool of water and a pool of ice. The chemical and atomic structures of these two bodies are the same. As we can see, water and ice will respond or behave in two different ways under the same external loads. Although their chemical and atomic structures are the same, the internal constitution of matter for water and ice is different. This constitution of matter governs the behavior of the body. The relationship between cause and effect can be called the *constitutive law* of the material for a given phenomenon.

THEORY OF CONSTITUTIVE EQUATIONS

A *constitutive equation* is a mathematical model that can permit reproduction of the observed response of a continuous medium. It is not easy to devise such a general constitutive law for all materials; even for a single class of materials, this can be a difficult task! Because, for instance, behavior of the same material under different excitations may or may not be related. As an example, consider the response of a piece of steel that is subjected to a voltage gradient. The flow of current in this medium can be assumed to be governed by a field law such as Ohm's law. There may exist a certain connection between the load–deformation properties and electrical properties; however, this may not always be the case, that is, the electrical properties are not always related to the deformation characteristics. Furthermore, these relationships may be different for different ranges of loads and voltages applied. Thus it becomes difficult to establish a general constitutive equation for a material to cover all possible ranges and modes of excitation and behavior. Hence it will be necessary to confine the considerations to ranges of specific interests.

Establishment of constitutive equations can be based on the experimental observations at a macroscopic level or from physical theories of molecular behavior. The second approach, which considers the properties at the microscopic level, can be quite complex. On the other hand, the establishment of constitutive equations based on observations made at the macroscopic level can impart physical significance in engineering and physical sciences. The identification of the relevant constitutive variables for a certain material is a difficult task. The selection of pertinent properties often may have to be done aided by experience and intuition.

Once the pertinent constitutive variables or parameters are identified, it is necessary to know the relationships among those variables. When there are many independent variables, the task of finding the functional relationships can become complicated. However, with modern theories of constitutive equations, certain restrictions can be imposed on the general functional relations (1,2). These restrictions will certainly reduce the rigor involved in the general functional forms. On the other hand, this reduces the effort required in experimental explorations of material characteristics, as well as explaining certain limitations on their applicability. Further discussions of such restrictions are given in subsequent chapters.

SIMPLE EXPLANATION

The important question is the determination of a formula, law, or device that defines the behavior of a material. For the rather simple problem involving the extension of a (linear) spring which is loaded axially [Fig. 1-2(a)], the quantity f is given by

$$fQ = u \tag{1-1}$$

Here, f describes (indirectly) the constitution of the spring. Although f does not include an explicit description of the physical constitution of the spring, it does it implicitly. Because if we know f, the response u can be found for a given excitation or load Q.

How do we find f? The usual procedure is to conduct a test in which a known value of Q is applied to the spring and the response u is measured. Then for linear material the ratio u/Q gives the value of f. For a simple spring, this is rather an easy task.

For multidimensional problems, however, it may no longer be possible to write such simple expressions connecting the load and the displacement. As indicated in Fig. 1-2(b), loads Q_i can cause responses (deformations) in all three directions. Moreover, in contrast to the single-element material in the spring [Fig. 1-2(a)], a body or substance [Fig. 1-2(b)] is composed of a continuum made up of interconnected particles. Hence, in addition to the effect of the load in one direction, there exist coupled effects that consist of responses in the three major directions. Here it becomes difficult to write explicit relations such as that in Eq. (1-1) in the individual direction.

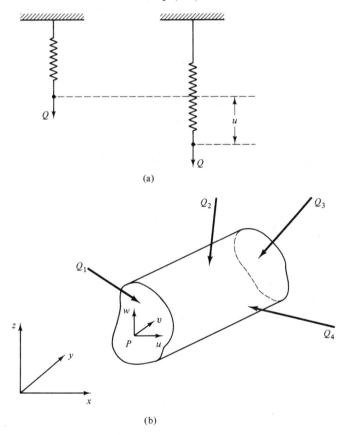

(a)

(b)

Figure 1-2 Load–deformation behavior: (a) extension of a spring under axial loading; (b) deformation of a continuous body.

For instance, consider the following equation for x-displacement of point P:

$$f_{x1}Q_1 = u_1 \tag{1-2}$$

where f_{x1} describes the constitutive behavior in the x-direction and u_1 is the displacement in the x-direction due to force Q_1. Is this expression sufficient? No, because Q_2 can also contribute to the displacement of point P in the x-direction, and similarly Q_3, and so on. Hence the total x-displacement u of point P can be written as

$$f_{x1}Q_1 + f_{x2}Q_2 + \cdots + f_{xn}Q_n = u_1 + u_2 + \cdots + u_n = u \tag{1-3}$$

In order to find the response u under the effects of $Q_1, Q_2, \ldots, Q_n$, it becomes necessary to know the constitutive parameters $f_{x1}, f_{x2}, \ldots, f_{xn}$, and similarly for displacements v and w in the y and z directions, respectively. Hence in matrix notation we can write

$$\begin{bmatrix} f_{x1} & f_{x2} & \cdots & f_{xn} \\ f_{y1} & f_{y2} & \cdots & f_{yn} \\ f_{z1} & f_{z2} & \cdots & f_{zn} \end{bmatrix} \begin{Bmatrix} Q_1 \\ Q_2 \\ Q_3 \\ \vdots \\ Q_n \end{Bmatrix} = \begin{Bmatrix} u \\ v \\ w \end{Bmatrix} \tag{1-4a}$$

or

$$[\mathbf{F}]\{\mathbf{Q}\} = \{\mathbf{u}\} \tag{1-4b}$$

A moment's reflection will show that determination of the constitutive parameters in $[F]$ can be a formidable task. Unlike the single spring element, the continuum is composed of an "infinite" number of springs interconnected in an arbitrary manner.

In continuum mechanics, the excitation (load) and the response (displacement) are usually not expressed explicitly as in Eqs. (1-4). They are expressed through quantities that are relevant to the internal reactions within the body. For instance, instead of force Q, a quantity called *stress*, σ, is defined as

$$\sigma = \lim_{A \to 0} \frac{Q}{A} \tag{1-5}$$

where A is the area on which Q is acting, and instead of displacement u, a quantity called *strain*, ϵ, is defined, in an average sense, as

$$\epsilon = \frac{u}{l} \tag{1-6}$$

where l is the length of the body (Fig. 1-3). Then Eq. (1-1) can be written as

$$f(A\sigma) = \epsilon l$$

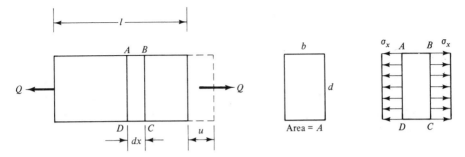

Figure 1-3 Axial extension of an elastic body.

or

$$\frac{fA}{l} = \frac{\epsilon}{\sigma} \tag{1-7}$$

or

$$D = \frac{\epsilon}{\sigma}$$

where $D(L^2/F)$ denotes flexibility or compliance of the body. A specimen of a body of known area A and length l can be tested with a given load Q, and ϵ and σ can be found; this can permit evaluation of D. At subsequent stages we shall generalize this approach for multidimensional bodies.

NEED FOR EXPERIMENTAL DATA

The only rational way to determine the parameters to define the constitutive laws is to conduct appropriate laboratory and/or field tests. For instance, to determine the parameter relevant to response in the x-direction, the test should simulate that response. For the response parameter relevant to the effect of load applied in the x-direction for response in the y-direction, the test must simulate this particular mode of deformation, and so on. The subject of identification of parameters and appropriate laboratory tests to determine them are the topics of interest in this book. They will be discussed subsequently.

Further Explanation

In order to present the concept of constitutive equations, let us again consider the following well-known example of uniaxial strain. Consider the small element $ABCD$ shown in Fig. 1-3. This body is in equilibrium under the applied load Q. Therefore, each element such as $ABCD$ of the body has to be in equilibrium. The stress system in the element is also shown in Fig. 1-3.

The equilibrium of the element can be expressed in the following way:

$$Q = \sigma_x A \tag{1-8}$$

where σ_x is the (uniaxial) stress and A is the cross-sectional area.

With the concept of strain presented in the preceding section, let us now define the state of strain in this one-dimensional element. The most simple definition of strain is given as the ratio of elongation u and the initial length of the rod:

$$\epsilon_x = \frac{u}{l} \tag{1-9}$$

In this problem, the known parameters are the geometry l and the externally applied load Q. The unknown parameters are the stress σ_x and the strain ϵ_x caused by the response, that is, the axial deformation u. However, we have only two equations, Eqs. (1-8) and (1-9), to solve the three unknown quantities, σ_x, ϵ_x, and u.

In Eqs. (1-8) and (1-9), no mention was made of the material properties. To solve for the three unknowns, an additional equation is required, and this third equation will be the constitutive equation of the material constituting the body.

As stated before, in this case, this relationship can be the well-known Hooke's law. Hooke's law gives a relationship between the stress and strain as

$$\sigma_x = E_x \epsilon_x \tag{1-10}$$

where E_x is called the *Young's modulus* or *modulus of elasticity*. This third equation includes a parameter E_x, which denotes the property of the material.

The example above shows clearly that for the solution of unknown quantities (deformation), the constitutive law of the material is an important component.

COMMENTS

A complete appreciation of constitutive behavior of materials is not possible without the use of the principles of physics and mathematics. Mathematics provides a concise way for expressing physical phenomenon. The physical phenomenon that we propose to consider in this text is the behavior of engineering materials under external loads.

We believe that the subject can be studied and understood at different levels. For instance, it can be studied strictly in terms of abstract mathematical considerations, on the one hand, and strictly intuitive and empirical considerations, on the other. Although a considerable amount of work toward development of constitutive relations has been done in continuum and solid mechanics, it has often remained essentially within the purview of one sophisticated in mathematics. In view of the growing importance of constitutive models, particularly with the availability of powerful computational methods, it becomes necessary to bring the subject to a level that the engineer can comprehend and apply. The purpose of this book is to provide a link between theoretical treatment and application, with emphasis on intuitive expositions. Also, particular attention is given to the importance of testing and determination of the constitutive parameters.

The subject of constitutive laws has developed under various subdisciplines of mechanics, such as theory of elasticity, hypoelasticity, and theory of plasticity. As a result, different constitutive laws based on different concepts have been proposed; often, proponents of a particular model proclaim its superiority over others. We believe that each model can be valid within its own local realm, and that no universal constitutive model has yet been developed that is valid for all materials under all conditions. We further believe that since the phenomenon described by a number of different models is unique, there may indeed be a common ground where many of those models may meet.

Finally, the study of constitutive laws of engineering materials belongs to the general subject of *mechanics*. Mechanics involves the study and understanding of motion. Hence our aim is to learn constitutive laws as they relate to engineering bodies in motion. In other words, we will study the behavior of materials as a body deforms under the influence of external and internal forces.

> Most fundamental ideas of science are essentially simple and may, as a rule, be expressed in a language comprehensible to everyone.
>
> Einstein and Infeld (7)*

NOTATION

In studying and understanding the theory of constitutive laws, it is necessary and useful to employ indicial and tensor notation. Since most readers may be familiar with matrix notation, an effort is made to use, wherever possible, both tensor and matrix notation. This allows easy understanding and physical interpretation of various quantities. The following notation is used commonly; other notation is defined wherever it occurs.

Tensor Notation	Corresponding Matrix Notation
Zero order, A	Scalar: A
First order, A_i	Vector
	Column: $\{A\}$
	Row: $[A]$ or $\{A\}^T$
Second order, A_{ij} or (A)	Matrix $[A]$

*Reprinted by permission from the Hebrew University of Jerusalem.

REFERENCES

1. Eringen, A. C., *Nonlinear Theory of Continuous Media*, McGraw-Hill Book Company, New York, 1962.

2. Fung, Y. C., *Foundations of Solid Mechanics*, Prentice-Hall, Inc., Englewood Cliffs, N.J., 1965.

3. Truesdell, C., *Continuum Mechanics—The Mechanical Foundations of Elasticity and Fluid Dynamics*, Vol. 1, Gordon and Breach, Science Publishers, Inc., New York, 1966.

4. Jaunzemis, W., *Continuum Mechanics*, Macmillan Publishing Co., Inc., New York, 1967.

5. Malvern, L. E., *Introduction to the Mechanics of a Continuous Medium*, Prentice-Hall, Inc., Englewood Cliffs, N.J., 1969.

6. Eringen, A. C., *Continuum Physics*, Vol. 2, Academic Press, Inc., New York, 1975.

7. Einstein, A., and Infeld, L., *The Evolution of Physics*, Simon and Schuster, New York, 1968.

2

STATEMENTS
OF AXIOMS
OF CONTINUUM MECHANICS

The main aim of this text is to present, in a simplified manner, the subject of constitutive laws for (engineering) materials with special attention to the determination of parameters and implementation. Since constitutive equations or laws relate to physical phenomena such as the response of structures to excitations or loads, they must obey certain principles or axioms that govern the physical phenomena. Details of mathematical and rigorous treatments and the relation of constitutive equations to these axioms are available in advanced treatments in continuum mechanics and mathematical physics (1–8). In this chapter we have simply stated some of these axioms in simple language essentially from an intuitive viewpoint; these statements are guided by the material available in the foregoing treatments.

An *axiom* is a well-established scientific principle that does not need proof because its implication has been accepted as truth for a long time. There are a number of axioms in continuum mechanics that are accepted universally and are based on principles derived from experience and rational facts.

Axiom of Determinism

Our current actions are influenced by our past deeds. The magnitude of stress and deformation in a body due to an external force are dependent on the past history of the external forces experienced by the body. These statements represent the principle of determinism, which in general states that every phenomenon is determined totally by the history or sequence of causes, and that its attributes are not free but destined by events that it experienced previously.

Determinism has always been, in one form or another, an integral part of the natural sciences. It is ordinarily understood to explain the fact that the *past determines the future*. In other words, future comes based on what happened in the past, and without the past, future is present.

In continuum physics, what we assume as determinism is much simpler, requiring that the past should determine the future, and that the material behavior of the body is independent of any point outside the body. In other words, all future outcome is determined by that state (or history) of the body up to the present time. That is, in continuum physics the history up to the present time is required to determine the future outcome. This phenomenon is sometimes referred to as the *principle of heredity*.

Axiom of Causality

Simply stated, *causality* means that effect and causes go together. In other words, things and phenomena cannot take place without a cause. There cannot occur deformation (effects) without an external force (cause).

Mathematical models to describe the internal constitution of matter are developed after selecting suitable constitutive variables. The axiom of causality provides a selection or identification rule to distinguish dependent constitutive variables from independent ones.

Axiom of Objectivity

To be objective means to be without bias or prejudice. The quality of being objective is objectivity. This axiom as stated in Refs. 4 and 8 is given below:

> The constitutive response functionals must be form-invariant under arbitrary rigid motions of spatial frame of reference and a constant shift of the origin of time.*

Physically, the axiom of objectivity means that the material properties cannot vary with the motion of the observer. For instance, the modulus of elasticity of a steel specimen found in a laboratory will not change even when the test is carried out in a moving aircraft. Furthermore, the axiom of objectivity mentions that the modulus of elasticity determined at two different times under the same conditions will not change if the material remains unchanged with time.

*From A. C. Eringen, ed., *Continuum Physics*, Vol. 2, Academic Press, Inc., New York, 1975.

Axiom of Neighborhood

This axiom, which was presented by Eringen (4, 8), is also known as the "axiom of local action." This axiom states that the values of response functions at a point are not affected by the conditions far away from that point. In other words, the axiom of neighborhood is used to exclude from constitutive equations the "actions at a distance."

Axiom of Memory

A detailed description of this axiom can be found elsewhere (4, 7, 8). It asserts that the values of present constitutive variables are not affected by the values of constitutive variables at distant past. This is the counterpart of the axiom of neighborhood in the time domain.

Axiom of Equipresence

This axiom, as stated in Refs. 4 and 8, is:

At the outset all constitutive response functionals are to be considered to depend on the same list of constitutive variables until the contrary is deduced.*

This principle says that once a constitutive variable has been identified for a material, it should be assumed to be present in all the constitutive equations of that material.

The principle of equipresence says that all the constitutive variables should be included in every constitutive equation, unless the presence of a certain variable violates a basic principle of mechanics or thermodynamics, or another axiom. However, when there are too many variables in the constitutive equation, the problem of retaining all of them becomes complicated. For this reason, the principle of equipresence is often not popular among engineers and scientists, who may not prefer to complicate a problem by having too many constitutive parameters or variables, particularly if those are not significant. However, for generality it is useful to adopt this principle and then simplify by making certain assumptions. This will make the engineer-analyst who develops the constitutive equations aware of the assumptions that were made and the associated limitations that entered the formulation.

Axiom of Admissibility

The constitutive laws will be different for different materials. The physical laws of nature, such as conservation of mass, linear and angular momen-

---

*From A. C. Eringen, ed., *Continuum Physics*, Vol. 2, Academic Press, Inc., New York, 1975.

tum, and laws of thermodynamics, should be satisfied by any system irrespective of the material type. These physical laws lead to the governing equations, such as the continuity equation, equation of motion, symmetry of stress tensor, energy balance, and the entropy inequality. Details of these laws are given in various references (1–8). The axiom of admissibility asserts that constitutive equations must be consistent with the physical laws.

The statements of various axioms that we have given are intended primarily to make the reader aware of certain principles as underlying foundations of constitutive modeling. Although we shall often study and use various laws in this book without direct reference or use of these axioms and principles, it is understood that they form the basis of such laws.

REFERENCES

1. Noll, W., "A Mathematical Theory of the Mechanical Behavior of Continuous Media," *Arch. Ration. Mech. Anal.*, Vol. 2, 1958, pp. 197–226.
2. Truesdell, C., and Noll, W., "The Nonlinear Field Theories of Mechanics," in *Encyclopedia of Physics*, S. Flugge (Ed.), Vol. 3/3, Springer-Verlag, Berlin, 1965.
3. Truesdell, C., *Continuum Mechanics—The Mechanical Foundations of Elasticity and Fluid Dynamics*, Vol. 1, Gordon and Breach, Science Publishers, Inc., New York, 1966.
4. Eringen, A. C., *Mechanics of Continua*, John Wiley & Sons, Inc., New York, 1967.
5. Jaunzemis, W., *Continuum Mechanics*, Macmillan Publishing Co., Inc., New York, 1967.
6. Malvern, L. E., *Introduction to the Mechanics of a Continuous Medium*, Prentice-Hall, Inc., Englewood Cliffs, N.J., 1969.
7. Coleman, B. D., and Mizel, V., "On the General Theory of Fading Memory," *Arch. Ration. Mech. Anal.*, Vol. 29, 1968, pp. 18–31.
8. Eringen, A. C., *Continuum Physics*, Vol. 2, Academic Press, Inc., New York, 1975.

3

STRAIN, COMPONENTS OF STRAIN, AND STRAIN TENSOR

When a body is subjected to an external force system, it experiences deformations in relation to the original configuration [Fig. 3-1(a)]. It is convenient to define strain as a relation (ratio) between the magnitude of deformations and the extents of the original or undeformed configurations.

There are a number of ways in which measures of strains are defined (1–3); often these are named after the person who defined them. The Cauchy, Green, Hencky, and Almansi are some of the measures of strains used commonly.

ONE-DIMENSIONAL IDEALIZATION

As a simple illustration, consider a circular bar loaded axially and approximated by a one-dimensional line element [Fig. 3-1(b)]. The original configuration (length) of the bar is l_0 and under an axial deformation, it extends or contracts to a new length equal to l. We have expressed below three of the measures of strain as uniform quantities over l_0.

Cauchy: This measure is based on the original configuration, that is,

$$\epsilon_x^c = \frac{l - l_0}{l_0} = \frac{l}{l_0} - 1 = \lambda - 1 \tag{3-1a}$$

where λ is equal to the ratio l/l_0.

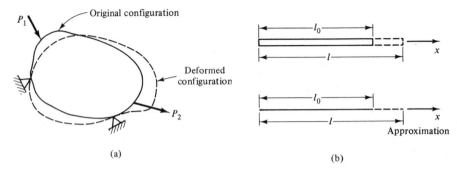

Figure 3-1　Schematic of deformation: (a) general body; (b) one-dimensional bar.

Green:　The Green strain is, again, based on the original configuration,

$$\epsilon_x^g = \frac{l^2 - l_0^2}{2l_0^2} \tag{3-1b}$$

Almansi: The Almansi strain is based on the current or deformed configuration of the bar, that is,

$$\epsilon_x^a = \frac{l^2 - l_0^2}{2l^2} \tag{3-1c}$$

THREE-DIMENSIONAL CASE

The foregoing concepts can be generalized for three-dimensional bodies. Here we need to consider nine components of strain, which under the absence of body couples, reduce to six strain components because $\epsilon_{ij} = \epsilon_{ji}$. Assuming small strains, the symmetric tensor of these components is written as

$$\epsilon_{ij} = \begin{pmatrix} \epsilon_{11} & \epsilon_{12} & \epsilon_{13} \\ \epsilon_{12} & \epsilon_{22} & \epsilon_{23} \\ \epsilon_{13} & \epsilon_{23} & \epsilon_{33} \end{pmatrix} \tag{3-2a}$$

In engineering use, components of strain are often expressed in vector form as

$$\{\epsilon\}^T = \begin{bmatrix} \epsilon_{11} & \epsilon_{22} & \epsilon_{33} & \epsilon_{12} & \epsilon_{23} & \epsilon_{13} \end{bmatrix} \tag{3-2b}$$

or*

$$\{\epsilon\}^T = \begin{bmatrix} \epsilon_{xx} & \epsilon_{yy} & \epsilon_{zz} & \gamma_{xy} & \gamma_{yz} & \gamma_{xz} \end{bmatrix} \tag{3-2c}$$

*Here γ_{xy}, γ_{yz}, and γ_{xz} are called engineering shear strains, and their magnitudes are twice those of their counterparts ϵ_{12}, ϵ_{23}, and ϵ_{13} in the tensor notation, respectively.

$\gamma_{xy} = 2\,\epsilon_{xy}$ etc,

or

$$\{\epsilon\}^{\mathrm{T}} = [\epsilon_x \quad \epsilon_y \quad \epsilon_z \quad \gamma_{xy} \quad \gamma_{yz} \quad \gamma_{xz}] \tag{3-2d}$$

where $\epsilon_{xx}, \epsilon_{yy}, \epsilon_{zz}$ or $\epsilon_x, \epsilon_y, \epsilon_z$ denote components of normal strains, whereas $\epsilon_{12}, \epsilon_{23},$ and ϵ_{13} or $\gamma_{xy}, \gamma_{yz},$ and γ_{xz} denote the components of shear strain.

Strain Tensor

In this text we consider essentially the case of small or infinitesimal strains; that is, external forces cause small deformations which result in strains that are much smaller compared to unity.

In Fig. 3-2(a) a particle or point P moves from position 1 to position 2 under the influence of an (external) force system. The two positions are denoted by P_1 and P_2 [Fig. 3-2(b)] and by position vectors a_i and x_i, respectively. The change in displacement of the particle is denoted by u_i $(i = 1, 2, 3)$. From Fig. 3-1 we can write

$$a_i + u_i = x_i \tag{3-3a}$$

or

$$u_i = x_i - a_i \tag{3-3b}$$

For small strains and rotations, it can be shown that (1, 3)

$$\tfrac{1}{2}(dx_i \, dx_i - da_i \, da_i) = \tfrac{1}{2}[u_{i,j} + u_{j,i}] \, da_i \, da_j \tag{3-4}$$

where $u_{i,j}$ denotes gradients of displacements u_i. The displacement gradient can be written as

$$u_{i,j} = \tfrac{1}{2}(u_{i,j} + u_{j,i}) + \tfrac{1}{2}(u_{i,j} - u_{j,i})$$
$$= \epsilon_{ij} + \omega_{ij} \tag{3-5}$$

The first term, ϵ_{ij}, is symmetrical and is called the *small* or *infinitesimal strain tensor*:

$$\epsilon_{ij} = \tfrac{1}{2}(u_{i,j} + u_{j,i}) \tag{3-6}$$

and the second term, ω_{ij}, which is skew symmetrical, is called the *small rotation tensor*:

$$\omega_{ij} = \tfrac{1}{2}(u_{i,j} - u_{j,i}) \tag{3-7}$$

For $i = j = 1,$ 2, or 3, Eq. (3-6) gives

$$\epsilon_{11} = u_{1,1} = \frac{\partial u_1}{\partial x_1} = \frac{\partial u}{\partial x} = \epsilon_{xx} = \epsilon_x$$

$$\epsilon_{22} = u_{2,2} = \frac{\partial u_2}{\partial x_2} = \frac{\partial v}{\partial y} = \epsilon_{yy} = \epsilon_y \tag{3-8a}$$

$$\epsilon_{33} = u_{3,3} = \frac{\partial u_3}{\partial x_3} = \frac{\partial w}{\partial z} = \epsilon_{zz} = \epsilon_z$$

and for $i \neq j$, we have

$$\epsilon_{12} = \frac{1}{2}(u_{1,2} + u_{2,1}) = \frac{1}{2}\left(\frac{\partial u_1}{\partial x_2} + \frac{\partial u_2}{\partial x_1}\right)$$

$$= \frac{1}{2}\left(\frac{\partial u}{\partial y} + \frac{\partial v}{\partial x}\right) = \frac{1}{2}\gamma_{xy} \qquad (3\text{-}8b)$$

$$\epsilon_{23} = \frac{1}{2}(u_{2,3} + u_{3,2}) = \frac{1}{2}\left(\frac{\partial u_2}{\partial x_3} + \frac{\partial u_3}{\partial x_2}\right)$$

$$= \frac{1}{2}\left(\frac{\partial v}{\partial z} + \frac{\partial w}{\partial y}\right) = \frac{1}{2}\gamma_{yz} \qquad (3\text{-}8c)$$

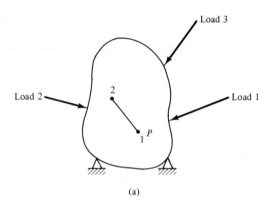

(a)

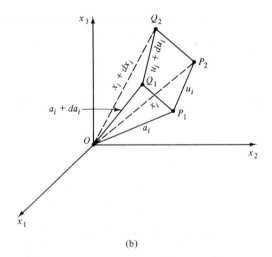

(b)

Figure 3-2 Deformations in a continuous medium: (a) deformation of a continuous body under external loading; (b) deformation of a line element PQ.

and

$$\epsilon_{13} = \frac{1}{2}(u_{1,3} + u_{3,1}) = \frac{1}{2}\left(\frac{\partial u_1}{\partial z} + \frac{\partial u_3}{\partial x}\right)$$

$$= \frac{1}{2}\left(\frac{\partial u}{\partial z} + \frac{\partial w}{\partial x}\right) = \frac{1}{2}\gamma_{xz} \qquad (3\text{-}8\text{d})$$

Here u, v, and w are the commonly used symbols in engineering practice and they correspond to u_1, u_2, and u_3, the components of displacements in the $x_1(x)$, $x_2(y)$, and $x_3(z)$ directions, respectively.

The rotation tensor can be expanded as

$$\omega_{11} = \omega_{22} = \omega_{33} = 0 \qquad (3\text{-}9\text{a})$$

and

$$\omega_{12} = \frac{1}{2}(u_{1,2} - u_{2,1}) = \frac{1}{2}\left(\frac{\partial u_1}{\partial x_2} - \frac{\partial u_2}{\partial x_1}\right)$$

$$= \frac{1}{2}\left(\frac{\partial u}{\partial y} - \frac{\partial v}{\partial x}\right) = -\omega_{21}$$

$$\omega_{23} = \frac{1}{2}(u_{2,3} - u_{3,2}) = \frac{1}{2}\left(\frac{\partial u_2}{\partial x_3} - \frac{\partial u_3}{\partial x_2}\right)$$

$$= \frac{1}{2}\left(\frac{\partial v}{\partial z} - \frac{\partial w}{\partial y}\right) = -\omega_{32} \qquad (3\text{-}9\text{b})$$

$$\omega_{13} = \frac{1}{2}(u_{1,3} - u_{3,1}) = \frac{1}{2}\left(\frac{\partial u_1}{\partial x_3} - \frac{\partial u_3}{\partial x_1}\right)$$

$$= \frac{1}{2}\left(\frac{\partial u}{\partial z} - \frac{\partial w}{\partial x}\right) = -\omega_{31}$$

INVARIANTS OF STRAIN TENSOR

The (small) strain tensor ϵ_{ij} in Eq. (3-6) is a second-order symmetric tensor. Hence, as discussed in Appendixes 1 and 2, it possesses the transformation and invariant properties of tensors.

There are two sets of invariants we shall define and use: one set associated with the characteristic equation of the (strain) tensor, and the other associated with the tensor through its property of trace and different orders of trace. Additional details on invariants of tensors are given in Appendix 2.

The characteristic equation of the tensor in Eq. (3-2a) can be written as

$$\epsilon^3 - I_{1\epsilon}\epsilon^2 + I_{2\epsilon}\epsilon - I_{3\epsilon} = 0 \qquad (3\text{-}10\text{a})$$

where $I_{1\epsilon}$, $I_{2\epsilon}$, and $I_{3\epsilon}$ are invariants with respect to the characteristic equation.

They are given by

$$I_{1\epsilon} = \epsilon_{ii} = \epsilon_{11} + \epsilon_{22} + \epsilon_{33} \tag{3-10b}$$

$$I_{2\epsilon} = \begin{vmatrix} \epsilon_{11} & \epsilon_{12} \\ \epsilon_{12} & \epsilon_{22} \end{vmatrix} + \begin{vmatrix} \epsilon_{22} & \epsilon_{23} \\ \epsilon_{23} & \epsilon_{33} \end{vmatrix} + \begin{vmatrix} \epsilon_{11} & \epsilon_{13} \\ \epsilon_{13} & \epsilon_{33} \end{vmatrix} \tag{3-10c}$$

$$I_{3\epsilon} = |\epsilon_{ij}| \tag{3-10d}$$

Here $|\cdot|$ denotes a determinant. $(\epsilon_{11})(\epsilon_{22}) - (\epsilon_{12})(\epsilon_{12})$

For engineering use, we can define an alternative form of the invariants based on the trace of ϵ_{ij}. One of the advantages of this alternative is that physical meanings can be assigned to these invariants. In the following definitions of invariants we have used a notation with an overbar and two subscripts, and also a simple notation without a bar and with only one subscript. The overbar notation is shown because sometimes it is used in other publications (1, 2); however, in this text, we generally use the simpler notation.

First invariant of the strain tensor:

$$\bar{I}_{1\epsilon} = I_1 = I_{1\epsilon} = \epsilon_{ii} = \epsilon_{11} + \epsilon_{22} + \epsilon_{33} = \text{tr}(\epsilon) \tag{3-10e}$$

This is the same as before and, for small strains, denotes volumetric strain caused by the applied load.

Second invariant of the strain tensor:

$$\bar{I}_{2\epsilon} = I_2 = \tfrac{1}{2}\epsilon_{ij}\epsilon_{ji} = \tfrac{1}{2}\,\text{tr}(\epsilon)^2$$
$$= \tfrac{1}{2}(I_{1\epsilon}^2 - 2I_{2\epsilon}) \tag{3-10f}$$

Third invariant of the strain tensor:

$$\bar{I}_{3\epsilon} = I_3 = \tfrac{1}{3}\epsilon_{ik}\epsilon_{km}\epsilon_{mi} = \tfrac{1}{3}\,\text{tr}(\epsilon)^3$$
$$= \tfrac{1}{3}(I_{1\epsilon}^3 - 3I_{1\epsilon}I_{2\epsilon} + 3I_{3\epsilon}) \tag{3-10g}$$

Example 3-1

For the strain tensor

$$\epsilon_{ij} = \begin{pmatrix} 4 & -1 & 0 \\ -1 & 2 & -0.5 \\ 0 & -0.5 & 3 \end{pmatrix} \times 10^{-2}$$

compute the invariants $I_{1\epsilon}, I_{2\epsilon}, I_{3\epsilon}$ and I_1, I_2, I_3.

Computation of $I_{1\epsilon}$:

$$I_{1\epsilon} = \epsilon_{11} + \epsilon_{22} + \epsilon_{33}$$
$$= (4 + 2 + 3) \times 10^{-2} = 9 \times 10^{-2}$$

Computation of $I_{2\epsilon}$:

$$I_{2\epsilon} = \epsilon_{11}\epsilon_{22} - \epsilon_{12}^2 + \epsilon_{22}\epsilon_{33} - \epsilon_{23}^2 + \epsilon_{11}\epsilon_{33} - \epsilon_{13}^2$$
$$= \left[4 \times 2 - (-1)^2 + 2 \times 3 - (-0.5)^2 + 3 \times 4 - (0)^2\right] \times 10^{-4}$$
$$= 24.75 \times 10^{-4}$$

Computation of $I_{3\epsilon}$:

$$I_{3\epsilon} = |\epsilon_{ij}|$$

Using the first row of the matrix, we can compute the determinant as

$$|\epsilon_{ij}| = (4 \times 10^{-2}) \times \left(\begin{vmatrix} 2 \times 10^{-2} & -0.5 \times 10^{-2} \\ -0.5 \times 10^{-2} & 3 \times 10^{-2} \end{vmatrix} \right)$$

$$+ (10^{-2}) \times \left(\begin{vmatrix} -1 \times 10^{-2} & -0.5 \times 10^{-2} \\ 0 & 3 \times 10^{-2} \end{vmatrix} \right)$$

$$- 0 \times \left(\begin{vmatrix} -1 \times 10^{-2} & 2 \times 10^{-2} \\ 0 & -0.5 \times 10^{-2} \end{vmatrix} \right)$$

$$= (4 \times 5.75 - 1 \times 3) \times 10^{-6} = 20 \times 10^{-6}$$

Computation of I_1:

$$I_1 = I_{1\epsilon} = \text{tr}(\epsilon) = \epsilon_{11} + \epsilon_{22} + \epsilon_{33} = 9 \times 10^{-2}$$

Computation of I_2:

$$I_2 = I_{2\epsilon} = \tfrac{1}{2}\epsilon_{ij}\epsilon_{ij}$$

Using Eq. (3-10f), we obtain

$$I_2 = \tfrac{1}{2}\left(I_{1\epsilon}^2 - 2I_{2\epsilon}\right)$$

$$= \tfrac{1}{2}\left[(9 \times 10^{-2})^2 - 2(24.75 \times 10^{-4})\right]$$

$$= 15.75 \times 10^{-4}$$

This quantity can also be computed by using the trace of $(\epsilon)^2$. That is,

$$I_2 = \tfrac{1}{2} \text{tr} \begin{pmatrix} 4 & -1 & 0 \\ -1 & 2 & -0.5 \\ 0 & -0.5 & 3 \end{pmatrix} \begin{pmatrix} 4 & -1 & 0 \\ -1 & 2 & -0.5 \\ 0 & -0.5 & 3 \end{pmatrix} \times 10^{-4}$$

By matrix multiplication, we obtain

$$I_2 = \tfrac{1}{2} \text{tr} \begin{pmatrix} 17 & -6 & 0.5 \\ -6 & 5.25 & -2.5 \\ -0.5 & -2.5 & 9.25 \end{pmatrix} \times 10^{-4}$$

$$= \tfrac{1}{2}(17 + 5.25 + 9.25) \times 10^{-4}$$

$$= 15.25 \times 10^{-4}$$

Computation of I_3:

$$I_3 = \bar{I}_{3\epsilon} = \tfrac{1}{3}\epsilon_{ik}\epsilon_{km}\epsilon_{mi}$$

Using Eq. (3-10g), we can write

$$I_3 = \tfrac{1}{3}\left(I_{1\epsilon}^3 - 3I_{1\epsilon}I_{2\epsilon} + 3I_{3\epsilon}\right)$$

$$= \tfrac{1}{3}\left[(9 \times 10^{-2})^3 - 3(9 \times 10^{-2} \times 24.75 \times 10^{-4}) + 3 \times 20 \times 10^{-6}\right]$$

$$= 40.25 \times 10^{-6}$$

This can also be computed by using the trace of $(\epsilon)^3$. That is,

$$I_3 = \frac{1}{3} \text{tr} \begin{pmatrix} 17 & -6 & 0.5 \\ -6 & 5.25 & -2.5 \\ 0.5 & -2.5 & 9.25 \end{pmatrix} \begin{pmatrix} 4 & -1 & 0 \\ -1 & 2 & 0.5 \\ 0 & -0.5 & 3 \end{pmatrix} \times 10^{-6}$$

$$= \frac{1}{3} \text{tr} \begin{pmatrix} 74 & -29.25 & 4.5 \\ 29.25 & 17.75 & -10.125 \\ 4.5 & -10.125 & 29 \end{pmatrix} \times 10^{-6}$$

$$= \frac{1}{3}(74 + 17.75 + 29) \times 10^{-6}$$

$$= 40.25 \times 10^{-6}$$

DECOMPOSITION OF STRAIN TENSOR

A tensor can be decomposed into two tensors. Accordingly, the strain tensor can be decomposed as

$$\epsilon_{ij} = E_{ij} + \tfrac{1}{3}\epsilon_{kk}\delta_{ij} \qquad (3\text{-}11)$$

The tensor E_{ij} is called the *strain deviation* tensor or *deviatoric strain* tensor, and $\frac{1}{3}\epsilon_{kk}\delta_{ij}$ is called the *spherical* or *volumetric* strain tensor. The former is given by

$$E_{ij} = \epsilon_{ij} - \tfrac{1}{3}\epsilon_{kk}\delta_{ij} \qquad (3\text{-}12)$$

For certain materials, it may be appropriate to consider the deviatoric and volumetric strain separately and superimpose their individual effects.

Consider E_{ij} first and set $i = j$; then

$$E_{ii} = \epsilon_{ii} - \tfrac{1}{3}\epsilon_{kk}\delta_{ii} \qquad (3\text{-}13\text{a})$$

Since $\delta_{ii} = 3$ [Eq. (A1-12)], we have

$$E_{ii} = \epsilon_{ii} - \epsilon_{kk} = 0 \qquad (3\text{-}13\text{b})$$

That is, the sum of the normal components of E_{ij}, $E_{ii} = E_{11} + E_{22} + E_{33}$ vanishes.

The volumetric strain tensor is given by

$$\frac{I_1}{3}\delta_{ij} = \frac{I_{1\epsilon}}{3}\delta_{ij} = \tfrac{1}{3}\left(\epsilon_{11}\delta_{ij} + \epsilon_{22}\delta_{ij} + \epsilon_{33}\delta_{ij}\right)$$

$$= \tfrac{1}{3}\left(\epsilon_{11} + \epsilon_{22} + \epsilon_{33}\right)\delta_{ij} \qquad (3\text{-}14)$$

Since $\delta_{ij} = 0$ for $i \neq j$, and $\delta_{ij} = 1$ for $i = j$, the volumetric strain tensor has nonzero terms on the diagonal only; for small strains the sum of the diagonal terms denotes mean volumetric strain.

Invariants of Deviatoric Strain Tensor

As indicated previously, the trace of E_{ij} vanishes; hence E_{ij} has only two nonzero invariants, which are described below.

Second invariant of deviatoric strain tensor

$$\bar{I}_{2E} = I_{2D} = \tfrac{1}{2}\,\text{tr}\,(E)^2 = \tfrac{1}{2}E_{ij}E_{ji}$$

$$= \bar{I}_{2\epsilon} - \frac{\bar{I}_{1\epsilon}^2}{6}$$

$$= I_2 - \frac{I_1^2}{6} \tag{3-15a}$$

It is important to note that the subscript D is used here simply to denote the deviatoric quantity. This quantity can also be expressed as

$$I_{2D} = \tfrac{1}{2}\left[\epsilon_{11}^2 + \epsilon_{22}^2 + \epsilon_{33}^2 + 2\epsilon_{12}^2 + 2\epsilon_{23}^2 + 2\epsilon_{13}^2 - 3\left(\frac{\epsilon_v}{3}\right)^2\right] \tag{3-15b}$$

or

$$I_{2D} = \tfrac{1}{6}\left[(\epsilon_{11} - \epsilon_{22})^2 + (\epsilon_{22} - \epsilon_{33})^2 + (\epsilon_{11} - \epsilon_{33})^2\right] + \epsilon_{12}^2 + \epsilon_{23}^2 + \epsilon_{13}^2 \tag{3-15c}$$

Here ϵ_v is the volumetric strain defined as $\epsilon_v = I_1 = \epsilon_{11} + \epsilon_{22} + \epsilon_{33}$.

In engineering use, often a quantity called *octahedral shear strain*, γ_{oct}, is defined. This strain is equal to the projection of the strain vector on an octahedral plane (Fig. 3-3), which makes equal angles with the three principal directions. The expression for γ_{oct} is given by

$$\gamma_{\text{oct}}^2 = \tfrac{1}{9}\left[(\epsilon_{11} - \epsilon_{22})^2 + (\epsilon_{22} - \epsilon_{33})^2 + (\epsilon_{11} - \epsilon_{33})^2\right]$$

$$+ \tfrac{2}{3}\left[(\epsilon_{12})^2 + (\epsilon_{23})^2 + (\epsilon_{13})^2\right] \tag{3-15d}$$

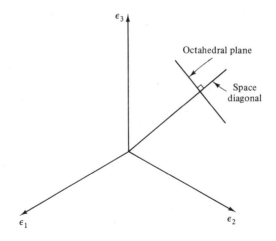

Figure 3-3 Octahedral plane in strain space.

The second invariant of the deviatoric strain tensor, I_{2D}, is directly proportional to γ_{oct} and is given by

$$I_{2D} = \tfrac{3}{2}\gamma_{\text{oct}}^2 \qquad (3\text{-}15\text{e})$$

Many yield criteria (Chapters 9 and 10) in the theory of plasticity are related to the deviatoric quantities, and hence octahedral quantities are frequently used in practice.

Third invariant of deviatoric strain tensor

$$\bar{I}_{3E} = I_{3D} = \tfrac{1}{3}\operatorname{tr}(E)^3 = \tfrac{1}{3}E_{ik}E_{km}E_{mi}$$
$$= \bar{I}_{3\epsilon} - \tfrac{2}{3}\bar{I}_{1\epsilon}\bar{I}_{2\epsilon} + \tfrac{2}{27}(\bar{I}_{1\epsilon})^3$$
$$= I_3 - \tfrac{2}{3}I_1 I_2 + \tfrac{2}{27}I_1^3 \qquad (3\text{-}15\text{f})$$

Example 3-2

Find the deviatoric strain tensor corresponding to the state of strain given in Example 3-1, and then evaluate I_{1D}, I_{2D}, and I_{3D}. From Example 3-1 we have

$$I_1 = 9 \times 10^{-2}$$
$$I_2 = 15.75 \times 10^{-4}$$
$$I_3 = 40.25 \times 10^{-6}$$

The deviatoric strain tensor, E_{ij}, is given by

$$E_{ij} = \epsilon_{ij} - \frac{I_1}{3}\delta_{ij}$$

$$= \begin{pmatrix} 0.04 & -0.01 & 0.0 \\ -0.01 & 0.02 & -0.005 \\ 0 & -0.005 & 0.03 \end{pmatrix} - \tfrac{1}{3}(0.09)\begin{pmatrix} 1 & 0 & 0 \\ 0 & 1 & 0 \\ 0 & 0 & 1 \end{pmatrix}$$

Therefore,

$$E_{ij} = \begin{pmatrix} 1.0 & -1.0 & 0.0 \\ -1.0 & -1.0 & -0.5 \\ 0 & -0.5 & 0.0 \end{pmatrix} \times 10^{-2}$$

It can be seen that $I_{1E} = 0.0$.

Computation of I_{2D}:   Using Eq. (3-15a), we obtain

$$I_{2D} = I_2 - \tfrac{1}{6}I_1^2$$

$$= \left(15.75 - \frac{9^2}{6}\right) \times 10^{-4} = 2.25 \times 10^{-4}$$

This can also be computed by finding the trace of $(E)^2$. That is,

$$I_{2D} = \tfrac{1}{2}\operatorname{tr}(E)^2$$
$$= \tfrac{1}{2}\left[E_{11}^2 + E_{22}^2 + E_{33}^2 + 2E_{12}^2 + 2E_{23}^2 + 2E_{13}^2\right]$$
$$= \tfrac{1}{2}[1.0 + 1.0 + 0.0 + 2 \times 1.0 + 2 \times 0.25 + 0.0] \times 10^{-4}$$
$$= 2.25 \times 10^{-4}$$

Computation of I_{3D}:   Using Eq. (3-15f), I_{3D} can be written as

$$I_{3D} = I_3 - \tfrac{2}{3}I_1I_2 + \tfrac{2}{27}I_1^3$$
$$= \left(40.25 - \tfrac{2}{3} \times 9 \times 15.75 + \tfrac{2}{27}9^3\right) \times 10^{-6}$$
$$= -0.25 \times 10^{-6}$$

This can be computed by finding the trace of $(E)^3$, and can be evaluated using matrix multiplication as

$$I_{3D} = \tfrac{1}{3}\operatorname{tr}(E)^3$$
$$= \tfrac{1}{3}\operatorname{tr}\begin{pmatrix} 2 & -2.25 & 0 \\ -2.25 & -2.5 & -1.125 \\ 0 & -1.125 & -0.25 \end{pmatrix} \times 10^{-6}$$
$$= \tfrac{1}{3}(2 - 2.5 - 0.25) \times 10^{-6} = -0.25 \times 10^{-6}$$

PRINCIPAL STRAINS

In this chapter we have confined ourselves to infinitesimal strains, and hence the relative displacement vector of two closely spaced points, P and Q (Fig. 3-4), can be expressed as (1,3)

$$du_i = (\epsilon_{ij} + \omega_{ij})\, da_j \tag{3-16a}$$

where da_j is the length of PQ. Here ϵ_{ij} and ω_{ij} are strain and rotation tensors at point P, respectively. Considering a pure state of strain, that is, $\omega_{ij} = 0$, the relative displacement vector du_i can be written as

$$du_i = \epsilon_{ij}\, da_j \tag{3-16b}$$

Note that $\omega_{ij} = 0$ does not necessarily imply that lines or planes in the body do not rotate. Hence the relative displacement of Q with respect to P per unit length of PQ can be expressed as

$$\frac{du_i}{|da|} = \epsilon_{ij}\frac{da_j}{|da|}$$

or

$$\epsilon_i = \epsilon_{ji}l_j \tag{3-16c}$$

where $l_j (j = 1, 2, 3)$ represent direction cosines of line of length PQ. In general, the strain vector ϵ_i [Eq. (3-16c)] does not have the same direction as the original line PQ (Fig. 3-4), nor the deformed direction of PQ. The change in length per unit length of PQ, ϵ, during deformation can be expressed in terms of the strain tensor at P, and the direction cosines of PQ as (1,3)

$$\epsilon = \epsilon_{ij}l_il_j \tag{3-17}$$

Although the strain vector in general does not coincide with the direction of the line element under consideration, there are three special directions along which the strain vector has the same orientation as the original line segment. These are called the *principal directions of the strain tensor* or the *principal axes of strain*. The magnitudes of normal strains (displacement per unit original length) along such directions are called *principal strains*.

Hence the strain vector along principal directions can be expressed as

$$\epsilon_i = \lambda l_i = \lambda \delta_{ij} l_j \tag{3-18}$$

where λ is a constant and ϵ_i denotes the three principal strains. Substituting Eq. (3-16c) into Eq. (3-18), we obtain

$$\left(\epsilon_{ji} - \lambda \delta_{ij}\right)l_j = 0 \tag{3-19a}$$

By using the symmetric property of the strain tensor, that is, $\epsilon_{ij} = \epsilon_{ji}$, we have

$$\left(\epsilon_{ij} - \lambda \delta_{ij}\right)l_j = 0 \tag{3-19b}$$

or in matrix notation,

$$\big[[\epsilon] - \lambda[\mathbf{I}]\big]\{l\} = 0 \tag{3-19c}$$

where [**I**] is the identity matrix given by

$$[\mathbf{I}] = \begin{bmatrix} 1 & 0 & 0 \\ 0 & 1 & 0 \\ 0 & 0 & 1 \end{bmatrix} \tag{3-20}$$

and

$$\{l\}^{\mathrm{T}} = \begin{bmatrix} l_1 & l_2 & l_3 \end{bmatrix}$$

Equation (3-19b) represents a set of three linear homogeneous algebraic equations corresponding to the direction cosines l_1, l_2, and l_3. Equation

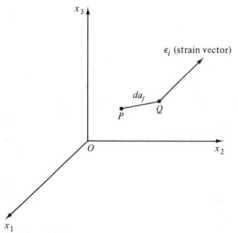

Figure 3-4 Relative displacement vector of two closely spaced points.

(3-19b) can be written in the algebraic form as

$$(\epsilon_{11} - \lambda)l_1 + \epsilon_{12}l_2 + \epsilon_{13}l_3 = 0 \tag{3-21a}$$

$$\epsilon_{12}l_1 + (\epsilon_{22} - \lambda)l_2 + \epsilon_{23}l_3 = 0 \tag{3-21b}$$

$$\epsilon_{13}l_1 + \epsilon_{23}l_2 + (\epsilon_{33} - \lambda)l_3 = 0 \tag{3-21c}$$

Furthermore, the direction cosines $l_i (i = 1, 2, 3)$ satisfy the following relationship:

$$l_i l_i = 1 \tag{3-22a}$$

which can be expressed as

$$l_1^2 + l_2^2 + l_3^2 = 1 \tag{3-22b}$$

Since Eq. (3-22b) has to be satisfied, all three components l_1, l_2, and l_3 cannot be zero. Therefore, for a nontrivial solution, the determinant of coefficient matrix in Eq. (3-19c) has to be zero, that is,

$$|\epsilon_{ij} - \lambda\delta_{ij}| = 0 \tag{3-23a}$$

or

$$\begin{vmatrix} \epsilon_{11} - \lambda & \epsilon_{12} & \epsilon_{13} \\ \epsilon_{12} & \epsilon_{22} - \lambda & \epsilon_{23} \\ \epsilon_{13} & \epsilon_{23} & \epsilon_{33} - \lambda \end{vmatrix} = 0 \tag{3-23b}$$

This reduces to a cubic equation in λ, and is known as the *characteristic equation* (see also Appendix 2). This can be simplified as

$$\lambda^3 - I_{1\epsilon}\lambda^2 + I_{2\epsilon}\lambda - I_{3\epsilon} = 0 \tag{3-23c}$$

It can be shown that this equation has three real roots; these three values are the principal strains. Their corresponding directions, which are mutually orthogonal, can be found by substituting the values of λ in Eq. (3-21) and solving Eqs. (3-21) and (3-22) for values of $l_i (i = 1, 2, 3)$.

When the principal directions are used as the axes of reference, the strain tensor takes the following form:

$$\epsilon_{ij} = \begin{pmatrix} \epsilon_1 & 0 & 0 \\ 0 & \epsilon_2 & 0 \\ 0 & 0 & \epsilon_3 \end{pmatrix} \tag{3-24}$$

It should be noted that Eq. (3-24) does not necessarily imply that $\epsilon_1 > \epsilon_2 > \epsilon_3$. The largest principal value is called the *major principal strain*, and the smallest value is called the *minor principal strain*. The other value is called the *intermediate principal strain*.

Example 3-3

Find the principal strains and principal strain directions at a point P where the strain tensor is given by

$$\epsilon_{ij} = \begin{pmatrix} 3 & -1 & 0 \\ -1 & 3 & 0 \\ 0 & 0 & 1 \end{pmatrix} \times 10^{-2}$$

Note: Since the factor 10^{-2} is common, it will not be carried out in the computations. First we compute the invariants:

$$I_{1\epsilon} = 7$$

$$I_{2\epsilon} = 8 + 3 + 3 = 14$$

$$I_{3\epsilon} = 8$$

The characteristic equation is

$$\epsilon^3 - 7\epsilon^2 + 14\epsilon - 8 = 0 \qquad (3\text{-}25\text{a})$$

or

$$\epsilon^3 - 6\epsilon^2 - \epsilon^2 + 6\epsilon + 8\epsilon - 8 = 0 \qquad (3\text{-}25\text{b})$$

or

$$\epsilon^2(\epsilon - 1) - 6\epsilon(\epsilon - 1) + 8(\epsilon - 1) = 0 \qquad (3\text{-}25\text{c})$$

or

$$(\epsilon - 1)(\epsilon - 4)(\epsilon - 2) = 0 \qquad (3\text{-}25\text{d})$$

Therefore, we obtain its roots as

$$\epsilon = 4.0, 2.0, \text{ and } 1.0$$

Hence

Major principal strain:	4.0×10^{-2}
Minor principal strain:	1.0×10^{-2}
Intermediate principal strain:	2.0×10^{-2}

In order to find principal strain directions, let us consider them separately.

Major principal strain direction: $\epsilon_1 = 4.0 \times 10^{-2}$. From Eq. (3-21), we have

$$(3 - 4)l_1 - l_2 \qquad\qquad = 0 \qquad (3\text{-}26\text{a})$$

$$- l_1 \qquad + (3 - 4)l_2 \qquad = 0 \qquad (3\text{-}26\text{b})$$

$$(1 - 4)l_3 = 0 \qquad (3\text{-}26\text{c})$$

and, from Eq. (3-22),

$$l_1^2 + l_2^2 + l_3^2 = 1.0 \qquad (3\text{-}27)$$

From Eq. (3-26a), we have

$$l_1 = -l_2 \qquad (3\text{-}28\text{a})$$

and from Eq. (3-26c),

$$l_3 = 0 \qquad (3\text{-}28\text{b})$$

Substituting Eqs. (3-28), we obtain

$$2l_1^2 = 1.0 \qquad (3\text{-}29\text{a})$$

Therefore,

$$l_1 = \pm \frac{1}{\sqrt{2}} \qquad (3\text{-}29b)$$

$$l_2 = \mp \frac{1}{\sqrt{2}} \qquad (3\text{-}29c)$$

Then the major principal strain direction is given by

$$l_1 = \frac{1}{\sqrt{2}}, \qquad l_2 = -\frac{1}{\sqrt{2}}, \qquad l_3 = 0$$

or

$$l_1 = -\frac{1}{\sqrt{2}}, \qquad l_2 = \frac{1}{\sqrt{2}}, \qquad l_3 = 0$$

Note that the second set of values refer to negative direction. Here we shall use only the first set (positive direction) for our illustration.

Minor principal strain direction: $\epsilon_3 = 1.0 \times 10^{-2}$. Substituting the value of ϵ_3 for λ in Eq. (3-21), we can write the following equations:

$$(3 - 1)l_1 - l_2 \qquad\qquad = 0 \qquad (3\text{-}30a)$$

$$-l_1 \qquad + (3 - 1)l_2 \qquad = 0 \qquad (3\text{-}30b)$$

$$(1 - 1)l_3 = 0 \qquad (3\text{-}30c)$$

From Eqs. (3-30a) and (3-30b), it can be seen that

$$l_1 = l_2 = 0.0 \qquad (3\text{-}31a)$$

Substituting values of l_1 and l_2 from Eq. (3-31a) into Eq. (3-22b), the value of l_3 can be found as

$$l_3 = \pm 1.0 \qquad (3\text{-}31b)$$

Hence, the minor principal strain directions are given by

$$l_1 = 0.0, \qquad l_2 = 0.0, \qquad l_3 = +1.0$$

or

$$l_1 = 0.0, \qquad l_2 = 0.0, \qquad l_3 = -1.0$$

Note that the second set of values refer to negative direction.

Intermediate principal strain direction: $\epsilon_2 = 2.0 \times 10^{-2}$. Substituting the value of ϵ_2 for λ in Eq. (3-21), we can write the following equations:

$$(3 - 2)l_1 - l_2 \qquad\qquad = 0 \qquad (3\text{-}32a)$$

$$-l_1 \qquad + (3 - 2)l_2 \qquad = 0 \qquad (3\text{-}32b)$$

$$(1 - 2)l_3 = 0 \qquad (3\text{-}32c)$$

From Eqs. (3-32a) and (3-32b), we can show that

$$l_1 = l_2 \qquad (3\text{-}33a)$$

and from Eq. (3-32c),

$$l_3 = 0.0 \tag{3-33b}$$

Substituting Eqs. (3-33a) and (3-33b) into Eq. (3-22b), we have

$$2l_1^2 = 1.0 \tag{3-33c}$$

Therefore,

$$l_1 = \pm \frac{1}{\sqrt{2}} \tag{3-33d}$$

$$l_2 = \pm \frac{1}{\sqrt{2}} \tag{3-33e}$$

Hence the intermediate principal direction is given by

$$l_1 = \frac{1}{\sqrt{2}}, \qquad l_2 = \frac{1}{\sqrt{2}}, \qquad l_3 = 0.0$$

or

$$l_1 = -\frac{1}{\sqrt{2}}, \qquad l_2 = -\frac{1}{\sqrt{2}}, \qquad l_3 = 0.0$$

Here the second set of values refer to the negative direction. Hence the direction cosines of principal strain directions can be arranged in the following way:

	x_1	x_2	x_3
Major x_1'	$\dfrac{1}{\sqrt{2}}$	$-\dfrac{1}{\sqrt{2}}$	0
Intermediate x_2'	$\dfrac{1}{\sqrt{2}}$	$\dfrac{1}{\sqrt{2}}$	0
Minor x_3'	0	0	1

In order to check the form of the strain tensor when principal strain directions are used as the reference axes, a coordinate transformation can be performed. The transformation tensor, a_{ij}, can be written as follows (see Appendix 2):

$$a_{ij} = \begin{pmatrix} \dfrac{1}{\sqrt{2}} & -\dfrac{1}{\sqrt{2}} & 0 \\ \dfrac{1}{\sqrt{2}} & \dfrac{1}{\sqrt{2}} & 0 \\ 0 & 0 & 1 \end{pmatrix} \tag{3-34}$$

The strain tensor, ϵ_{ij}', in the new reference frame is given as follows:

$$\epsilon_{ij}' = a_{im}a_{jn}\epsilon_{mn} \tag{3-35a}$$

or in the vector notation,

$$[\epsilon]' = [a][\epsilon][a]^T$$

$$= \begin{bmatrix} \dfrac{1}{\sqrt{2}} & -\dfrac{1}{\sqrt{2}} & 0 \\ \dfrac{1}{\sqrt{2}} & \dfrac{1}{\sqrt{2}} & 0 \\ 0 & 0 & 1 \end{bmatrix} \begin{bmatrix} 3 & -1 & 0 \\ -1 & 3 & 0 \\ 0 & 0 & 1 \end{bmatrix} \begin{bmatrix} \dfrac{1}{\sqrt{2}} & \dfrac{1}{\sqrt{2}} & 0 \\ -\dfrac{1}{\sqrt{2}} & \dfrac{1}{\sqrt{2}} & 0 \\ 0 & 0 & 1 \end{bmatrix} \times 10^{-2}$$

(3-35b)

or

$$[\epsilon]' = \begin{bmatrix} 4 & 0 & 0 \\ 0 & 2 & 0 \\ 0 & 0 & 1 \end{bmatrix} \times 10^{-2} \qquad (3\text{-}35c)$$

This gives the same values as obtained by solving the characteristic equation. It can be noted that in Eq. (3-35c), the principal values are arranged in the same way as the principal directions were arranged in the transformation tensor, a_{ij}.

STATES OF STRAIN FOR VARIOUS MODES OF DEFORMATION

When a deformable body is subjected to external forces, the general coupled state of its deformation can be expressed through the six components in the strain tensor ϵ_{ij}. It is often convenient to consider specific states of the general deformation; as we shall see subsequently, this approach would provide a basis for determination of constitutive parameters relevant to the specific mode under consideration.

Uniaxial state of strain. Figure 3-5(a) shows a line bar of uniform cross section subjected to a purely axial strain while other strains are constrained. Under these conditions, strains can be assumed to occur only in one direction, along the axis of the body, and the strain tensor reduces to

$$\epsilon_{ij} = \begin{pmatrix} \epsilon_{11} & 0 & 0 \\ 0 & 0 & 0 \\ 0 & 0 & 0 \end{pmatrix} \qquad (3\text{-}36a)$$

or

$$= \begin{pmatrix} \epsilon_1 & 0 & 0 \\ 0 & 0 & 0 \\ 0 & 0 & 0 \end{pmatrix}$$

Hydrostatic state of strain [Fig. 3-5(b)]. Here there is no deviatoric component of strain and the body is assumed to experience only equal volumetric strains (deformation) in all directions. Hence the strain tensor becomes

$$\epsilon_{ij} = \begin{pmatrix} \epsilon_{11} & 0 & 0 \\ 0 & \epsilon_{11} & 0 \\ 0 & 0 & \epsilon_{11} \end{pmatrix} = \begin{pmatrix} \epsilon_1 & 0 & 0 \\ 0 & \epsilon_1 & 0 \\ 0 & 0 & \epsilon_1 \end{pmatrix} \qquad (3\text{-}36b)$$

An isotropic element of material subjected to all-round hydrostatic stress (pressure), discussed in Chapter 4, undergoes purely volumetric strain as given in Eq. (3-36b).

Triaxial state of strain [Fig. 3-5(c)].   This state is caused in a cubical specimen of a material in which the three (independent) strains applied to its faces are the principal strains, thus

$$\epsilon_{ij} = \begin{pmatrix} \epsilon_{11} & 0 & 0 \\ 0 & \epsilon_{22} & 0 \\ 0 & 0 & \epsilon_{33} \end{pmatrix} = \begin{pmatrix} \epsilon_1 & 0 & 0 \\ 0 & \epsilon_2 & 0 \\ 0 & 0 & \epsilon_3 \end{pmatrix} \qquad (3\text{-}36c)$$

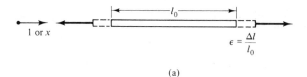

(a)

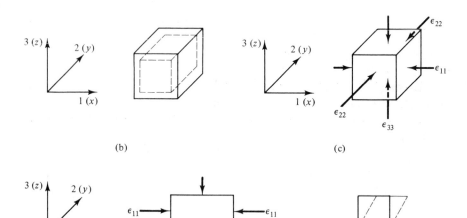

(b)   (c)

(d)   (e)

Figure 3-5   Various states of strain: (a) uniaxial strain; (b) hydrostatic strain; (c) triaxial strain; (d) cylindrical state of strain; (e) simple shear strain.

A special case of the triaxial state of strain is the *cylindrical state of strain*, in which two of the applied principal strains are equal [Fig. 3-5(d)]; for this case, the strain tensor takes the form

$$\epsilon_{ij} = \begin{pmatrix} \epsilon_{11} & 0 & 0 \\ 0 & \epsilon_{22} & 0 \\ 0 & 0 & \epsilon_{22} \end{pmatrix} = \begin{pmatrix} \epsilon_1 & 0 & 0 \\ 0 & \epsilon_2 & 0 \\ 0 & 0 & \epsilon_2 \end{pmatrix} \tag{3-36d}$$

Simple shear strain [Fig. 3-5(e)].   Here an element of material is subjected to shearing without any volume change. Hence

$$\epsilon_{ij} = \begin{pmatrix} 0 & \epsilon_{12} & 0 \\ \epsilon_{12} & 0 & 0 \\ 0 & 0 & 0 \end{pmatrix} \tag{3-36e}$$

Consideration and analysis of individual states of strain are important. As we shall see subsequently, they allow development and use of (laboratory) test devices (Chapter 5) in order to simulate such states, and then determine appropriate parameters to define various constitutive laws.

PROBLEMS

3-1. Find the values of $I_{1\epsilon}$, $I_{2\epsilon}$, $I_{3\epsilon}$, I_1, I_2, I_3, I_{2D}, and I_{3D} for the following (strain) tensor:

$$\epsilon_{ij} = \begin{pmatrix} 3 & 2 & 1 \\ 2 & 4 & 2 \\ 1 & 2 & 5 \end{pmatrix} \times 10^{-3}$$

Answers: $I_{1\epsilon} = 12 \times 10^{-3}$ cm/cm; $I_{2\epsilon} = 38 \times 10^{-6}$ (cm/cm)^2; $I_{3\epsilon} = 32 \times 10^{-9}$ (cm/cm)^3; $I_1 = 12 \times 10^{-3}$ cm/cm; $I_2 = 34 \times 10^{-6}$ (cm/cm)^2; $I_3 = 152 \times 10^{-9}$ (cm/cm)^3; $I_{2D} \doteq 10 \times 10^{-6}$ (cm/cm)^2; $I_{3D} = 8 \times 10^{-9}$ (cm/cm)^3.

3-2. Derive the expressions for (a) $\bar{I}_{2\epsilon} = I_{2D}$ in Eq. (3-15a), and (b) $\bar{I}_{3\epsilon} = I_{3D}$ in Eq. (3-15f).

Partial Steps:

(a) $I_{2D} = \dfrac{1}{2} E_{ij} E_{ij} = \dfrac{1}{2} \left[\epsilon_{ij} \epsilon_{ij} - 2 \dfrac{I_1}{3} \epsilon_{ij} \delta_{ij} + \dfrac{I_1}{9} \delta_{ij} \delta_{ij} \right]$

$\quad = I_2 - \dfrac{I_1^2}{3} + \dfrac{I_1^2}{6} = I_2 - \dfrac{I_1^2}{6}$

(b) $I_{3D} = \dfrac{1}{3} E_{im} E_{mn} E_{in}$

$\quad = \dfrac{1}{3} \left[\left(\epsilon_{im} \epsilon_{mn} - \dfrac{I_1}{3} \epsilon_{in} - \dfrac{I_1}{3} \epsilon_{in} + \dfrac{I_1^2}{9} \delta_{in} \right) \left(\epsilon_{in} - \dfrac{I_1}{3} \delta_{in} \right) \right]$

$\quad = \dfrac{1}{3} \left[\epsilon_{im} \epsilon_{mn} \epsilon_{in} - \dfrac{2 I_1 I_2}{3} - \dfrac{4 I_1 I_2}{3} + \dfrac{3 I_1^3}{9} - \dfrac{3 I_1^3}{27} \right]$

$\quad = I_3 - \dfrac{2}{3} I_1 I_2 + \dfrac{2}{27} I_1^3$

3-3. For the strain tensor:

$$\epsilon_{ij} = \begin{pmatrix} \epsilon_{11} & 0 & 0 \\ 0 & \epsilon_{22} & 0 \\ 0 & 0 & \epsilon_{33} \end{pmatrix}$$

derive expressions for (a) I_1, (b) I_{2D}, and (c) I_{3D}.

Partial Steps and Answers:

(a) $I_1 \;\; = \epsilon_{11} + \epsilon_{22} + \epsilon_{33}$

(b) $I_{2D} = I_2 - \dfrac{I_1^2}{6}$

$= \frac{1}{6}[3\epsilon_{11}^2 + 3\epsilon_{22}^2 + 3\epsilon_{33}^2 - \epsilon_{11}^2 - \epsilon_{22}^2 - \epsilon_{33}^2$

$\qquad - 2\epsilon_{11}\epsilon_{22} - 2\epsilon_{22}\epsilon_{33} - 2\epsilon_{33}\epsilon_{11}]$

$= \frac{1}{6}\left[\left(\epsilon_{11} - \epsilon_{22} \right)^2 + \left(\epsilon_{22} - \epsilon_{33} \right)^2 + \left(\epsilon_{33} - \epsilon_{11} \right)^2 \right]$

(c) $I_{3D} = I_3 - \frac{2}{3} I_1 I_2 + \frac{2}{27} I_1^3$

$= \frac{1}{3}\left[\left(\epsilon_{11}^3 + \epsilon_{22}^3 + \epsilon_{33}^3 \right) - \epsilon_{11}^3 - \epsilon_{22}^3 - \epsilon_{33}^3 - \epsilon_{11}\epsilon_{22}^2 - \epsilon_{11}\epsilon_{33}^2 \right.$

$\qquad \left. - \epsilon_{22}\epsilon_{11}^2 - \epsilon_{22}\epsilon_{33}^2 - \epsilon_{33}\epsilon_{11}^2 - \epsilon_{33}\epsilon_{22}^2 \right]$

$\qquad + \frac{2}{27}\left[\epsilon_{11}^3 + \epsilon_{22}^3 + \epsilon_{33}^3 + 3\epsilon_{11}\epsilon_{22}^2 + 3\epsilon_{22}\epsilon_{11}^2 + 3\epsilon_{22}\epsilon_{33}^2 \right.$

$\qquad \left. + 3\epsilon_{22}^2\epsilon_{33} + 3\epsilon_{11}\epsilon_{33}^2 + 3\epsilon_{11}^2\epsilon_{33} + 6\epsilon_{11}\epsilon_{22}\epsilon_{33} \right]$

$= \frac{1}{27}\left[3\left(\epsilon_{11}^3 + \epsilon_{22}^3 + \epsilon_{33}^3 \right) - \left(\epsilon_{11} + \epsilon_{22} + \epsilon_{33} \right)^3 + 18\epsilon_{11}\epsilon_{22}\epsilon_{33} \right]$

REFERENCES

1. Fung, Y. C., *Foundations of Solid Mechanics*, Prentice-Hall, Inc., Englewood Cliffs, N.J., 1965.

2. Eringen, A. C., *Nonlinear Theory of Continuous Media*, McGraw-Hill Book Company, New York, 1962.

3. Malvern, L. E., *Introduction to the Mechanics of a Continuous Medium*, Prentice-Hall, Inc., Englewood Cliffs, N.J., 1969.

4

STRESS, COMPONENTS OF STRESS, AND STRESS TENSOR

The components of the stress tensor are shown in Fig. 4-1. The components of the tensor in the Cartesian coordinate system, x_i ($i = 1, 2, 3$), can be defined as follows:

$$\sigma_{ij} = \lim_{A_i \to 0} \left(\frac{F_j}{A_i} \right) \tag{4-1}$$

where F_j is the force in the coordinate direction j ($j = 1, 2, 3$) and A_i is the area normal to the direction i on which the force F_j acts.

Sign Convention

Different sign conventions are used in conjunction with solid and geo-mechanics. In certain applications compressive normal stresses have been taken as positive, while in some other applications tensile normal stresses have been assumed as positive. Although a sign convention does not influence the mechanics of deformations, it is important in interpreting information with a physical insight. In this text, components σ_{11}, σ_{22}, and σ_{33} are assumed positive when they are compressive. The positive quantities of shear and normal stresses are shown in Fig. 4-1.

In absence of body moments, the stress tensor σ_{ij} can be shown to be symmetric; that is,

$$\sigma_{ij} = \sigma_{ji} \tag{4-2a}$$

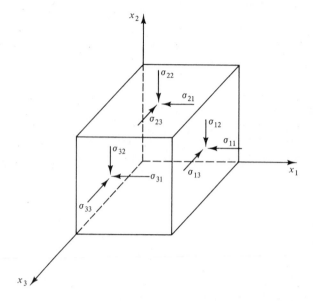

Figure 4-1   Components of stress tensor.

This implies that

$$\sigma_{12} = \sigma_{21}$$

$$\sigma_{13} = \sigma_{31} \tag{4-2b}$$

$$\sigma_{23} = \sigma_{32}$$

The stress tensor σ_{ij}, like any other second-order tensor, possesses three independent invariants.

Important Comment

The presentations in this text are based on effective stresses, that is, the stresses carried by the (soil) skeleton (1); the stress tensor σ_{ij} is considered as the effective stress tensor. For porous geologic materials, the existence and development of pressures in the fluid (water) in the pores can be highly important. However, if the pore water pressures are known, the effective stresses can be evaluated and become the primary quantities for constitutive models.

INVARIANTS OF STRESS TENSOR

The invariants of a stress tensor can be defined in a number of ways; additional details on invariants of tensors are given in Appendix 2. Let us consider the "characteristic equation" of the stress tensor σ_{ij}. The characteristic equation can be written as

$$\sigma^3 - I_{1\sigma}\sigma^2 + I_{2\sigma}\sigma - I_{3\sigma} = 0 \tag{4-3a}$$

where

$$I_{1\sigma} = \sigma_{ii} = \sigma_{11} + \sigma_{22} + \sigma_{33} \tag{4-3b}$$

$$I_{2\sigma} = \begin{vmatrix} \sigma_{11} & \sigma_{12} \\ \sigma_{12} & \sigma_{22} \end{vmatrix} + \begin{vmatrix} \sigma_{22} & \sigma_{23} \\ \sigma_{23} & \sigma_{33} \end{vmatrix} + \begin{vmatrix} \sigma_{11} & \sigma_{13} \\ \sigma_{13} & \sigma_{33} \end{vmatrix} \tag{4-3c}$$

$$I_{3\sigma} = |\sigma_{ij}| \tag{4-3d}$$

These three quantities, $I_{1\sigma}$, $I_{2\sigma}$, and $I_{3\sigma}$, are invariants of the stress tensor. There are alternative ways to define invariants of the stress tensor. The invariants in Eqs. (4-3) derived from the characteristic equation of the tensor are denoted by $I_{1\sigma}$, $I_{2\sigma}$, and $I_{3\sigma}$ (without an overbar), while in an alternative method, they are denoted by $\bar{I}_{1\sigma}$, $\bar{I}_{2\sigma}$, and $\bar{I}_{3\sigma}$ (with an overbar) and are derived from the tensor itself.

First invariant of stress tensor:

$$\bar{I}_{1\sigma} = J_1 = \sigma_{ii} = \sigma_{11} + \sigma_{22} + \sigma_{33} = \operatorname{tr}(\sigma)$$

$$= I_{1\sigma} \tag{4-4a}$$

Second invariant of stress tensor:

$$\bar{I}_{2\sigma} = J_2 = \tfrac{1}{2}\sigma_{ij}\sigma_{ji} = \tfrac{1}{2}\operatorname{tr}(\sigma)^2$$

$$= \tfrac{1}{2}\left(I_{1\sigma}^2 - 2I_{2\sigma}\right) \tag{4-4b}$$

Third invariant of stress tensor:

$$\bar{I}_{3\sigma} = J_3 = \tfrac{1}{3}\sigma_{ik}\sigma_{km}\sigma_{mi} = \tfrac{1}{3}\operatorname{tr}(\sigma)^3$$

$$= \tfrac{1}{3}\left(I_{1\sigma}^3 - 3I_{1\sigma}I_{2\sigma} + 3I_{3\sigma}\right) \tag{4-4c}$$

Note that we have used the general symbols for the invariants ($\bar{I}_{i\sigma}$, $i = 1, 2, 3$), and have also introduced the symbols J_1, J_2, and J_3, which we use in this text.

Example 4-1

Consider the state of stress represented by the stress tensor

$$\sigma_{ij} = \begin{pmatrix} 2 & 1 & 0 \\ 1 & 3 & 2 \\ 0 & 2 & 4 \end{pmatrix} \qquad \text{units: kN/m}^2$$

First invariant:

$$J_1 = \sigma_{mm} = 2 + 3 + 4 = 9 \text{ kN/m}^2$$

Second invariant:

$$J_2 = \tfrac{1}{2}\left(I_{1\sigma}^2 - 2I_{2\sigma} \right)$$
$$= \tfrac{1}{2}\left[9^2 - 2(5 + 8 + 8) \right]$$
$$= 19.5 \left(kN/m^2 \right)^2$$

Third invariant:

$$J_3 = \tfrac{1}{3}\left[I_{1\sigma}^3 - 3I_{1\sigma}I_{2\sigma} + 3I_{3\sigma} \right]$$
$$= \tfrac{1}{3}\left[9^3 - 3(9)(21) + 3(12) \right]$$
$$= 66 \left(kN/m^2 \right)^3$$

DECOMPOSITION OF STRESS TENSOR

The symmetric stress tensor σ_{ij} can be decomposed into two symmetric tensors, the *deviatoric stress tensor* and the *hydrostatic or spherical stress tensor*. The deviatoric stress tensor is sometimes referred to as the *stress deviation tensor*. The decomposition is given by

$$\sigma_{ij} = S_{ij} + \tfrac{1}{3}\sigma_{nn}\delta_{ij} \tag{4-5a}$$

where $= S_{ij}$ is the deviatoric stress tensor, $\sigma_{nn} = \sigma_{11} + \sigma_{22} + \sigma_{33}$ is the hydrostatic stress, and δ_{ij} is the Kronecker delta. Equation (4-5a) can be written in matrix notation as

$$
\begin{bmatrix} \sigma_{11} & \sigma_{12} & \sigma_{13} \\ \sigma_{21} & \sigma_{22} & \sigma_{23} \\ \sigma_{31} & \sigma_{32} & \sigma_{33} \end{bmatrix} =
\begin{bmatrix} S_{11} & \sigma_{12} & \sigma_{13} \\ \sigma_{21} & S_{22} & \sigma_{23} \\ \sigma_{31} & \sigma_{32} & S_{33} \end{bmatrix} +
\begin{bmatrix} \dfrac{\sigma_{nn}}{3} & 0 & 0 \\ 0 & \dfrac{\sigma_{nn}}{3} & 0 \\ 0 & 0 & \dfrac{\sigma_{nn}}{3} \end{bmatrix}
$$

or

$$[\sigma] = [S] + [p] \tag{4-5b}$$

Here $S_{12} = \sigma_{12}$, $S_{13} = \sigma_{13}$, and so on.

The tensor S_{ij} or matrix $[S]$ is that part of the stress state which represents the shear or deviatoric state of stress and excludes the hydrostatic state of stress. The tensor $\tfrac{1}{3}\sigma_{nn}\delta_{ij}$ or the matrix $[p]$ refers strictly to the spherical or hydrostatic stress. In the matrix $[p]$ above, each component on the diagonal, that is, $\sigma_{nn}/3$, denotes *mean pressure* $p = (\sigma_{11} + \sigma_{22} + \sigma_{33})/3$, and can be expressed as

$$p\delta_{ij} = \begin{pmatrix} p & 0 & 0 \\ 0 & p & 0 \\ 0 & 0 & p \end{pmatrix} \tag{4-5c}$$

The deviatoric stress tensor S_{ij} can be expressed as follows:

$$S_{ij} = \sigma_{ij} - \tfrac{1}{3}\sigma_{nn}\delta_{ij} = \sigma_{ij} - p\delta_{ij} \qquad (4\text{-}5\text{d})$$

where p is the mean pressure and is equal to $J_1/3$.

Invariants of Deviatoric Stress Tensor

First invariant of deviatoric stress tensor. Deviatoric stress tensor is a second-order symmetric tensor. It is important to note that the trace of this tensor is zero; that is,

$$S_{ii} = S_{11} + S_{22} + S_{33}$$

$$= \sigma_{ii} - \tfrac{1}{3}\sigma_{nn}\delta_{ii}$$

$$= \sigma_{ii} - \sigma_{nn}$$

$$= 0 \qquad (4\text{-}6\text{a})$$

Hence S_{ij} has only two independent nonzero invariants.

Second invariant of deviatoric stress tensor. We follow one of the common symbols and denote the second invariant by J_{2D}:

$$J_{2D} = \bar{I}_{2s} = \frac{1}{2}S_{ij}S_{ij} = \frac{1}{2}\operatorname{tr}(S)^2$$

$$= \frac{1}{2}\left(\sigma_{ij} - \frac{\sigma_{mm}}{3}\delta_{ij}\right)\left(\sigma_{ij} - \frac{\sigma_{mm}}{3}\delta_{ij}\right)$$

$$= \frac{1}{2}\left(\sigma_{ij}\sigma_{ij} - 2\sigma_{ij}\frac{\sigma_{mm}}{3}\delta_{ij} + \frac{\sigma_{mm}^2}{9}\delta_{ij}\delta_{ij}\right)$$

$$= J_2 - \frac{\sigma_{mm}^2}{3} + \frac{\sigma_{mm}^2}{6}$$

$$= J_2 - \frac{J_1^2}{6} \qquad (4\text{-}6\text{b})$$

In matrix notation, since $J_{2D} = \tfrac{1}{2}\operatorname{tr}(S)^2$, we have

$$J_{2D} = \tfrac{1}{2}\operatorname{tr}\begin{pmatrix} S_{11} & S_{12} & S_{13} \\ S_{12} & S_{22} & S_{23} \\ S_{13} & S_{23} & S_{33} \end{pmatrix}\begin{pmatrix} S_{11} & S_{12} & S_{13} \\ S_{12} & S_{22} & S_{23} \\ S_{13} & S_{23} & S_{33} \end{pmatrix}$$

$$= \tfrac{1}{2}\left[S_{11}^2 + S_{12}^2 + S_{13}^2 + S_{12}^2 + S_{22}^2 + S_{23}^2 + S_{13}^2 + S_{23}^2 + S_{33}^2\right]$$

$$= \tfrac{1}{2}\left[(\sigma_{11} - p)^2 + (\sigma_{22} - p)^2 + (\sigma_{33} - p)^2 + 2S_{12}^2 + 2S_{23}^2 + 2S_{13}^2\right]$$

$$\qquad (4\text{-}6\text{c})$$

From the definition, we have $S_{12} = \sigma_{12}$, $S_{23} = \sigma_{23}$, and $S_{13} = \sigma_{13}$; hence the foregoing expression is simplified to

$$J_{2D} = \tfrac{1}{2}\left[\sigma_{11}^2 + \sigma_{22}^2 + \sigma_{33}^2 + 2\sigma_{12}^2 + 2\sigma_{23}^2 + 2\sigma_{13}^2 - 3p^2\right] \qquad (4\text{-}6\text{d})$$

It can also be shown that

$$J_{2D} = \tfrac{1}{6}\left[(\sigma_{11} - \sigma_{22})^2 + (\sigma_{22} - \sigma_{33})^2 + (\sigma_{11} - \sigma_{33})^2\right] + \sigma_{12}^2 + \sigma_{23}^2 + \sigma_{13}^2$$
$$(4\text{-}6\text{e})$$

In terms of principal stresses,

$$J_{2D} = \tfrac{1}{6}\left[(\sigma_1 - \sigma_2)^2 + (\sigma_2 - \sigma_3)^2 + (\sigma_1 - \sigma_3)^2\right] \qquad (4\text{-}6\text{f})$$

Using the expression for J_2 given by Eq. (4-4b), J_{2D} can be written as

$$J_{2D} = J_2 - \frac{J_1^2}{6} \qquad (4\text{-}6\text{g})$$

This expression is the same as that given before [Eq. (4-6b)].

In engineering practice, often a quantity called *octahedral shear stress*, τ_{oct}, is defined and used. This shear stress is equal to the magnitude of the projection of the stress vector on an octahedral plane (Fig. 4-2) which makes equal angles with the three principal directions. The mean pressure, $J_1/3$, remains constant on an octahedral plane. The expression for τ_{oct} is given by

$$\tau_{\text{oct}}^2 = \tfrac{1}{9}\left[(\sigma_{11} - \sigma_{22})^2 + (\sigma_{22} - \sigma_{33})^2 + (\sigma_{11} - \sigma_{33})^2\right]$$
$$+ \tfrac{2}{3}\left[(\sigma_{12})^2 + (\sigma_{23})^2 + (\sigma_{13})^2\right] \qquad (4\text{-}6\text{h})$$

The second invariant of the deviatoric stress tensor, J_{2D}, is directly proportional to τ_{oct}, and given by

$$J_{2D} = \tfrac{3}{2}\tau_{\text{oct}}^2 \qquad (4\text{-}6\text{i})$$

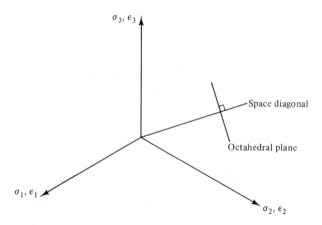

Figure 4-2   Octahedral plane in principal stress space.

Third invariant of deviatoric stress tensor. This is denoted as J_{3D}, and can be expressed as

$$J_{3D} = \bar{I}_{3S} = \frac{1}{3} S_{ij} S_{jm} S_{mi} = \frac{1}{3} \operatorname{tr}(S)^3$$

$$= \frac{1}{3} \left(\sigma_{ij} - \frac{J_1}{3} \delta_{ij} \right) \left(\sigma_{jm} - \frac{J_1}{3} \delta_{jm} \right) \left(\sigma_{mi} - \frac{J_1}{3} \delta_{mi} \right)$$

$$= \frac{1}{3} \left(\sigma_{ij} \sigma_{jm} \sigma_{mi} - \frac{J_1}{3} \sigma_{ij} \sigma_{ij} - \frac{2}{3} J_1 \sigma_{im} \sigma_{im} \right.$$

$$\left. + \frac{2}{9} J_1^2 \sigma_{mm} + \frac{J_1^2}{9} \sigma_{mm} - \frac{J_1^3}{27} \delta_{mm} \right)$$

$$= J_3 - \frac{2}{3} J_1 J_2 + \frac{2}{27} J_1^3 \tag{4-6j}$$

Example 4-2

Consider the same state of stress given in Example 4-1.

$$\sigma_{ij} = \begin{pmatrix} 2 & 1 & 0 \\ 1 & 3 & 2 \\ 0 & 2 & 4 \end{pmatrix} \quad \text{kN/m}^2$$

From Example 4-1, we have

$$J_1 = 9 \left(\text{kN/m}^2 \right)$$

$$J_2 = 19.5 \left(\text{kN/m}^2 \right)^2$$

$$J_3 = 66 \left(\text{kN/m}^2 \right)^3$$

The deviatoric stress tensor S_{ij} is given by

$$S_{ij} = \sigma_{ij} - \frac{J_1}{3} \delta_{ij}$$

$$= \begin{pmatrix} 2 & 1 & 0 \\ 1 & 3 & 2 \\ 0 & 2 & 4 \end{pmatrix} - \begin{pmatrix} 3 & 0 & 0 \\ 0 & 3 & 0 \\ 0 & 0 & 3 \end{pmatrix}$$

$$S_{ij} = \begin{pmatrix} -1 & 1 & 0 \\ 1 & 0 & 2 \\ 0 & 2 & 1 \end{pmatrix}$$

Hence

$$J_{1D} = -1 + 0 + 1 = 0$$

By using Eq. (4-4b) or (A2-13) we can express the second invariant as

$$J_{2D} = \bar{I}_{2s} = \tfrac{1}{2} \left(I_{1s}^2 - 2 I_{2s} \right)$$

$$= \tfrac{1}{2} [0 - 2(-1 - 4 - 1)]$$

$$= 6.0 \left(\text{kN/m}^2 \right)^2$$

This can also be evaluated from the expression derived for J_{2D} in the preceding section [Eq. (4-6g)] as

$$J_{2D} = J_2 - \tfrac{1}{6}J_1^2$$

$$= 19.5 - \tfrac{1}{6}(9^2) = 6.0 \left(kN/m^2\right)^2$$

The third invariant can be computed by using Eq. (4-4c) or (A2-14e) as

$$J_{3D} = \bar{I}_{3s} = \tfrac{1}{3}\left(I_{1s}^3 - 3I_{1s}I_{2s} + 3I_{3s}\right)$$

$$= \tfrac{1}{3}[0 - 0 + 3(4 - 1)]$$

$$= 3.0 \left(kN/m^2\right)^3$$

The value of J_{3D} can also be found from Eq. (4-6j) as

$$J_{3D} = J_3 - \tfrac{2}{3}J_1 J_2 + \tfrac{2}{27}J_1^3$$

$$= 66 - \tfrac{2}{3}(9)(19.5) + \tfrac{2}{27}(9)^3$$

$$= 3.0 \left(kN/m^2\right)^3$$

Stress on a Plane

Consider a state of stress at a point represented by the stress tensor σ_{ij}.

$$\sigma_{ij} = \begin{pmatrix} \sigma_{11} & \sigma_{12} & \sigma_{13} \\ \sigma_{21} & \sigma_{22} & \sigma_{23} \\ \sigma_{31} & \sigma_{32} & \sigma_{33} \end{pmatrix} \tag{4-7a}$$

The stress vector on any plane passing through this point can be found by considering the equilibrium of forces [Fig. 4-3(a)]. The stress vector on the plane ABC is given by the expression

$$\sigma_i^P = \sigma_{ji} l_j \tag{4-7b}$$

where l_j ($j = 1, 2, 3$) are the direction cosines corresponding to the unit normal of the plane.

The components of stresses on the plane ABC according to Eq. (4-7b) can be written in expanded form as

$$\sigma_1^P = \sigma_{11} l_1 + \sigma_{12} l_2 + \sigma_{13} l_3$$
$$\sigma_2^P = \sigma_{12} l_1 + \sigma_{22} l_2 + \sigma_{23} l_3 \tag{4-7c}$$
$$\sigma_3^P = \sigma_{13} l_1 + \sigma_{23} l_2 + \sigma_{33} l_3$$

The superscript denotes the plane ABC. Later, we have used σ_1, σ_2, and σ_3 to denote principal stresses. In the foregoing section, σ_i ($i = 1, 2, 3$) denote the components of stresses on plane ABC in the x_1, x_2, and x_3 directions. They are expressed in terms of the components of the stress tensor σ_{ij}, or the stress vector

$$\{\sigma\}^T = \begin{bmatrix} \sigma_{11} & \sigma_{22} & \sigma_{33} & \sigma_{12} & \sigma_{23} & \sigma_{13} \end{bmatrix} \tag{4-7d}$$

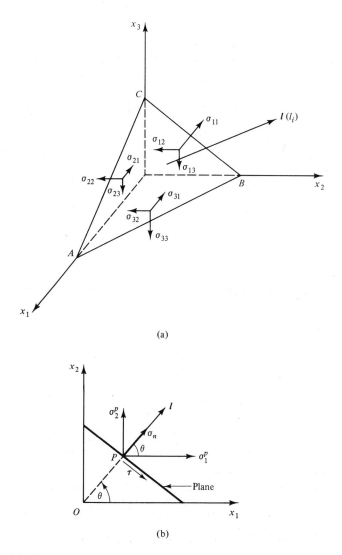

(a)

(b)

Figure 4-3 Stresses on a tetrahedral element: (a) stresses on a plane; (b) stresses on a two-dimensional plane.

Example 4-3

Consider the two-dimensional state of stress at point P [Fig. 4-3(b)].

$$\sigma_{ij} = \begin{pmatrix} 2 & 1 & 0 \\ 1 & 4 & 0 \\ 0 & 0 & 0 \end{pmatrix} \qquad (4\text{-}8a)$$

Let us find the stress vector on a plane whose normal makes an angle of θ with the

x-axis. The direction cosines are given by

$$l_1 = \cos \theta$$
$$l_2 = \sin \theta \qquad\qquad (4\text{-}8b)$$
$$l_3 = 0.0$$

Therefore, by substituting the values in Eq. (4-8a) in Eq. (4-7c), we have the following relationships:

$$\sigma_1^P = \sigma_{11} \cos \theta + \sigma_{12} \sin \theta = 2 \cos \theta + \sin \theta$$
$$\sigma_2^P = \sigma_{21} \cos \theta + \sigma_{22} \sin \theta = \cos \theta + 4 \sin \theta \qquad (4\text{-}8c)$$
$$\sigma_3^P = 0.0$$

The normal stress on the plane can be found by solving the foregoing stresses in the normal direction; that is,

$$\sigma_n^P = \sigma_1^P \cos \theta + \sigma_2^P \sin \theta$$
$$= \sigma_{11} \cos^2 \theta + \sigma_{22} \sin^2 \theta + 2\sigma_{12} \cos \theta \sin \theta \qquad (4\text{-}8d)$$

In a similar way, the shear stress τ^P on the plane can be calculated; its expression can also be obtained by using the concept of the Mohr circle of stress:

$$\tau^P = \sigma_1^P \sin \theta - \sigma_2^P \cos \theta$$
$$= \sigma_{12}(\sin^2 \theta - \cos^2 \theta) + \cos \theta \sin \theta (\sigma_{11} - \sigma_{22}) \qquad (4\text{-}8e)$$

PRINCIPAL STRESSES

When the stress tensor is known at a point, it is always possible to find a set of planes on which the stress vectors are normal to them. On these planes the shear stresses are zero. These are called the *principal planes*, and the direction of their normal vectors are called the *principal directions* or *principal axes* of stress. It is possible to find three such planes. The normal stresses on those three principal planes are called *principal stresses* and can be expressed as follows by introducing a scalar quantity λ:

$$\sigma_i = \lambda l_i, \qquad i = 1, 2, 3 \qquad (4\text{-}9a)$$

where l_i is the unit normal vector to the principal plane. By using Eq. (4-7b), σ_i can also be expressed as

$$\sigma_i = \lambda l_i = \lambda \delta_{ij} l_j = \sigma_{ji} l_j \qquad (4\text{-}9b)$$

Therefore,

$$(\sigma_{ji} - \lambda \delta_{ij}) l_j = 0 \qquad (4\text{-}9c)$$

By using the symmetric property of the stress tensor, that is, $\sigma_{ij} = \sigma_{ji}$, we have

$$(\sigma_{ij} - \lambda \delta_{ij}) l_j = 0 \qquad (4\text{-}10a)$$

or in matrix notation,

$$[[\sigma] - \lambda[\mathbf{I}]]\{l\} = 0 \qquad (4\text{-}10b)$$

where I is the identity matrix given by

$$[\mathbf{I}] = \begin{bmatrix} 1 & 0 & 0 \\ 0 & 1 & 0 \\ 0 & 0 & 1 \end{bmatrix} \qquad (4\text{-}10c)$$

and

$$\{l\}^{\mathrm{T}} = \begin{bmatrix} l_1 & l_2 & l_3 \end{bmatrix}$$

Equation (4-10a) represents a set of three linear homogeneous algebraic equations corresponding to the direction cosines l_1, l_2, and l_3. Equation (4-10b) can be written in algebraic form as

$$\begin{aligned} (\sigma_{11} - \lambda)l_1 + \sigma_{12}l_2 + \sigma_{13}l_3 &= 0 \\ \sigma_{12}l_1 + (\sigma_{22} - \lambda)l_2 + \sigma_{23}l_3 &= 0 \\ \sigma_{13}l_1 + \sigma_{23}l_2 + (\sigma_{33} - \lambda)l_3 &= 0 \end{aligned} \qquad (4\text{-}11a)$$

This can be expressed in matrix form as

$$\begin{bmatrix} \sigma_{11} - \lambda & \sigma_{12} & \sigma_{13} \\ \sigma_{12} & \sigma_{22} - \lambda & \sigma_{23} \\ \sigma_{13} & \sigma_{23} & \sigma_{33} - \lambda \end{bmatrix} \begin{Bmatrix} l_1 \\ l_2 \\ l_3 \end{Bmatrix} = 0 \qquad (4\text{-}11b)$$

It should be noted that the following relation also holds:

$$l_i l_i = 1 \qquad (4\text{-}12a)$$

which can be expanded as

$$l_1^2 + l_2^2 + l_3^2 = 1 \qquad (4\text{-}12b)$$

Since Eq. (4-12) has to be satisfied, all three components l_1, l_2, and l_3 cannot be zero. Therefore, for a nontrivial solution, the determinant of coefficients of the matrix in Eq. (4-11b) has to be zero; that is,

$$|\sigma_{ij} - \lambda\delta_{ij}| = 0 \qquad (4\text{-}13a)$$

or

$$\begin{vmatrix} \sigma_{11} - \lambda & \sigma_{12} & \sigma_{13} \\ \sigma_{12} & \sigma_{22} - \lambda & \sigma_{23} \\ \sigma_{13} & \sigma_{23} & \sigma_{33} - \lambda \end{vmatrix} = 0 \qquad (4\text{-}13b)$$

This reduces to a cubic equation in λ, and is known as the characteristic equation (see also Appendix 2). This can be written as

$$\lambda^3 - I_{1\sigma}\lambda^2 + I_{2\sigma}\lambda - I_{3\sigma} = 0 \qquad (4\text{-}14)$$

It can be shown that this equation has three real roots. These three values are the three principal stresses, σ_i $(i = 1, 2, 3)$. Their corresponding directions can be found by back substituting the values of λ in Eqs. (4-11) and (4-12) for l_i, that is, l_1, l_2, and l_3.

If the principal directions are used as the axes of reference, the stress tensor takes the form

$$\sigma_{ij} = \begin{pmatrix} \sigma_1 & 0 & 0 \\ 0 & \sigma_2 & 0 \\ 0 & 0 & \sigma_3 \end{pmatrix} \tag{4-15}$$

It should be noted that Eq. (4-15) does not necessarily imply that $\sigma_1 > \sigma_2 > \sigma_3$. The largest principal value is called the *major principal stress*, and the smallest value is called the *minor principal stress*. The other value is known as the *intermediate principal stress*.

Example 4-4

Find the principal stresses and the principal stress directions at point P [Fig. 4-2(b)] if the stress tensor at the point is given by

$$\sigma_{ij} = \begin{pmatrix} 4 & -2 & 0 \\ -2 & 4 & 0 \\ 0 & 0 & 5 \end{pmatrix} \quad kN/m^2$$

First we compute the invariants:

$$I_{1\sigma} = 13 \ (kN/m^2)$$

$$I_{2\sigma} = 12 + 20 + 20 = 52 \ (kN/m^2)^2$$

$$I_{3\sigma} = 60 \ (kN/m^2)^3$$

The characteristic equation is

$$\sigma^3 - 13\sigma^2 + 52\sigma - 60 = 0 \tag{4-16a}$$

or

$$\sigma^3 - 5\sigma^2 - 8\sigma^2 + 40\sigma + 12\sigma - 60 = 0 \tag{4-16b}$$

or

$$\sigma^2(\sigma - 5) - 8\sigma(\sigma - 5) + 12(\sigma - 5) = 0 \tag{4-16c}$$

or

$$(\sigma - 5)(\sigma - 6)(\sigma - 2) = 0 \tag{4-16d}$$

Therefore, we obtain its roots as

$$\sigma = 6.0, 5.0, \text{ and } 2.0$$

Hence

　　　Major principal stress:　　　　6.0 (kN/m^2)
　　　Minor principal stress:　　　　2.0 (kN/m^2)
　　　Intermediate principal stress:　5.0 (kN/m^2)

In order to find principal stress directions, let us consider them separately.

Major principal stress direction: $\sigma_1 = 6.0$: From Eq. (4-11a), we have

$$(4-6)l_1 - 2l_2 = 0 \tag{4-17a}$$

$$-(2)l_1 + (4-6)l_2 = 0 \tag{4-17b}$$

$$(5-6)l_3 = 0 \tag{4-17c}$$

and also

$$l_1^2 + l_2^2 + l_3^2 = 1.0 \tag{4-17d}$$

From Eq. (4-17c), $l_3 = 0$. From Eq. (4-17a) or (4-17b), $l_1 = -l_2$. Substituting in Eq. (4-17d), we have

$$2l_1^2 = 1.0$$

Therefore,

$$l_1 = \pm \frac{1}{\sqrt{2}}$$

$$l_2 = \mp \frac{1}{\sqrt{2}}$$

Then the major principal stress direction is given by

$$l_1 = \frac{1}{\sqrt{2}}, \qquad l_2 = -\frac{1}{\sqrt{2}}, \qquad l_3 = 0$$

or

$$l_1 = -\frac{1}{\sqrt{2}}, \qquad l_2 = \frac{1}{\sqrt{2}}, \qquad l_3 = 0$$

The second set of values refer to negative directions of the principal stresses. Here we shall use only the first for our illustration.

Minor principal stress direction: $\sigma_3 = 2.0$: Equations (4-11a) and (4-12b) can be simplified as

$$2l_1 - 2l_2 = 0 \tag{4-18a}$$

$$-2l_1 + 2l_2 = 0 \tag{4-18b}$$

$$3l_3 = 0 \tag{4-18c}$$

$$l_1^2 + l_2^2 + l_3^2 = 1 \tag{4-18d}$$

Hence $l_3 = 0$ and $l_1 = l_2$. Substituting these values in Eq. (4-18d), we have

$$2l_1^2 = 1$$

$$l_1 = \pm \frac{1}{\sqrt{2}}$$

$$l_2 = \pm \frac{1}{\sqrt{2}}$$

The minor principal stress direction is given by

$$l_1 = \frac{1}{\sqrt{2}}, \qquad l_2 = \frac{1}{\sqrt{2}}, \qquad l_3 = 0$$

or

$$l_1 = -\frac{1}{\sqrt{2}}, \qquad l_2 = -\frac{1}{\sqrt{2}}, \qquad l_3 = 0$$

Intermediate principal stress direction: $\sigma_2 = 5.0$: For this case, Eqs. (4-11a) and (4-12b) can be simplified as

$$-l_1 - 2l_2 = 0 \tag{4-19a}$$

$$-2l_1 - l_2 = 0 \tag{4-19b}$$

$$0.l_3 = 0 \tag{4-19c}$$

$$l_1^2 + l_2^2 + l_3^2 = 1.0 \tag{4-19d}$$

Here $l_1 = l_2 = 0$ and $l_3 = \pm 1.0$. The intermediate principal stress direction is given by

$$l_1 = 0 = l_2, \qquad l_3 = +1$$

or

$$l_1 = l_2 = 0, \qquad l_3 = -1$$

Therefore, the direction cosines of the principal stress directions can be arranged in the following way:

	x_1	x_2	x_3
Major x_1'	$\dfrac{1}{\sqrt{2}}$	$-\dfrac{1}{\sqrt{2}}$	0
Minor x_2'	$\dfrac{1}{\sqrt{2}}$	$\dfrac{1}{\sqrt{2}}$	0
Intermediate x_3'	0	0	1

To check the form of the stress tensor, when the principal stress directions are used as the reference axes, a transformation can be performed. The transformation tensor a_{ij} can be written as follows:

$$a_{ij} = \begin{pmatrix} \dfrac{1}{\sqrt{2}} & -\dfrac{1}{\sqrt{2}} & 0 \\ \dfrac{1}{\sqrt{2}} & \dfrac{1}{\sqrt{2}} & 0 \\ 0 & 0 & 1 \end{pmatrix}$$

The stress tensor, σ_{ij}', in the new reference frame is given as follows:

$$\sigma_{ij}' = a_{im} a_{jn} \sigma_{mn} \tag{4-20a}$$

or in matrix notation

$$[\sigma]' = [a][\sigma][a]^T$$

$$= \begin{bmatrix} \dfrac{1}{\sqrt{2}} & -\dfrac{1}{\sqrt{2}} & 0 \\[2mm] \dfrac{1}{\sqrt{2}} & \dfrac{1}{\sqrt{2}} & 0 \\[2mm] 0 & 0 & 1 \end{bmatrix} \begin{bmatrix} 4 & -2 & 0 \\ -2 & 4 & 0 \\ 0 & 0 & 5 \end{bmatrix} \begin{bmatrix} \dfrac{1}{\sqrt{2}} & \dfrac{1}{\sqrt{2}} & 0 \\[2mm] -\dfrac{1}{\sqrt{2}} & \dfrac{1}{\sqrt{2}} & 0 \\[2mm] 0 & 0 & 1 \end{bmatrix}$$

$$= \begin{bmatrix} \dfrac{6}{\sqrt{2}} & -\dfrac{6}{\sqrt{2}} & 0 \\[2mm] \dfrac{2}{\sqrt{2}} & \dfrac{2}{\sqrt{2}} & 0 \\[2mm] 0 & 0 & 5 \end{bmatrix} \begin{bmatrix} \dfrac{1}{\sqrt{2}} & \dfrac{1}{\sqrt{2}} & 0 \\[2mm] -\dfrac{1}{\sqrt{2}} & \dfrac{1}{\sqrt{2}} & 0 \\[2mm] 0 & 0 & 1 \end{bmatrix} \tag{4-20b}$$

$$[\sigma_{ij}]' = \begin{bmatrix} 6 & 0 & 0 \\ 0 & 2 & 0 \\ 0 & 0 & 5 \end{bmatrix} = \begin{bmatrix} \sigma_1 & 0 & 0 \\ 0 & \sigma_3 & 0 \\ 0 & 0 & \sigma_2 \end{bmatrix} \tag{4-20c}$$

This result is the same as that obtained by solving the characteristic equation. It is interesting to note that the principal stresses are arranged in the same way as the principal directions were arranged.

Shear stresses: As described earlier, the shear stresses on the principal planes are zero. By considering the stresses on the tetrahedron [Fig. 4-3(a)], we have

$$\sigma_i^P = \sigma_{ji} l_j \tag{4-7b}$$

or

$$\sigma_x^P = \sigma_{xx} l_x + \sigma_{xy} l_y + \sigma_{xz} l_z$$

$$\sigma_y^P = \sigma_{xy} l_x + \sigma_{yy} l_y + \sigma_{yz} l_z$$

$$\sigma_z^P = \sigma_{xz} l_x + \sigma_{yz} l_y + \sigma_{zz} l_z$$

where l_x, l_y, and l_z, are the direction cosines of the unit normal vector of the plane, in x, y, and z directions, respectively. The normal stress on the plane considered can be written as

$$\sigma_n = l_x \sigma_x^P + l_y \sigma_y^P + l_z \sigma_z^P \tag{4-21a}$$

Substituting for σ_x^P, σ_y^P, and σ_z^P, we obtain

$$\sigma_n = l_x^2 \sigma_{xx} + l_y^2 \sigma_{yy} + l_z^2 \sigma_{zz} + 2\left(l_x l_y \sigma_{xy} + l_y l_z \sigma_{yz} + l_z l_x \sigma_{zx}\right) \tag{4-21b}$$

The resultant shear stress σ_s on this surface can be obtained as follows:

$$\sigma_n^2 + \sigma_s^2 = \left(\sigma_x^P\right)^2 + \left(\sigma_y^P\right)^2 + \left(\sigma_z^P\right)^2 \tag{4-22a}$$

Therefore,

$$\sigma_s^2 = \left(\sigma_x^p\right)^2 + \left(\sigma_y^p\right)^2 + \left(\sigma_z^p\right)^2 - \sigma_n^2 \qquad (4\text{-}22\text{b})$$

Let us consider a special case where the coordinate axes and principal axes coincide. The shear components of the stress tensor will be zero for this case. Therefore,

$$
\begin{array}{lll}
\sigma_x^p = l_x \sigma_{xx} & \text{or} & \sigma_1^p = l_1 \sigma_{11} \\
\sigma_y^p = l_y \sigma_{yy} & \text{or} & \sigma_2^p = l_2 \sigma_{22} \\
\sigma_z^p = l_z \sigma_{zz} & \text{or} & \sigma_3^p = l_3 \sigma_{33}
\end{array}
\qquad (4\text{-}23\text{a})
$$

and

$$\sigma_n = l_1^2 \sigma_{11} + l_2^2 \sigma_{22} + l_3^2 \sigma_{33} \qquad (4\text{-}23\text{b})$$

Substituting the values above in Eq. (4-22b) for shear stress,

$$\sigma_s^2 = l_1^2 \sigma_{11}^2 + l_2^2 \sigma_{22}^2 + l_3^2 \sigma_{33}^2 - \left(l_1^2 \sigma_{11} + l_2^2 \sigma_{22} - l_3^2 \sigma_{33}\right)^2 \qquad (4\text{-}23\text{c})$$

It can be shown that the maximum shear stress occurs on certain planes on which l_1, l_2, and l_3 will have the following values:

	Plane 1	Plane 2	Plane 3
l_1	0	$\pm\sqrt{\dfrac{1}{2}}$	$\pm\sqrt{\dfrac{1}{2}}$
l_2	$\pm\sqrt{\dfrac{1}{2}}$	0	$\pm\sqrt{\dfrac{1}{2}}$
l_3	$\pm\sqrt{\dfrac{1}{2}}$	$\pm\sqrt{\dfrac{1}{2}}$	0

These planes bisect the angles between the coordinate axes. Substituting the values above in Eq. (4-23c), we have

$$
\begin{aligned}
\sigma_{s1} &= \pm \tfrac{1}{2}\left(\sigma_{22} - \sigma_{33}\right) \\
\sigma_{s2} &= \pm \tfrac{1}{2}\left(\sigma_{11} - \sigma_{33}\right) \\
\sigma_{s3} &= \pm \tfrac{1}{2}\left(\sigma_{11} - \sigma_{22}\right)
\end{aligned}
\qquad (4\text{-}24)
$$

where σ_{s1}, σ_{s2}, and σ_{s3} represent the maximum shear stress values on the three planes above, respectively.

The normal stresses on the planes above can be computed by using Eq. (4-23b):

$$\sigma_{n1} = \tfrac{1}{2}\left(\sigma_{22} + \sigma_{33}\right) \qquad (4\text{-}25\text{a})$$

$$\sigma_{n2} = \tfrac{1}{2}\left(\sigma_{11} + \sigma_{33}\right) \qquad (4\text{-}25\text{b})$$

$$\sigma_{n3} = \tfrac{1}{2}\left(\sigma_{11} + \sigma_{22}\right) \qquad (4\text{-}25\text{c})$$

so that the normal stress on each of these planes is equal to the average of the principal stresses on the two planes whose angle it bisects.

STATES OF STRESS FOR VARIOUS MODES
OF DEFORMATION

One of the important aspects of constitutive modeling is the determination of parameters from laboratory tests. As will be seen in subsequent chapters, it is often convenient to consider specific states of stress in the determination of parameters. These states of stress contain information regarding the stress path followed (Chapter 5). In Fig. 4-4, some specific stress states are shown; further details are given in Chapter 5.

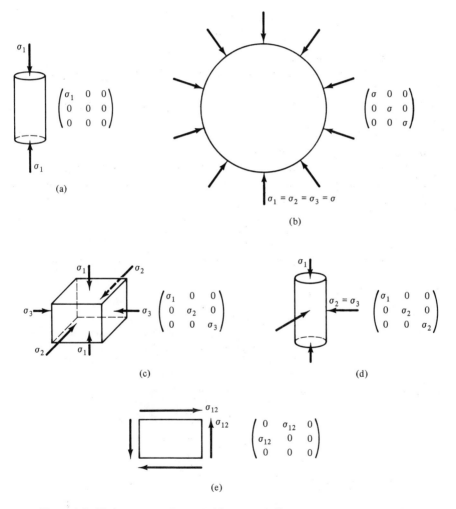

Figure 4-4 Various states of stress with corresponding stress tensor: (a) uniaxial state of stress; (b) hydrostatic state of stress; (c) triaxial state of stress; (d) cylindrical state of stress; (e) pure shear state of stress.

In this chapter we have described briefly stress and relevant quantities that are used in this text. For additional information, the reader may consult available literature, such as Refs. 2 to 6.

PROBLEMS

4-1. Find the values of $I_{1\sigma}, I_{2\sigma}, I_{3\sigma}, J_1, J_2, J_3, J_{2D}, J_{3D}, J_{1p}, J_{2p},$ and J_{3p} for the (stress) tensor

$$\sigma_{ij} = \begin{pmatrix} 4 & 3 & 2 \\ 3 & 5 & 1 \\ 2 & 1 & 6 \end{pmatrix} \quad \text{kPa}$$

Here subscript p denotes hydrostatic stress tensor.
Answers: $I_{1\sigma} = 15$ kPa; $I_{2\sigma} = 60$ (kPa)^2; $I_{3\sigma} = 54$ (kPa)^3; $J_1 = 15$ kPa; $J_2 = 52.5$ (kPa)^2; $J_3 = 275$ (kPa)^3; $J_{2D} = 15$ (kPa)^2; $J_{3D} = 0.0$ (kPa)^3; $J_{1p} = 15$ kPa; $J_{2p} = 37.5$ (kPa)^2; $J_{3p} = 125$ (kPa)^3.

REFERENCES

1. Terzaghi, K., and Peck, R. B., *Soil Mechanics in Engineering Practice*, 2nd ed., John Wiley & Sons, Inc., New York, 1967.

2. Timoshenko, S. P., and Goodier, J. N., *Theory of Elasticity*, 3rd ed., McGraw-Hill Book Company, New York, 1970.

3. Frederick, D., and Chang, T. S., *Continuum Mechanics*, Allyn and Bacon, Inc., Boston, 1963.

4. Fung, Y. C., *Foundations of Solid Mechanics*, Prentice-Hall, Inc., Englewood Cliffs, N.J., 1965.

5. Malvern, L. E., *Introduction to the Mechanics of a Continuous Medium*, Prentice-Hall, Inc., Englewood Cliffs, N.J., 1969.

6. Desai, C. S., and Christian, J. T. (Eds.), *Numerical Methods in Geotechnical Engineering*, McGraw-Hill Book Company, New York, 1977.

5

NONLINEAR BEHAVIOR, IMPORTANCE OF TESTING, AND TEST DATA

NONLINEAR BEHAVIOR

Very often, when we talk of nonlinear behavior we tend to visualize it as represented by a stress–strain curve (Fig. 5-1) that is *not linear* or *nonlinear*. Although such symbolic interpretations are always useful in a limited sense, nonlinearity generally implies something more. Nonlinearity may be defined as the failure of the magnitude of response to be proportional to the magnitude of excitation. There can be a number of factors that can contribute to nonlinearity, which for engineering bodies is usually manifested through large changes in the geometry and changes in material properties; the first is called *geometric* and the second *material* nonlinearity. Often, nonlinearity occurs as a mixture of both categories. In this text we are concerned primarily with *material nonlinearity*. Before we discuss various models for material nonlinearity in detail, let us first consider an explanation of nonlinear behavior.

Explanation of Nonlinear Behavior

Figure 5-2 shows a symbolic representation of the relation between a forcing function or excitation (load, Q) and a response (displacement, q). In incremental form, an approximation for this relation can be expressed as (1)

$$[\mathbf{k}_t]\{d\mathbf{q}\} = \{d\mathbf{Q}\} \tag{5-1}$$

where $[\mathbf{k}_t]$ denotes the material property (stiffness) matrix of the system, $\{d\mathbf{q}\}$ the incremental response (displacement) vector, and $\{d\mathbf{Q}\}$ the incremental

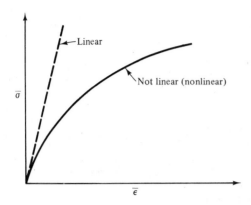

Figure 5-1 Symbolic stress–strain relation.

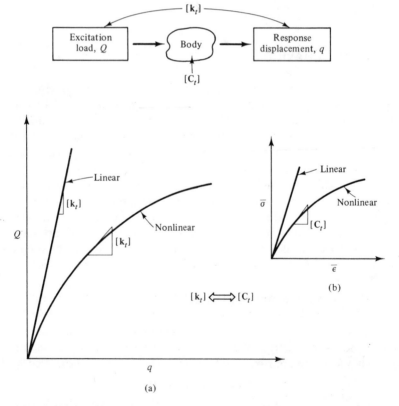

Figure 5-2 General representation of nonlinear behavior: (a) load–displacement behavior; (b) stress–strain behavior.

vector of forcing functions (loads). If the material were linear,

$$[\mathbf{k}_t] = [\mathbf{k}_i] \tag{5-2}$$

where $[\mathbf{k}_i]$ is the constant initial stiffness matrix at the initial stage of the body. For a linear material, $[\mathbf{k}_i]$ remains the same irrespective of the magnitude of stress (load).

In numerical (finite element) procedures, the stiffness matrix $[\mathbf{k}_t]$ is often expressed as (1, 2)

$$[\mathbf{k}_t] = \iiint_V [\mathbf{B}^T][\mathbf{C}_t][\mathbf{B}] \, dV \tag{5-3}$$

where $[\mathbf{B}]$ is the strain–displacement transformation matrix, and V is the volume of the region or element. Matrix $[\mathbf{C}_t]$ in Eq. (5-3) is called the *stress–strain* or *constitutive matrix*, and represents material behavior. It can be expressed as

$$\{d\boldsymbol{\sigma}\} = [\mathbf{C}_t]\{d\boldsymbol{\epsilon}\} \tag{5-4}$$

or

$$d\sigma_{ij} = C_{ijkl}^t \, d\epsilon_{kl} \tag{5-5}$$

where $\{d\boldsymbol{\sigma}\}$ and $\{d\boldsymbol{\epsilon}\}$ are vectors of incremental stress and strain, and C_{ijkl}^t is (tangent) constitutive tensor.

The implication in Eq. (5-4) is that the nonlinear load–displacement behavior in Eq. (5-1) contains as an integral part the material behavior expressed through the constitutive matrix $[\mathbf{C}]$. This is indicated in Fig. 5-2 in symbolic form. If the material is linear [Fig. 5-2(b)], the load–displacement relation is usually linear. If the material behavior is nonlinear, it renders the load–displacement behavior nonlinear. Thus when talking of nonlinear behavior, it may be understood that usually it is exhibited through a forcing function (load)–response (displacement) relationship. Hence when we talk of nonlinear behavior, it is appropriate to express it through a nonlinear load–displacement relation in conjunction with nonlinear constitutive law (Fig. 5-2).

Factors Causing Nonlinearity

One of the major factors causing nonlinearity is the state of stress (Figs. 5-1 and 5-2). As the level of loading increases or decreases, the material experiences continuous changes in its physical characteristics. At any instant during the changes in the load, the material is different than it was in the previous instant. In other words, the moduli or parameters that define the material stiffness experience continuous changes as the loading is changed. This can be expressed by writing

$$[\mathbf{k}_t] = [\mathbf{k}_t(C_t)] \tag{5-6}$$

$$[\mathbf{C}_t] = [\mathbf{C}_t(\sigma)] \tag{5-7}$$

It is the matrix $[\mathbf{C}_t]$ that is the main subject of this text. As we shall see later, the major aim of constitutive modeling will be to define $[\mathbf{C}_t]$ in such a way that it can be used in (numerical) solution techniques such as the finite element method.

Other factors.   In addition to the state of load or stress, there are a number of other factors that can cause nonlinearity. Chief among these are the initial or in situ stresses, physical state of material represented by density, water content or void ratio, type of loading: -static, dynamic, or repetitive, rate of loading, path of load or stress and existence, and/or occurrence of discontinuities such as joints, interfaces, fractures, and cracks; the emphasis in this text is on continuous media. Furthermore, although many of the constitutive models described here can be applied to cyclic loading, the main attention is given to static loading.

GENERAL STATEMENT OF CONSTITUTIVE LAWS

A general mathematical form, f, for constitutive equations or laws can be expressed as (3, 4)

$$f(\sigma, \dot{\sigma}, \epsilon, \dot{\epsilon}) = 0 \qquad (5\text{-}8)$$

where σ denotes stress, ϵ denotes (infinitesimal) strain, and the overdot denotes rate of change with respect to time.

Most useful constitutive laws or models for complex (geological) materials are obtained by using the incremental forms. Then relations between increments of stress and strain can be expressed as follows:

Stress–strain form:

$$\dot{\sigma}_{ij} = f_{ij}(\sigma_{kl})\dot{\epsilon}_{mn} \qquad (5\text{-}9a)$$

or

$$\dot{\sigma}_{ij} = C_{ijkl}(\sigma_{mn})\dot{\epsilon}_{kl} \qquad (5\text{-}9b)$$

Strain–stress or inverse form:

$$\dot{\epsilon}_{ij} = g_{ij}(\epsilon_{kl})\dot{\sigma}_{mn} \qquad (5\text{-}10a)$$

or

$$\dot{\epsilon}_{ij} = D_{ijkl}(\epsilon_{mn})\dot{\sigma}_{kl} \qquad (5\text{-}10b)$$

In this book we consider the foregoing equations homogeneous in time. Hence the quantities $\dot{\sigma}$ and $\dot{\epsilon}$ can be replaced by increments $d\sigma$ and $d\epsilon$, respectively. Then Eq. (5-9) can be expressed in the matrix notation as in Eq. (5-4):

$$\{d\sigma\} = [\mathbf{C}]\{d\epsilon\} \qquad (5\text{-}11)$$

or inversely from Eq. (5-10),

$$\{d\boldsymbol{\epsilon}\} = [\mathbf{D}]\{d\boldsymbol{\sigma}\} \qquad (5\text{-}12)$$

where C_{ijkl} or [C] and D_{ijkl} or [D] denote constitutive tensors or matrices, f_{ij} and g_{ij} are constitutive functions, σ_{ij} and ϵ_{ij} are the stress and strain tensors, respectively, and the overdot denotes an increment.

As stated earlier, formulation, determination, and implementation of C_{ijkl} or [C] and D_{ijkl} or [D] are the subjects of this text.

STEPS IN DEVELOPMENT OF A CONSTITUTIVE LAW

Development of a viable constitutive law for successful implementation in (numerical) solution techniques can be considered to consist of five main steps (5):

1. Mathematical formulation
2. Identification of significant parameters
3. Determination of parameters from laboratory tests, and verification, which can involve the following two additional steps:
4. Successful prediction of a majority of observed data from which the parameters were determined, and of other test data under different stress paths
5. Satisfactory comparisons between predictions from a solution scheme in which the constitutive law is introduced, and observations or closed-form solutions for relevant practical boundary value problems

A principal purpose of this text is to consider these five steps and illustrate them for various laws.

DETERMINATION OF CONSTITUTIVE PARAMETERS

Once a mathematical constitutive law consistent with the physical laws (Chapter 2) is derived, it is necessary to identify and choose all significant parameters that are needed to define it. Then it is essential to perform appropriate (laboratory) tests to evaluate the parameters. This is an extremely important step which is often not given the attention it deserves. It is our intention to emphasize and give comprehensive consideration to this step.

As we shall see in subsequent chapters, the most common way to identify parameters is to consider specialized forms of a general three-dimensional law, and then ascertain and devise testing configurations that are relevant to the specialization. For instance, hydrostatic, simple shear stress and strain, pure shear stress, one-dimensional stress and strain, and conventional triaxial states are some of the specializations that allow physical interpretations of the model(s).

In order to define each of the special states of stress or strain, we must consider tests with appropriate specimens and loading conditions for a given material. This brings us to the topic of laboratory testing, which is a vital step toward the development of a consistent constitutive law.

LABORATORY TEST DEVICES

Laboratory and/or field testing play a significant role in the development of constitutive laws of materials. Mathematical basis and development are indeed important; however, unless the parameters in the models are determined from appropriate tests, there will always be a gap between theory and practice. Hence development of appropriate models for the behavior of materials must include realistic tests that can simulate all significant factors that govern the behavior of the intended boundary value problem. In this section we describe a few of the conventional and recently developed test devices for solid (geological) materials.

Solid Materials

Uniaxial tests (6) [Fig. 5-3(a)]. A specimen of circular or square cross section is tested by applying axial stress symmetrically about the center line. The axial state of stress causes an extension (or compression) strain in the direction of the load and compression or extensional strain, respectively, in the

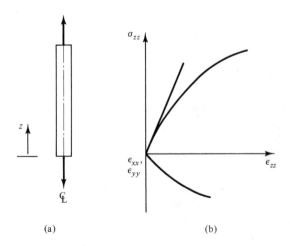

(a)   (b)

Figure 5-3   Uniaxial test.

other two directions. Figure 5-3(b) shows plots of measured axial stress, σ_{zz}, versus axial strain, ϵ_{zz}, and lateral strain $\epsilon_{xx}(\epsilon_{yy})$ versus ϵ_{zz}.

Cylindrical triaxial device (7) [Fig. 5-4(a)].   This device permits testing of a cylindrical specimen of a material. A confining stress (pressure) equal to $\sigma_3(=\sigma_2)$ is applied, usually by using a fluid in the chamber. In the case of geologic media, this confining pressure simulates initial or in situ stresses that exist at the site before a load increment is applied. The axial (σ_1) or deviator stress, $(\sigma_1 - \sigma_3)$, is then applied, which causes shearing of the sample.

Measurements are obtained in terms of the two stresses, σ_1, $\sigma_2 = \sigma_3$, axial deformation (strain), lateral or radial deformation (strain), and in the case of drained tests by allowing expulsion of fluid from the pores of the medium, the volume of the exiting liquid. These measurements allow plotting of results in various forms; typical schematic plots in terms of stress difference $(\sigma_1 - \sigma_3)$ versus axial strain (ϵ_1) and radial strain (ϵ_3) versus (ϵ_1) are shown in Fig.

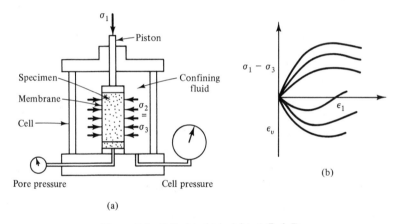

(a)

(b)

Figure 5-4   Cylindrical triaxial test (Ref. 7).

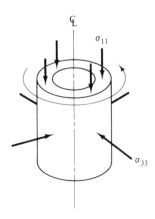

Figure 5-5   Hollow cylinder test (Ref. 8).

5-4(b). In the case of undrained tests, measurements of fluid pressure in the pores are also obtained.

Hollow cylinder tests (8, 9) (Fig. 5-5).   In this test (pure) shear stress is induced in a hollow cylindrical specimen subjected to a torque. Measurements are made for shear stress τ and shear strain γ.

Simple shear test (Fig. 5-6).   This test, which is often used for shear testing of soils, was devised by Roscoe (10). Here it is possible to obtain stress–strain response in terms of shear stress versus shear strain.

Direct shear device (11).   Figure 5-7 shows a schematic diagram of the direct shear device. A specimen with a square or circular cross section and thickness of about 1.0 in. (2.54 cm) is sheared along the plane at midsection. The results are plotted in terms of the shear stress on the plane and the relative movement of top and bottom halves of the box. The tests can be conducted under various normal stresses on the specimen.

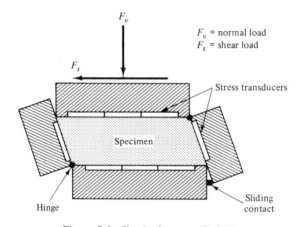

Figure 5-6   Simple shear test (Ref. 10).

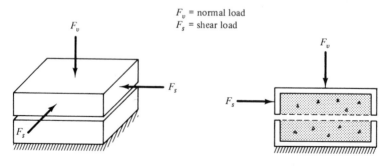

Figure 5-7   Direct shear test.

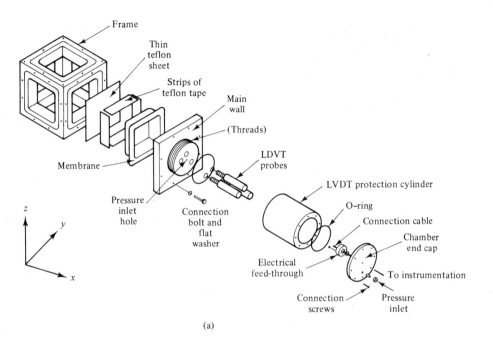

(a)

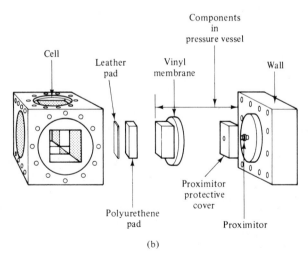

(b)

Figure 5-8 Truly triaxial or multiaxial devices: (a) capacity, 200 psi (1.38 MPa); (b) capacity, 20,000 psi (138 MPa) (Ref. 15, 16). (Copyright ASTM, reprinted with permission.)

Truly triaxial (TT) or multiaxial testing (12–16). The foregoing testing devices allow application of load and measurement of stress only in limited directions and modes of deformation. In reality, materials are subjected to three-dimensional states of stress and strain. The truly triaxial device (Fig. 5-8) permits application of three independent (principal) stresses, σ_1, σ_2, and σ_3, on six faces of a cubical specimen of a material. The application of the stresses can be such that any path of loading in the three-dimensional stress space can be followed. Testing under some of the common stress paths is discussed subsequently.

Development and use of truly triaxial testing is expected to play a significant role in research for development of constitutive laws of engineering materials. The test data for a soil included in this chapter and the test data for other materials in subsequent chapters were obtained by using the truly triaxial devices shown in Fig. 5-8. Reviews and details of truly triaxial testing are presented in Refs. 15 and 16.

Interfaces and Joints

Behavior at interfaces between two (dissimilar) media, such as concrete and soil or rock joints can be significantly different from that for the individual "solid" media. Under loading, an interface or joint can experience different modes, such as relative slip or debonding between two media, loss of contact, and then reestablishment of the lost contact. Influence of these modes in the behavior of an engineering structure can be important. Since this book is intended for materials assumed to be "continuous," details of modeling for interfaces and joints is beyond its scope. For such details the reader can consult various references (17–23).

TESTING PROGRAM

In order to evolve a constitutive model valid for all (significant) generalized states of loading and deformation, we must conduct a comprehensive series of tests with as many relevant test modes as possible. Here factors such as density, water content, (initial) anisotropy, stress path and type of loading—static, dynamic, and repetitive—can be important depending on the expected use of the model for practical boundary value problems. Hence it becomes necessary to perform appropriate tests on specimens of materials by including and by varying the foregoing factors. One of the significant factors is the stress path, which is discussed next.

Stress Paths

Figure 5-9 shows a practical boundary value problem of a structure embedded in a nonlinear half-space. Under axial and lateral load, different elements of the soil undergo different paths of loading or stress paths. As explained in subsequent chapters, one of the ways to describe stress path is to represent stress history on the shear (deviatoric) stress–mean pressure space.

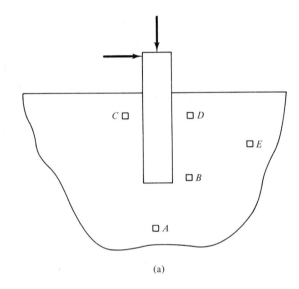

(a)

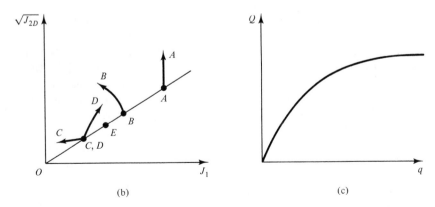

(b)   (c)

Figure 5-9 Typical stress paths followed by different points in the medium: (a) laterally loaded structure; (b) stress paths; (c) load–displacement curve.

In Fig. 5-9(b) we have used the stress space in terms of $\sqrt{J_{2D}}$, the second invariant of the deviatoric stress tensor S_{ij}, and J_1, the first invariant of the stress tensor σ_{ij}; the latter denotes a quantity that is related to (three times) the mean pressure.

Figure 5-9(b) shows a symbolic representation of the stress paths for a selected number of points in the half-space. Under the loading, point A may load in shear while its initial mean pressure may remain essentially constant. Point B may first load and then essentially unload in mean pressure with a small increase or decrease in the value of $\sqrt{J_{2D}}$; this stress relief can cause the phenomenon known as *arching*, particularly at higher loads [Fig. 5-9(c)]. Point C can unload in $\sqrt{J_{2D}}$ and J_1, and in fact, may experience tensile stress condition. The stress path followed by point D may indicate an arbitrary direction of loading, while point E, far away from the load, may not show any substantial change from the initial stresses.

The foregoing problem indicates that if we want to simulate and predict the behavior of the structure–soil system accurately, it is imperative that the constitutive law be valid for all the major stress paths that the system can experience in reality. Thus it is required that the specimens of the medium be tested under different stress paths.

The influence of the stress paths or stress history on the behavior of a material, and hence on the behavior of the relevant practical problems, is illustrated later in this chapter.

Details of Stress Paths

Figure 5-10 shows a schematic representation of various commonly considered stress paths in the principal stress space. They are shown in the triaxial and octahedral planes, and their definitions are given in the following section.

In practice, it is customary to use conventional (cylindrical) triaxial devices in studying the stress–strain behavior of geologic materials. In this apparatus two of the major principal stresses are always equal: that is, $\sigma_2 = \sigma_3$. Therefore, all possible stress paths followed in cylindrical triaxial device will fall on a plane in the stress space. This plane is called the *triaxial plane* and is shown in Fig. 5-10(a).

Figure 5-10(a) shows a projection of the space diagonal on the triaxial plane. Definitions and explanations of various forms, such as stress space, space diagonal, and octahedral planes, are given in Chapters 4 and 9.

In this text we include comprehensive test data for a number of materials tested with the foregoing devices; here the truly triaxial device for solid media is used extensively. In the following paragraphs explanations are given of the major stress paths as used with the truly triaxial device.

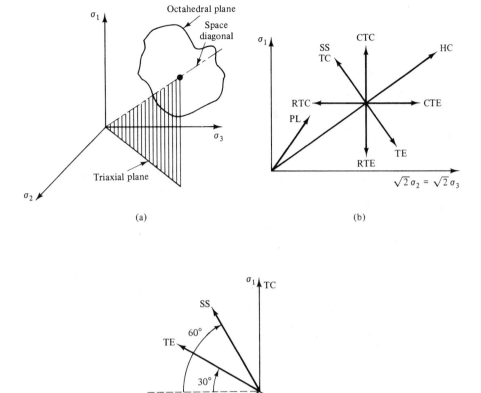

Figure 5-10 Representation of stress paths: (a) principal stress space; (b) projections of stress paths on triaxial plane; (c) stress paths in octahedral plane.

Hydrostatic compression (HC) stress path. A specimen starts from an initial hydrostatic or isotropic state of stress, $p_0 = \sigma_0$. Then it is subjected to increments of hydrostatic or mean pressure, $p = J_1/3$. Thus the loading occurs along the space diagonal [Fig. 5-10(a)].

The hydrostatic stress test provides information on the volumetric or bulk behavior of a medium; in this test for isotropic materials no shear stresses are induced. The observed test data from this test provide evaluation of the bulk modulus K and other information, such as the hardening parameter for various constitutive laws (Chapters 6 to 12).

HC and other tests are conducted with various initial physical states defined by initial density γ_0 and/or initial void ratio e_0. Measurements are obtained in terms of loads (stresses) and deformations (strains) as the loading progresses. In the case of porous media with fluid in the pores, measurements can also be made for pore water pressures.

Conventional triaxial compression (CTC).   This stress path is very often followed in the conventional cylindrical triaxial device (Fig. 5-4). Here the sample tested is in the form of a cylinder; hence two of the principal stresses, σ_2 and σ_3, are always equal. In the CTC stress path, the sample is subjected to an initial confining stress, $\sigma_0 = \sigma_1 = \sigma_2 = \sigma_3$. Then σ_2 and σ_3 are kept constant while σ_1 is increased [Fig. 5-10(b)]. Here σ_1 is the major principal stress and σ_2 and σ_3 are the intermediate and minor principal stresses, respectively. In the truly triaxial device (Fig. 5-8), the initial values of σ_2 and σ_3 are maintained at σ_0, and σ_1 is increased. The states of both shear stress and hydrostatic pressure change during loading. The increments of octahedral shear, $\Delta\tau_{oct}$, and octahedral normal or mean pressure, $\Delta\sigma_{oct} = \Delta p = \Delta J_1/3$, are given as

$$\Delta\tau_{oct} = \frac{\sqrt{2}}{3}\Delta\sigma_1 \tag{5-13a}$$

$$\Delta\sigma_{oct} = p = \frac{\Delta J_1}{3} = \frac{\Delta\sigma_1}{3} \tag{5-13b}$$

Here Δ denotes an increment. Equations (5-13) are special cases of the general definition of τ_{oct} and σ_{oct} given in Chapter 4:

$$\tau_{oct} = \frac{1}{3}\sqrt{\left(\sigma_1 - \sigma_2\right)^2 + \left(\sigma_2 - \sigma_3\right)^2 + \left(\sigma_1 - \sigma_3\right)^2} \tag{5-14a}$$

$$\sigma_{oct} = \frac{\sigma_1 + \sigma_2 + \sigma_3}{3} \tag{5-14b}$$

Reduced triaxial extension (RTE).   In the truly triaxial device, the sample is subjected to a given initial stress $\sigma_0 = \sigma_1 = \sigma_2 = \sigma_3$. Then σ_2 and σ_3 are held constant, whereas σ_1 is reduced. Here σ_2 and σ_3 are the major and intermediate principal stresses, and σ_1 becomes the minor principal stress.

The changes in the octahedral shear and normal stresses are essentially given by Eqs. (5-13a) and (5-13b); the former increases while the latter decreases. Since there is an increase in the shear stress, the path can be considered as loading rather than unloading in shear.

Conventional triaxial extension (CTE).   In the truly triaxial device, CTE is simulated by holding one of the stresses, say σ_1 constant while σ_2 and σ_3 are increased. It may be noted that here σ_1 becomes minor principal stress. The increase in the octahedral shear stress is given by Eq. (5-13a) with $\Delta\sigma_1$ replaced by $\Delta\sigma_2$, while the increase in the octahedral normal stress is given by

$$\Delta\sigma_{\text{oct}} = \tfrac{2}{3}\Delta\sigma_2 \qquad (5\text{-}15)$$

Reduced triaxial compression (RTC).   For this test with the truly triaxial device, σ_2 and σ_3 are reduced while σ_1 is held constant. As a consequence, the octahedral shear stress increases essentially according to Eq. (5-13a), while the octahedral normal stress decreases as

$$\Delta\sigma_{\text{oct}} = -\tfrac{2}{3}\Delta\sigma_3 \qquad (5\text{-}16)$$

Triaxial compression (TC) and triaxial extension (TE).   In TC and TE tests, the stress is applied such that we remain always on an octahedral plane [Fig. 5-10(a)]. In other words, during loading (unloading), the octahedral normal stress remains constant.

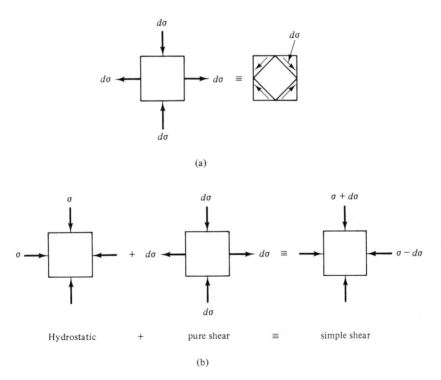

(a)

Hydrostatic   +   pure shear   ≡   simple shear

(b)

Figure 5-11   States of shear stress: (a) pure shear stress state; (b) simple shear stress state.

In the TC test, σ_1 is increased, whereas σ_2 and σ_3 are reduced such that σ_{oct} remains constant. In other words, if there is an increase in σ_1 of $\Delta\sigma_1$, σ_2 and σ_3 are decreased by equal amounts of $\Delta\sigma_1/2$. In the case of a triaxial extension (TE) test, σ_2 and σ_3 are increased but σ_1 is decreased such that σ_{oct} remains the same.

As noted earlier, the HC test isolates the volumetric behavior. The TC and TE tests, on the other hand, isolate the effects of shear or deviatoric behavior. Hence TC and TE tests, among other things, permit evaluation of shear moduli for a material.

The increment of the octahedral shear stress for TC and TE is given by

$$\Delta\tau_{oct} = \frac{1}{\sqrt{2}}\Delta\sigma_2 \qquad (5\text{-}17)$$

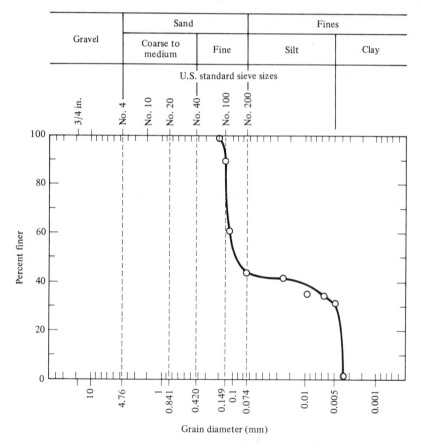

Figure 5-12   Grain size distribution for artificial soil (Refs. 24–28).

Simple shear. The SS stress path [Fig. 5-10(c)] is conducted in an octahedral plane; that is, the mean pressure is kept constant. This path is similar in principle to TC and TE paths except for the direction of the paths followed. In the SS test, one of the stresses, say σ_2, is held constant, while the other two, σ_1 and σ_3, are increased and decreased, respectively, by the same amount (Fig. 5-11). Hence $\Delta\sigma_1 = -\Delta\sigma_3$ and $\Delta\sigma_2 = 0$. The increase in the octahedral shear stress is given as

$$\Delta\tau_{\text{oct}} = \frac{\sqrt{2}}{\sqrt{3}}\Delta\sigma_1 \tag{5-18}$$

Proportional loading (PL). The proportional loading or constant obliquity tests are conducted such that the ratio of σ_1, σ_2, and σ_3 remain the same during the change in loading. Often, in order to identify the influence of the intermediate stress, say σ_2, tests are conducted under paths defined by the following ratio:

$$b = \frac{\sigma(\text{intermedia}) - \sigma(\text{minor})}{\sigma(\text{major}) - \sigma(\text{minor})} \tag{5-19}$$

This value indicates the relative magnitude of the intermediate principal stress. The value of b can vary from zero for the CTC path to unity for the CTE path.

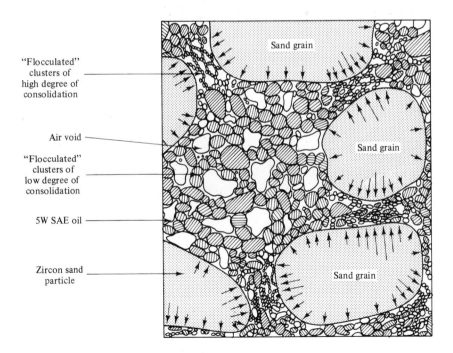

"Flocculated" clusters of high degree of consolidation

Air void

"Flocculated" clusters of low degree of consolidation

5W SAE oil

Zircon sand particle

Sand grain

Sand grain

Sand grain

Figure 5-13 Schematic representation of structure of artificial soil (Refs. 24, 28).

"STANDARD" TEST DATA

In order to define the constitutive laws and their parameters, test data for various materials are considered in subsequent chapters while discussing specific constitutive laws; then determination of the parameters for those laws are illustrated. In this chapter a set of "standard" data is included; the word "standard" is used in the sense that the same data are used as common for a number of constitutive models in various chapters. The standard test pertains to a material made of a sand, clay, and oil, called *artificial soil*.

Important note.   In this text we have employed laboratory test data from various devices described previously; however, a majority of data used are derived from the truly triaxial devices.

Very often the constitutive parameters are determined from conventional tests such as cylindrical triaxial and consolidation. In subsequent chapters we illustrate the use of data from all such devices for various constitutive laws.

TABLE 5-1   Tests for Artificial Soil

Stress Path	Initial Density, γ_0, lb/ft^3 (g/cm^3)	Initial Pressure, σ_0, psi (kPa)	Relevant Figure
HC	110 (1.76)	2.00 (13.80)	5-14(a)
	114 (1.82)	2.00 (13.80)	5-14(b)
	141 (2.25)	4.00 (27.60)	5-14(c)
CTC	101 (1.62)	1.00 (6.90)	
	103 (1.65)	2.00 (13.80)	5-15
	109 (1.75)	5.00 (34.50)	
RTE	117 (1.87)	10.00 (69.00)	5-16
CTE	123 (1.96)	10.00 (69.00)	5-17
RTC	125 (2.00)	10.00 (69.00)	5-18
TC	128 (2.04)	10.00 (69.00)	5-19(a)
	128 (2.04)	28.00 (193.20)	5-19(b)
TE	118 (1.89)	10.00 (69.00)	5-20
SS	109 (1.75)	15.00 (103.50)	5-21(a)
	128 (2.05)	10.00 (69.00)	5-21(b)
	141 (2.25)	10.00 (69.00)	5-21(c)
$b = 0.20$	125 (2.00)	10.00 (69.00)	5-22(a)
$= 0.80$	125 (2.00)	10.00 (69.00)	5-22(b)

For some cases, parameters determined from limited conventional test configurations are compared with those found from test results from a greater number of testing modes obtained by using TT devices. From a practical viewpoint, the philosophy should be to use a minimum number of testing modes consistent with the required accuracy. At the same time, for many boundary value problems involving a variety of stress paths, parameters found from limited number of stress paths may not be appropriate, and recourse must be made to truly triaxial devices that permit a wide range of stress paths.

Data for a Soil

Artificial soil is used for a number of recent research projects (24–28), and has been tested extensively by using truly triaxial, cylindrical triaxial, and direct shear test devices. The artificial soil consists of a mixture of fire clay (50%), and Florida zircon sand (50%). Ten percent of No. 5 SAE mineral oil is added to the mixture; oil is used as a binder to reduce the influence of "moisture loss" during tests for practical boundary value problems. The soil can be classified as sandy silt with a small amount of cohesion. It exhibits highly compressive characteristics with large changes in volume under shear stress. The soil has maximum and minimum densities of 2.65 and 1.00 g/cm^3, respectively. Figure 5-12 shows the grain size distribution for the artificial soil. A schematic representation of the fabric of the soil is shown in Fig. 5-13.

Table 5-1 shows the details of tests conducted with different stress paths. For many stress paths, more than one test was performed with different initial densities, γ_0, and mean pressure or confining stress, σ_0.

Figures 5-14 to 5-23 show results of typical tests under different stress paths obtained by using the truly triaxial device (15, 16). These test data are used to derive appropriate parameters for the soil for various hyperelastic, hypoelastic, classical plasticity, and advanced plasticity models (Chapters 6 to 12).

In the stress–strain data used in this and other chapters from three-dimensional tests on cubical specimens, the curves are plotted in terms of a measure of stress such as $(\sigma_1 - \sigma_3)$ and τ_{oct} versus principal strains labeled as $\epsilon_1, \epsilon_2, \epsilon_3$ or $\epsilon_x, \epsilon_y, \epsilon_z$. Depending on the stress path followed, the nature of major, minor, and intermediate strains will change. Table 5-2 shows the relation between $\epsilon_1, \epsilon_2, \epsilon_3$ and $\epsilon_x, \epsilon_y, \epsilon_z$ with respect to various stress paths.

Illustrations of the Influence of Stress Paths

Most geological materials are influenced by stress history or stress path. If such a medium is subjected to a change in state of stress, and brought to the same state of stress but through different stress paths, the (final) deformation can be different for the same state of stress. Figure 5-23 shows a number of

TABLE 5-2   Information on Testing Configuration

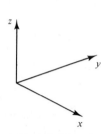

Physical coordinate
directions of testing
device

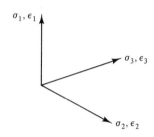

Indicial coordinate
directions of principal
stresses and strains*

Stress Path (Fig. 5-10)	Figure Showing Data	Corresponding Physical Direction*		
		1	2	3
HC	5-14(a)	z	x	y
HC	5-14(b)	z	x	y
HC	5-14(c)	z	x	y
CTC	5-15	z	x	y
RTE	5-16	z	x	y
CTE	5-17	z	x	y
RTC	5-18	z	x	y
TC	5-19(a)	z	x	y
TC	5-19(b)	z	x	y
TE	5-20	z	x	y
SS	5-21(a)	z	x	y
SS	5-21(b)	z	x	y
SS	5-21(c)	z	x	y
$b = 0.2$	5-22(a)	z	x	y

*Here, the directions denoted by 1, 2, 3 *do not* imply major, intermediate, and minor principal directions; they simply indicate reference axes used for presenting data in Figs. 5-14 to 5-22.

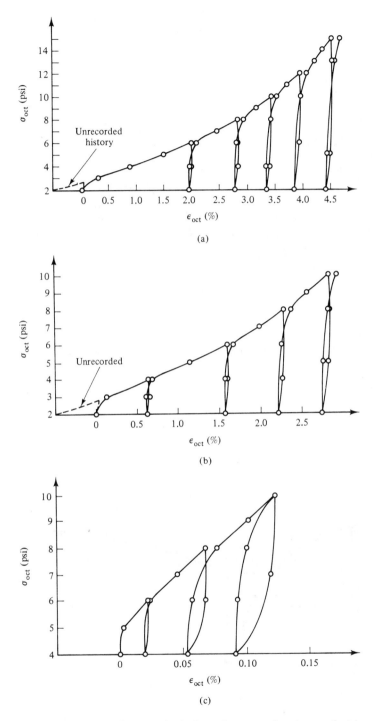

Figure 5-14 Stress–strain curves for hydrostatic compression stress path: (a) $\gamma_0 = 110 \text{ lb/ft}^3$, $\sigma_0 = 2$ psi; (b) $\gamma_0 = 114 \text{ lb/ft}^3$, $\sigma_0 = 2$ psi; (c) $\gamma_0 = 141 \text{ lb/ft}^3$, $\sigma_0 = 4$ psi.

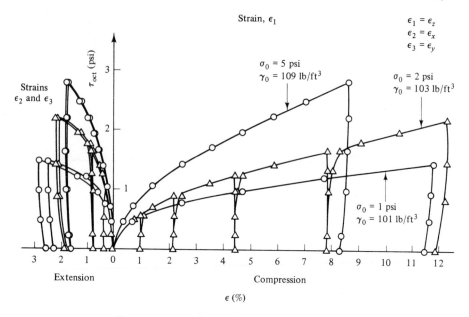

Figure 5-15   Stress–strain curves for CTC stress path.

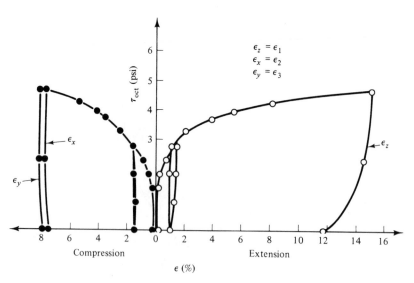

Figure 5-16   Stress–strain curves for reduced triaxial extension stress path: $\gamma_0 = 117$ lb/ft³, $\sigma_0 = 10$ psi.

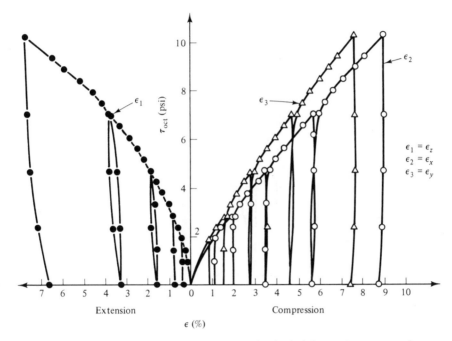

Figure 5-17 Stress–strain curves for conventional triaxial extension stress path: $\gamma_0 = 123 \ \mathrm{lb/ft^3}$, $\sigma_0 = 10$ psi.

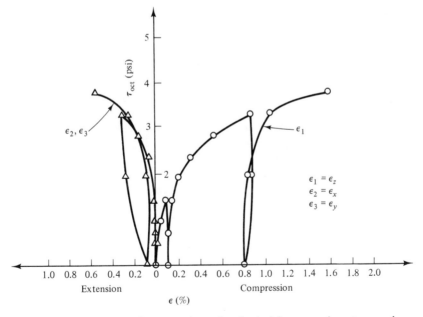

Figure 5-18 Stress–strain curves for reduced triaxial compression stress path: $\gamma_0 = 125 \ \mathrm{lb/ft^3}$, $\sigma_0 = 10$ psi.

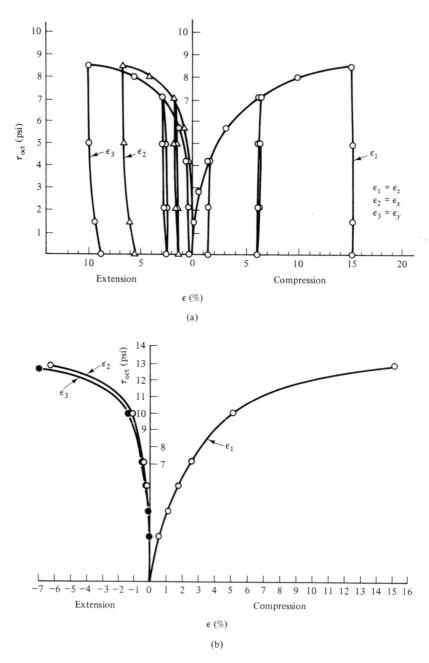

Figure 5-19 Stress–strain curves for triaxial compression stress path: (a) $\gamma_0 = 128$ lb/ft³, $\sigma_0 = 10$ psi; (b) $\gamma_0 = 128$ lb/ft³, $\sigma_0 = 28$ psi.

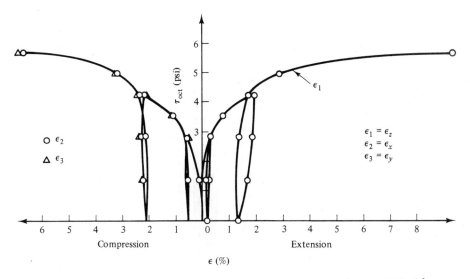

Figure 5-20   Stress–strain curves for triaxial extension stress path: $\gamma_0 = 118$ lb/ft^3, $\sigma_0 = 10$ psi.

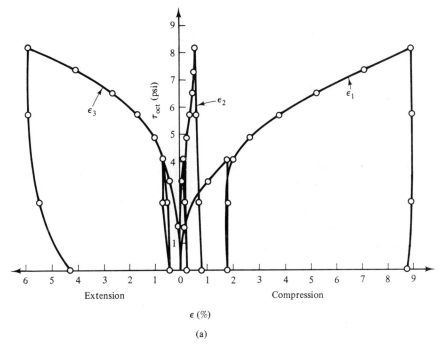

(a)

Figure 5-21   Stress–strain curves for simple shear stress path: (a) $\gamma_0 = 109$ lb/ft^3, $\sigma_0 = 15$ psi; (b) $\gamma_0 = 128$ lb/ft^3, $\sigma_0 = 10$ psi; (c) $\gamma_0 = 141$ lb/ft^3, $\sigma_0 = 10$ psi.

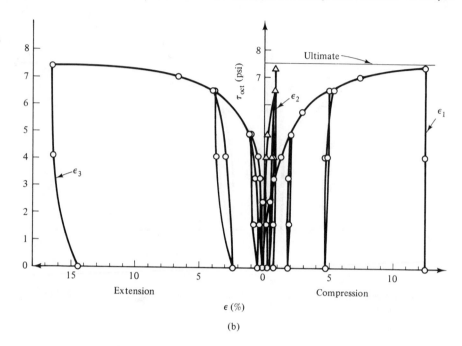

(b)

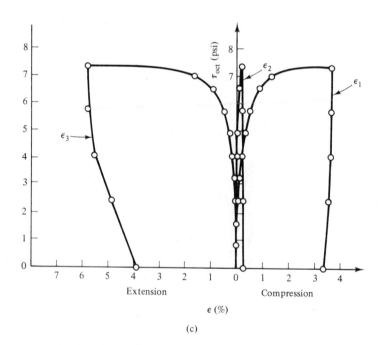

(c)

Figure 5-21 (*continued*)

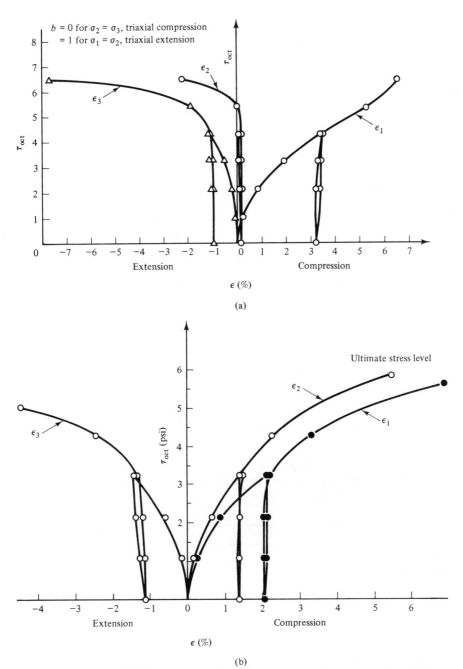

Figure 5-22 Stress–strain curves from PL stress path: (a) $b = 0.2$, $\gamma_0 = 125$ lb/ft^3, $\sigma_0 = 10$ psi; (b) $b = 0.8$, $\gamma_0 = 125$ lb/ft^3, $\sigma_0 = 10$ psi (Refs. 24, 27, 28).

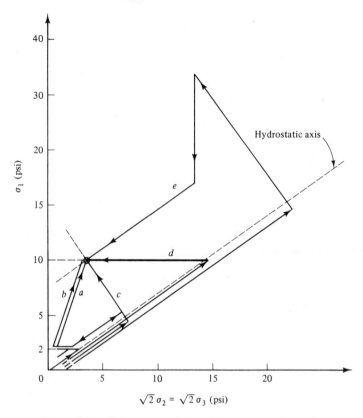

Figure 5-23   Some stress paths in triaxial plane (Refs. 27, 28).

stress paths under which the artificial soil ($\gamma_0 = 2.0$ g/cm^3) was tested (27, 28): the final state of stress reached in each case was $\sigma_1 = 10$ psi (69 kPa). The stress paths followed were:

Path	Symbol
PL	a, b
TC	c
RTC	d
Combination	e

In each test, the zero state of strain correspond to the hydrostatic condition immediately preceding application of the first increment of shear stress. The initial hydrostatic state of stress was $\sigma_0 = 2.00$ psi (13.80 kPa). It was found that the states of strains at the end of different stress paths were significantly different.

REFERENCES

1. Desai, C. S., and Abel, J. F., *Introduction to the Finite Element Method*, Van Nostrand Reinhold Company, New York, 1972.

2. Desai, C. S., and Christian, J. T. (Eds.), *Numerical Methods in Geotechnical Engineering*, McGraw-Hill Book Company, New York, 1977.

3. Eringen, A. C., *Nonlinear Theory of Continuous Media*, McGraw-Hill Book Company, New York, 1962.

4. Truesdell, C., *Continuum Mechanics—The Mechanical Foundations of Elasticity and Fluid Dynamics*, Vol. 1, Gordon and Breach, Science Publishers, Inc., New York, 1966.

5. Desai, C. S., Phan, H. V., and Sture, S., "Procedure, Selection and Application of Plasticity Models for a Soil," *Int. J. Numer. Anal. Methods Geomech.*, Vol. 5, 1981, pp. 295–311.

6. Timoshenko, S., and Young, D. H., *Elements of Strength of Materials*, D. Van Nostrand Company, Princeton, N.J., 1963.

7. Bishop, A., and Henkel, D. J., *The Triaxial Test*, Edward Arnold Ltd., London, 1962.

8. Hvorslev, M. J., and Kaufman, R. I., "Torsion Shear Apparatus and Testing Procedures," *Bull. No. 38*, U.S. Army Corps of Engineers Waterways Exp. Stn., Vicksburg, Miss., 1952.

9. Hvorslev, M. J., "Physical Components of the Shear Strength of Saturated Clays," *Proc. ASCE Res. Conf. Shear Strength Cohesive Soils*, Boulder, Colo., 1960, pp. 169–273.

10. Roscoe, K. H., "An Apparatus for the Application of Simple Shear to Soil Samples," *Proc. 3rd Int. Conf. Soil Mech. Found. Eng.*, Vol. I, 1953, p. 186.

11. "Laboratory Shear Testing of Soils," *ASTM Spec. Tech. Publ. No. 36*, ASTM, Philadelphia, 1963.

12. Ko, H. Y., and Scott, R. F., "A New Soil Testing Apparatus," *Geotechnique*, Vol. 17, No. 1, 1967, pp. 40–57.

13. Arthur, J. R. F., and Menzies, B. K., "Inherent Anisotropy in a Sand," *Geotechnique*, Vol. 22, No. 1, 1972, pp. 115–128.

14. Lade, P. V., "Cubical Triaxial Apparatus for Soil Testing," *Geotech. Test. J.*, Vol. 1, No. 2, June 1978, pp. 93–101.

15. Sture, S., and Desai, C. S., "Fluid Cushion Truly Triaxial or Multiaxial Testing Device," *Geotech. Test. J., ASTM*, Vol. 2, No. 1, Mar. 1979, pp. 20–33.

16. Desai, C. S., Janardhanam, R., and Sture, S., "High Capacity Truly Triaxial Multiaxial Device," *Geotech. Test. J., ASTM*, Vol. 5, No. 1/2, Mar. 1982, pp. 26–33.

17. Brummund, W. F., and Leonards, G. A., "Experimental Study of Static and Dynamic Friction between Sand and Typical Construction Materials," *ASTM J. Test. Eval.*, Vol. 1, No. 2, 1973, pp. 162–165.

18. Huck, P. J., et al., "Dynamic Response of Soil/Concrete Interfaces at High Pressure," *Rep. 1 AFWL-TR-73-264*, by IITRI, Chicago, to Defense Nuclear Agency, Washington, D.C., 1974.

19. Desai, C. S., "Soil–Structure Interaction and Simulation Problems," in *Finite Elements in Geomechanics*, G. Gudehus (Ed.), John Wiley & Sons, Ltd., Chichester, England, 1977, Chapter 7.

20. Desai, C. S., "Behavior of Interfaces between Structural and Geologic Media," State-of-the-Art Paper, *Int. Conf. Recent Adv. Geotech. Earthquake Eng. Soil Dyn.*, St. Louis, Mo., Apr., 1981.

21. Desai, C. S., "A Cyclic Multi-Degree-of-Freedom Shear Device," Report, Dept. of Civil Eng., Va. Polytech. Inst. and State Univ., Blacksburg, 1980.

22. Desai, C. S., Zaman, M. M., Lightner, J. G., and Siriwardane, H. J., "Thin-Layer Element for Interfaces and Joints," *Int. J. Numer. Anal. Methods Geomech.*, in press.

23. Desai, C. S., Zaman, M. M., and Drumm, E. C., "Cyclic Testing and Modelling of Interfaces," *Report*, Dept. Civil Eng. & Eng. Mech., Univ. of Arizona, Tucson, AZ, 1982.

24. Desai, C. S., Sture, S., and Perumpral, J. V., "Soil–Tool Interaction in Tillage," *Reports*, Dept. of Civil Eng., Va. Polytech. Inst. and State Univ., Blacksburg, 1978–1980.

25. Desai, C. S., "Some Aspects of Constitutive Laws of Geologic Media," *Proc. 3rd Int. Conf. Numer. Methods Geomech.*, Aachen, W. Germany, Vol. 1, W. Wittke (Ed.), Balkema Press, Rotterdam, 1979, pp. 299–308.

26. Sture, S., Desai, C. S., and Janardhanam, R., "Development of a Constitutive Law for an Artificial Soil," *Proc. 3rd Int. Conf. Numer. Methods Geomech.*, Aachen, W. Germany, W. Wittke (Ed.), Balkema Press, Rotterdam, 1979.

27. Mould, J., "Multiaxial Testing and Analytical Constitutive Characterization of Granular Materials," M.S. thesis, Va. Polytech. Inst. and State Univ., Blacksburg, 1979.

28. Sture, S., Desai, C. S., and Perumpral, J. V., "Laboratory Behavior of an Artificial Soil," *Report No. ENG VOI-E80.12*, Va. Polytech Inst. and State Univ. Blacksburg, 1980.

6

ELASTICITY: FIRST- AND SECOND-ORDER MODELS

In this chapter we consider two elasticity models: the Green and the Cauchy elastic. The existence of a strain energy function is assumed for the Green elastic materials; these are called *hyperelastic* materials. On the other hand, for Cauchy elastic materials, the stress is assumed to be a function of strain. Literally, "hyper" means over or beyond; hence hyperelastic can imply beyond or higher-order elastic. Here we consider only the first- and second-order models, and describe the determination of their material parameters based on experimental data.

ELASTIC MATERIAL

The internal mechanical response of a material can be expressed in terms of stresses and strains. For an elastic material, the state of stress is a function of the current state of deformation only. An elastic medium returns to its initial state after a cycle of loading and unloading. In other words, it retains no permanent strain (Fig. 6-1).

An elastic material in general can be nonlinear [Fig. 6-1(a)]. A special case is that of linear elastic behavior [Fig. 6-1(b)]. The linear elastic Hooke's law is the simplest example of a constitutive law; for uniaxial loading it can be expressed as

$$\sigma = E\epsilon \tag{6-1}$$

where σ is the stress, ϵ the strain, and E the response parameter commonly known as the Young's modulus.

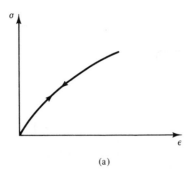

(a)

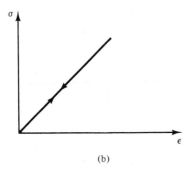

(b)

Figure 6-1 Elastic models: (a) nonlinear elastic; (b) linear elastic.

For three-dimensional bodies, the generalized Hooke's law can be expressed as (1, 2)

$$\sigma_{11} = C_{11}\epsilon_{11} + C_{12}\epsilon_{22} + C_{13}\epsilon_{33} + C_{14}\epsilon_{12} + C_{15}\epsilon_{23} + C_{16}\epsilon_{13}$$
$$\sigma_{22} = C_{21}\epsilon_{11} + C_{22}\epsilon_{22} + C_{23}\epsilon_{33} + C_{24}\epsilon_{12} + C_{25}\epsilon_{23} + C_{26}\epsilon_{13} \qquad \text{(6-2a)}$$
$$\cdots\cdots\cdots\cdots\cdots\cdots\cdots\cdots\cdots\cdots\cdots\cdots\cdots\cdots$$
$$\sigma_{13} = C_{61}\epsilon_{11} + C_{62}\epsilon_{22} + C_{63}\epsilon_{33} + C_{64}\epsilon_{12} + C_{65}\epsilon_{23} + C_{66}\epsilon_{13}$$

or

$$\sigma_{ij} = C_{ijkl}\epsilon_{kl} \qquad \text{(6-2b)}$$

or

$$\{\sigma\} = [\mathbf{C}]\{\epsilon\} \qquad \text{(6-2c)}$$

In general, the relation for nonlinear elastic law can be expressed as a unique relation between stress and strain as

$$\sigma_{ij} = f_{ij}(\epsilon_{kl}) \qquad \text{(6-3)}$$

where the f_{ij} are response functions.

Very often, basic laws and equations that govern problems in engineering are derived on the basis of the energy stored in a body. We talk of the potential of a body to perform work with reference to a datum configuration, and in the same sense, we talk of the potential of a force to perform work when it acts on a body. As the body deforms, it stores internal energy due to the strain or deformation. There is a definite relation between the potential of the force spent and the internal strain energy. The latter is usually related to the constitution of the medium, and is often used to derive constitutive or stress–strain laws. In this chapter we consider two such laws based on nonlinear elasticity. Eringen (3) has discussed them as the Cauchy and Green elastic models.

CAUCHY ELASTIC MODELS

This approach is based on the assumption that for an elastic material, the stress is a function of strain. Here the functional relationship of Eq. (6-3) is expressed as

$$\sigma_{ij} = \alpha_0 \delta_{ij} + \alpha_1 \epsilon_{ij} + \alpha_2 \epsilon_{im} \epsilon_{mj} + \alpha_3 \epsilon_{im} \epsilon_{mn} \epsilon_{nj} + \cdots \tag{6-4}$$

where $\alpha_0, \alpha_1, \alpha_2, \ldots, \alpha_n$ represent response functions or coefficients or parameters. In order to express Eq. (6-4) in terms of the three invariants of strain, we can make use of the Cayley–Hamilton (C-H) theorem (see Ref. 4 and Appendix 2). Then we can write a function expressed with terms including n powers of [a], simply in terms of [a] and $[a]^2$. Thus if the function is given by

$$f([a]) = \alpha_0 [I] + \alpha_1 [a] + \alpha_2 [a]^2 + \cdots + \alpha_n [a]^n \tag{6-5a}$$

it can be written as

$$f([a]) = \beta_0 [I] + \beta_1 [a] + \beta_2 [a]^2 \tag{6-5b}$$

where β_0, β_1, and β_2 are polynomial functions of the invariants $\bar{I}_{1a}$, $\bar{I}_{2a}$, and $\bar{I}_{3a}$, and [I] is the identity matrix.

Use of Eq. (6-5b) now allows writing the expression of σ_{ij} in Eq. (6-4) as

$$\sigma_{ij} = \phi_0 \delta_{ij} + \phi_1 \epsilon_{ij} + \phi_2 \epsilon_{im} \epsilon_{mj} \tag{6-6}$$

where ϕ_0, ϕ_1, and ϕ_2 are the elastic response functions dependent on the strain invariants $\bar{I}_{1\epsilon}$, $\bar{I}_{2\epsilon}$, and $\bar{I}_{3\epsilon}$, or I_1, I_2, I_3 (see Chapter 3).

We can also write an expression for strain in terms of invariants of stress $\bar{I}_{1\sigma}$, $\bar{I}_{2\sigma}$, and $\bar{I}_{3\sigma}$ through response functions ψ_0, ψ_1, and ψ_2 as

$$\epsilon_{ij} = \psi_0 \delta_{ij} + \psi_1 \sigma_{ij} + \psi_2 \sigma_{im} \sigma_{mj} \tag{6-7}$$

For $\epsilon_{ij} = 0$, that is, for zero strain levels, Eq. (6-6) becomes

$$\sigma_{ij} = \phi_0 \delta_{ij} \tag{6-8}$$

The expressions (6-6) and (6-7) are invariants with respect to coordinate transformation. For instance, consider

$$\sigma'_{ab} = l_{ai} l_{bj} \sigma_{ij} \tag{6-9a}$$

Substituting Eq. (6-6) into Eq. (6-9a), we obtain

$$\sigma'_{ab} = \phi_0 l_{ai} l_{bj} \delta_{ij} + \phi_1 l_{ai} l_{bj} \epsilon_{ij} + \phi_2 l_{ai} l_{bj} \epsilon_{ik} \epsilon_{kj} \tag{6-9b}$$

or

$$\sigma'_{ab} = \phi_0 \delta'_{ab} + \phi_1 \epsilon'_{ab} + \phi_2 \epsilon'_{ak} \epsilon'_{kb} \tag{6-9c}$$

This has the same form as Eq. (6-6), showing that the expression above is invariant with respect to coordinate transformations; this is in compliance with the axiom of objectivity discussed in Chapter 2. This shows that the quantities ϕ_0, ϕ_1, and ϕ_2 are functions of material constants and invariants.

First-Order Cauchy Elastic Model

For linear elastic behavior, the second-order strain terms vanish, and hence $\phi_2 = 0$. Also, ϕ_1 is constant and ϕ_0 is a linear function of the first strain invariant. Then Eq. (6-6) can be written as

$$\sigma_{ij} = (\alpha_0) \delta_{ij} + \alpha_1 I_1 \delta_{ij} + \alpha_2 \epsilon_{ij} \tag{6-10a}$$

where $(\alpha_0)\delta_{ij}$ is initial (isotropic) stress. In the absense of initial stress, Eq. (6-10a) becomes

$$\sigma_{ij} = \alpha_1 I_1 \delta_{ij} + \alpha_2 \epsilon_{ij} \tag{6-10b}$$

where α_1 and α_2 are material constants. This represents the generalized Hooke's law.

In order to identify the meanings of α_1 and α_2 in terms of the familiar notion of Hooke's law, let us first consider the case of simple shear deformations, which implies that there is no volumetric strain. Thus, there is only one nonzero shear strain component $\epsilon_{12} = \epsilon_{21}$, and Eq. (6-10b) can be written as

$$\sigma_{ij} = \alpha_2 \epsilon_{ij} \tag{6-11a}$$

or in expanded form

$$\sigma_{ij} = \begin{pmatrix} 0 & \alpha_2 \epsilon_{12} & 0 \\ \alpha_2 \epsilon_{12} & 0 & 0 \\ 0 & 0 & 0 \end{pmatrix} \tag{6-11b}$$

where ϵ_{12} is equal to one-half the engineering shear strain, γ_{12}. Hence

$$\sigma_{12} = \alpha_2 \left(\frac{\gamma_{12}}{2} \right)$$

or

$$\frac{\sigma_{12}}{\gamma_{12}} = \frac{\alpha_2}{2} = G \tag{6-11c}$$

That is, $\alpha_2 = 2G$. Here G is the shear modulus of the material.

Now consider uniform dilatation or volume change. For this case the strain tensor can be expressed as

$$\epsilon_{ij} = \left(\frac{I_1}{3}\right)\delta_{ij} = \begin{pmatrix} \dfrac{I_1}{3} & 0 & 0 \\ 0 & \dfrac{I_1}{3} & 0 \\ 0 & 0 & \dfrac{I_1}{3} \end{pmatrix} \tag{6-12}$$

Substitution of Eq. (6-12) and $\alpha_2 = 2G$ in Eq. (6-10b) leads to

$$\sigma_{ij} = \left(\alpha_1 I_1 + 2G\frac{I_1}{3}\right)\delta_{ij} \tag{6-13}$$

For hydrostatic compression, we have mean pressure p as

$$p = \frac{\sigma_{ii}}{3} = \frac{\sigma_{11} + \sigma_{22} + \sigma_{33}}{3} = \frac{J_1}{3} \tag{6-14a}$$

Hence Eq. (6-13) reduces to

$$\frac{J_1}{3} = \left(\alpha_1 + \frac{2G}{3}\right)I_1 \tag{6-14b}$$

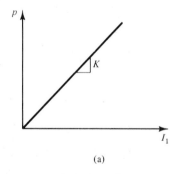

(a)

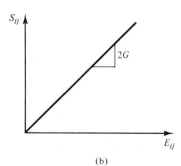

Figure 6-2 Linear behavior: (a) hydrostatic compression; (b) shear.

(b)

This relation can be plotted as shown in Fig. 6-2, in which the slope of the hydrostatic stress p versus volumetric strain I_1 curve denotes the bulk modulus K, given by

$$K = \alpha_1 + \frac{2G}{3} \tag{6-14c}$$

Therefore,

$$\alpha_1 = K - \frac{2G}{3}$$

$$= \lambda \tag{6-14d}$$

where λ is known as Lamé's constant in the classical theory of elasticity (3). We can now express Eq. (6-10) as

$$\sigma_{ij} = \left(K - \frac{2G}{3} \right) I_1 \delta_{ij} + 2G\epsilon_{ij}$$

$$= K I_1 \delta_{ij} + 2G \left(\epsilon_{ij} - \frac{I_1}{3} \delta_{ij} \right) \tag{6-15a}$$

As defined in Chapter 3, the term

$$E_{ij} = \epsilon_{ij} - \frac{I_1}{3} \delta_{ij} \tag{6-15b}$$

denotes the strain deviation tensor. Equation (6-15a) can also be written as

$$\sigma_{ij} = K I_1 \delta_{ij} + 2G E_{ij}$$

$$= \frac{J_1}{3} \delta_{ij} + S_{ij} \tag{6-16}$$

This is an interesting and useful relation that we use often in the classical theory of elasticity for linear (isotropic) materials. It says that the stress tensor is decomposed in two parts: volumetric stress or mean pressure tensor

$$p\delta_{ij} = \frac{J_1}{3} \delta_{ij} = K I_1 \delta_{ij} \tag{6-17a}$$

and deviatoric stress tensor

$$S_{ij} = 2G E_{ij} \tag{6-17b}$$

These are well-known relations and the values of K and G are defined as slopes in Fig. 6-2. As in the case of the bulk modulus K, the slope of the shear stress–shear strain curve gives a measure of shear modulus G [Fig. 6-2(b)]. Here the slope is expressed as $2G$ because of the tensor definition of the components of shear strain.

Equation (6-16) can be expressed in matrix notation as

$$
\begin{Bmatrix} \sigma_{11} \\ \sigma_{22} \\ \sigma_{33} \\ \sigma_{12} \\ \sigma_{23} \\ \sigma_{13} \end{Bmatrix} =
\begin{bmatrix}
K & K & K & 0 & 0 & 0 \\
K & K & K & 0 & 0 & 0 \\
K & K & K & 0 & 0 & 0 \\
0 & 0 & 0 & 0 & 0 & 0 \\
0 & 0 & 0 & 0 & 0 & 0 \\
0 & 0 & 0 & 0 & 0 & 0
\end{bmatrix}
\begin{Bmatrix} \epsilon_{11} \\ \epsilon_{22} \\ \epsilon_{33} \\ \epsilon_{12} \\ \epsilon_{23} \\ \epsilon_{13} \end{Bmatrix}
$$

$$
+
\begin{bmatrix}
\dfrac{4G}{3} & -\dfrac{2G}{3} & -\dfrac{2G}{3} & 0 & 0 & 0 \\
-\dfrac{2G}{3} & \dfrac{4G}{3} & -\dfrac{2G}{3} & 0 & 0 & 0 \\
-\dfrac{2G}{3} & -\dfrac{2G}{3} & \dfrac{4G}{3} & 0 & 0 & 0 \\
0 & 0 & 0 & 2G & 0 & 0 \\
0 & 0 & 0 & 0 & 2G & 0 \\
0 & 0 & 0 & 0 & 0 & 2G
\end{bmatrix}
\begin{Bmatrix} \epsilon_{11} \\ \epsilon_{22} \\ \epsilon_{33} \\ \epsilon_{12} \\ \epsilon_{23} \\ \epsilon_{13} \end{Bmatrix}
\quad (6\text{-}18a)
$$

or

$$
\begin{Bmatrix} \sigma_{11} \\ \sigma_{22} \\ \sigma_{33} \\ \sigma_{12} \\ \sigma_{23} \\ \sigma_{13} \end{Bmatrix} =
\begin{bmatrix}
K+\dfrac{4G}{3} & K-\dfrac{2G}{3} & K-\dfrac{2G}{3} & 0 & 0 & 0 \\
K-\dfrac{2G}{3} & K+\dfrac{4G}{3} & K-\dfrac{2G}{3} & 0 & 0 & 0 \\
K-\dfrac{2G}{3} & K-\dfrac{2G}{3} & K+\dfrac{4G}{3} & 0 & 0 & 0 \\
0 & 0 & 0 & 2G & 0 & 0 \\
0 & 0 & 0 & 0 & 2G & 0 \\
0 & 0 & 0 & 0 & 0 & 2G
\end{bmatrix}
\begin{Bmatrix} \epsilon_{11} \\ \epsilon_{22} \\ \epsilon_{33} \\ \epsilon_{12} \\ \epsilon_{23} \\ \epsilon_{13} \end{Bmatrix}
$$

$$(6\text{-}18b)$$

This expression is the same as the generalized Hooke's law in the theory of elasticity. Thus the lowest order of the Cauchy elastic law is the same as the generalized Hooke's law.

Inverse relation. Equations (6-10) and (6-16) express relation between stress and strain. The inverse relation between strain and stress can be written as

$$\epsilon_{ij} = \frac{S_{ij}}{2G} + \frac{J_1}{9K}\delta_{ij}$$

$$= E_{ij} + \frac{I_1}{3}\delta_{ij} \tag{6-19a}$$

where

$$S_{ij} = 2GE_{ij} \tag{6-19b}$$

$$J_1 = 3KI_1 \tag{6-19c}$$

Equation (6-19a) can be expressed in matrix notation as

$$
\begin{Bmatrix} \epsilon_{11} \\ \epsilon_{22} \\ \epsilon_{33} \\ \epsilon_{12} \\ \epsilon_{23} \\ \epsilon_{13} \end{Bmatrix} =
\begin{bmatrix}
\frac{1}{2G} & 0 & 0 & 0 & 0 & 0 \\
0 & \frac{1}{2G} & 0 & 0 & 0 & 0 \\
0 & 0 & \frac{1}{2G} & 0 & 0 & 0 \\
0 & 0 & 0 & \frac{1}{2G} & 0 & 0 \\
0 & 0 & 0 & 0 & \frac{1}{2G} & 0 \\
0 & 0 & 0 & 0 & 0 & \frac{1}{2G}
\end{bmatrix}
\begin{Bmatrix} \sigma_{11} \\ \sigma_{22} \\ \sigma_{33} \\ \sigma_{12} \\ \sigma_{23} \\ \sigma_{13} \end{Bmatrix}
$$

$$
+
\begin{bmatrix}
\frac{1}{9K}-\frac{1}{6G} & \frac{1}{9K}-\frac{1}{6G} & \frac{1}{9K}-\frac{1}{6G} & 0 & 0 & 0 \\
\frac{1}{9K}-\frac{1}{6G} & \frac{1}{9K}-\frac{1}{6G} & \frac{1}{9K}-\frac{1}{6G} & 0 & 0 & 0 \\
\frac{1}{9K}-\frac{1}{6G} & \frac{1}{9K}-\frac{1}{6G} & \frac{1}{9K}-\frac{1}{6G} & 0 & 0 & 0 \\
0 & 0 & 0 & 0 & 0 & 0 \\
0 & 0 & 0 & 0 & 0 & 0 \\
0 & 0 & 0 & 0 & 0 & 0
\end{bmatrix}
\begin{Bmatrix} \sigma_{11} \\ \sigma_{22} \\ \sigma_{33} \\ \sigma_{12} \\ \sigma_{23} \\ \sigma_{13} \end{Bmatrix}
$$

$$\tag{6-20a}$$

or

$$
\begin{Bmatrix} \epsilon_{11} \\ \epsilon_{22} \\ \epsilon_{33} \\ \epsilon_{12} \\ \epsilon_{23} \\ \epsilon_{13} \end{Bmatrix} =
\begin{bmatrix}
\dfrac{1}{3G}+\dfrac{1}{9K} & \dfrac{1}{9K}-\dfrac{1}{6G} & \dfrac{1}{9K}-\dfrac{1}{6G} & 0 & 0 & 0 \\
\dfrac{1}{9K}-\dfrac{1}{6G} & \dfrac{1}{3G}+\dfrac{1}{9K} & \dfrac{1}{9K}-\dfrac{1}{6G} & 0 & 0 & 0 \\
\dfrac{1}{9K}-\dfrac{1}{6G} & \dfrac{1}{9K}-\dfrac{1}{6G} & \dfrac{1}{3G}+\dfrac{1}{9K} & 0 & 0 & 0 \\
0 & 0 & 0 & \dfrac{1}{2G} & 0 & 0 \\
0 & 0 & 0 & 0 & \dfrac{1}{2G} & 0 \\
0 & 0 & 0 & 0 & 0 & \dfrac{1}{2G}
\end{bmatrix}
\begin{Bmatrix} \sigma_{11} \\ \sigma_{22} \\ \sigma_{33} \\ \sigma_{12} \\ \sigma_{23} \\ \sigma_{13} \end{Bmatrix}
$$

$$(6\text{-}20\text{b})$$

Various Idealizations

Equation (6-16) represents the stress–strain law for three-dimensional bodies. For two-dimensional idealizations such as plane stress and plane strain, it can be simplified as shown below (1).

Plane strain. Figure 6-3(a) shows various problems that can be approximated as plane strain. The loading is uniform in the longitudinal direction (z). Because of the loading and geometry, the strain in the z-direction is assumed to be zero or constant. Then the nonzero strains occur only in the 1–2 directions (1).

The stress–strain relation can be derived for the plane strain case as a special case of Eq. (6-15) as

$$
\begin{aligned}
\sigma_{11} &= KI_1 + 2G\left(\epsilon_{11} - \frac{I_1}{3}\right) \\
&= \left(K - \frac{2G}{3}\right)I_1 + 2G\epsilon_{11} \\
&= \left(\frac{3K+4G}{3}\right)\epsilon_{11} + \left(K - \frac{2G}{3}\right)\epsilon_{22}
\end{aligned}
\qquad (6\text{-}21\text{a})
$$

Similarly,

$$
\sigma_{22} = \left(\frac{3K+4G}{3}\right)\epsilon_{22} + \left(K - \frac{2G}{3}\right)\epsilon_{11} \qquad (6\text{-}21\text{b})
$$

$$
\sigma_{33} = \left(K - \frac{2G}{3}\right)\epsilon_{11} + \left(K - \frac{2G}{3}\right)\epsilon_{22} \qquad (6\text{-}21\text{c})
$$

$$
\sigma_{12} = 2G\epsilon_{12} \qquad (6\text{-}21\text{d})
$$

$$
\sigma_{23} = \sigma_{13} = 0 \qquad (6\text{-}21\text{e})
$$

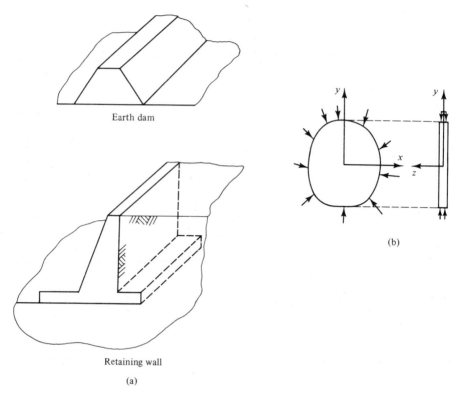

Earth dam

Retaining wall

(a)

(b)

Figure 6-3   (a) Examples of plane strain problems; (b) plane stress problem, thin plate with in-plane loading.

or in matrix form,

$$
\begin{Bmatrix} \sigma_{11} \\ \sigma_{22} \\ \sigma_{12} \end{Bmatrix} = \begin{bmatrix} \dfrac{3K+4G}{3} & \dfrac{3K-2G}{3} & 0 \\ \dfrac{3K-2G}{3} & \dfrac{3K+4G}{3} & 0 \\ 0 & 0 & 2G \end{bmatrix} \begin{Bmatrix} \epsilon_{11} \\ \epsilon_{22} \\ \epsilon_{12} \end{Bmatrix}
\tag{6-21f}
$$

and

$$
\sigma_{33} = \left(K - \frac{2G}{3} \right)(\epsilon_{11} + \epsilon_{22})
\tag{6-21g}
$$

Plane stress.   Figure 6-3(b) shows an example of plane stress case. The plate is subjected to load in the plane of the plate, causing in-plane or membrane effects. If the plate is thin, that is, if the thickness, h, is much smaller than the width of the plate, the normal stress σ_{33} is assumed to be zero and so are the shear stresses σ_{23} and σ_{13}. Hence the stress tensor and vector are

$$\sigma_{ij} = \begin{pmatrix} \sigma_{11} & \sigma_{12} & 0 \\ \sigma_{12} & \sigma_{22} & 0 \\ 0 & 0 & 0 \end{pmatrix} \tag{6-22a}$$

and

$$\{\sigma\}^T = [\sigma_{11} \quad \sigma_{22} \quad \sigma_{12}]$$

The stress–strain relationship for the plane stress case can be derived as a special case of Eq. (6-19a) as

$$\epsilon_{11} = \frac{S_{11}}{2G} + \frac{J_1}{9K}$$

$$= \frac{1}{2G}\left(\sigma_{11} - \frac{\sigma_{11} + \sigma_{22}}{3}\right) + \frac{1}{9K}(\sigma_{11} + \sigma_{22})$$

$$= \sigma_{11}\left(\frac{1}{3G} + \frac{1}{9K}\right) - \sigma_{22}\left(\frac{1}{6G} - \frac{1}{9K}\right) \tag{6-22b}$$

Similarly,

$$\epsilon_{22} = -\sigma_{11}\left(\frac{1}{6G} - \frac{1}{9K}\right) + \sigma_{22}\left(\frac{1}{3G} + \frac{1}{9K}\right) \tag{6-22c}$$

and

$$\epsilon_{12} = \frac{\sigma_{12}}{2G}$$

Hence in matrix form

$$\begin{Bmatrix} \epsilon_{11} \\ \epsilon_{22} \\ \epsilon_{12} \end{Bmatrix} = \begin{bmatrix} \frac{1}{3G} + \frac{1}{9K} & -\left(\frac{1}{6G} - \frac{1}{9K}\right) & 0 \\ -\left(\frac{1}{6G} - \frac{1}{9K}\right) & \frac{1}{3G} + \frac{1}{9K} & 0 \\ 0 & 0 & \frac{1}{2G} \end{bmatrix} \begin{Bmatrix} \sigma_{11} \\ \sigma_{22} \\ \sigma_{12} \end{Bmatrix} \tag{6-22d}$$

The normal strain ϵ_{33} can be derived as

$$\epsilon_{33} = \frac{S_{33}}{2G} + \frac{J_1}{9K} = -\frac{1}{2G}\left(\frac{\sigma_{11} + \sigma_{22}}{3}\right) + \frac{1}{9K}(\sigma_{11} + \sigma_{22})$$

$$= (\sigma_{11} + \sigma_{22})\left(\frac{1}{9K} - \frac{1}{6G}\right) \tag{6-22e}$$

Uniaxial stress. For the uniaxial state of stress, we have

$$\sigma_{ij} = \begin{pmatrix} \sigma_{11} & 0 & 0 \\ 0 & 0 & 0 \\ 0 & 0 & 0 \end{pmatrix} \tag{6-23a}$$

Substitution of this specialized relation in Eq. (6-19a) leads to

$$
\epsilon_{ij} = \begin{pmatrix} \left(\dfrac{3K+G}{9KG}\right)\sigma_{11} & 0 & 0 \\[2ex] 0 & \left(\dfrac{2G-3K}{18KG}\right)\sigma_{11} & 0 \\[2ex] 0 & 0 & \left(\dfrac{2G-3K}{18KG}\right)\sigma_{11} \end{pmatrix} \qquad (6\text{-}23b)
$$

which implies that

$$
\sigma_{11} = \left(\frac{9KG}{3K+G}\right)\epsilon_{11} = E\epsilon_{11} \qquad (6\text{-}23c)
$$

$$
\epsilon_{22} = \epsilon_{33} = -\left(\frac{3K-2G}{6K+2G}\right)\epsilon_{11}
$$

$$
= \nu\epsilon_{11} \qquad (6\text{-}23d)
$$

where E and ν are the well-known Young's modulus and Poisson's ratio, respectively. The latter defines a relation between the lateral strains ϵ_{22} and ϵ_{33}, and the axial strain ϵ_{11}. Thus

$$
E = \frac{9KG}{3K+G} \qquad (6\text{-}23e)
$$

$$
\nu = \frac{3K-2G}{6K+2G} \qquad (6\text{-}23f)
$$

Physically, the foregoing implies that a uniaxial state of stress causes strains in the axial and lateral directions.

Uniaxial strain. If a bar is subjected to simply an uniaxial state of strain given by

$$
\epsilon_{ij} = \begin{pmatrix} \epsilon_{11} & 0 & 0 \\ 0 & 0 & 0 \\ 0 & 0 & 0 \end{pmatrix} \qquad (6\text{-}24a)
$$

it will cause the stresses as given in Eq. (6-16) as

$$
\sigma_{ij} = \begin{pmatrix} \left(K+\dfrac{4G}{3}\right)\epsilon_{11} & 0 & 0 \\[2ex] 0 & \left(K-\dfrac{2G}{3}\right)\epsilon_{11} & 0 \\[2ex] 0 & 0 & \left(K-\dfrac{2G}{3}\right)\epsilon_{11} \end{pmatrix} \qquad (6\text{-}24b)
$$

Then

$$\sigma_{11} = \left(K + \frac{4G}{3} \right)\epsilon_{11} = \frac{E(1-\nu)}{(1+\nu)(1-2\nu)}\epsilon_{11} \qquad (6\text{-}24c)$$

$$\sigma_{22} = \sigma_{33} = \left(\frac{3K-2G}{3K+4G} \right)\sigma_{11} = \left(\frac{\nu}{1-\nu} \right)\sigma_{11} \qquad (6\text{-}24d)$$

Physically, this implies that if the strain is applied only in the axial direction and the lateral strain ϵ_{22} and ϵ_{33} are restrained, we need $\sigma_{22} = \sigma_{33}$ as in Eq. (6-24d) to prevent such lateral strains.

The ratio M between the stress σ_{11} and ϵ_{11} is often referred to as *constrained modulus*, given by

$$\frac{\sigma_{11}}{\epsilon_{11}} = M = K + \frac{4G}{3} = \frac{E(1-\nu)}{(1+\nu)(1-2\nu)} \qquad (6\text{-}24e)$$

Stress Path

As described in Chapter 5, the path of applied stress is an important parameter that influences the material behavior. The same material can exhibit different behavior under different stress paths. Its effect can be very important in characterizing the behavior of geological media, because they are quite sensitive to the stress path. In view of this, we give considerable attention to it in this text.

For the uniaxial strain state, the stress path on $\sqrt{J_{2D}}$ versus $J_1/3$ will have a slope given by

$$\sqrt{J_{2D}} = \sqrt{3}\left(\frac{1-2\nu}{1+\nu} \right)\frac{J_1}{3} \qquad (6\text{-}24f)$$

GREEN ELASTIC MODELS

As mentioned previously, for an elastic body, the state of stress is a function of the state of strain only. For elastic bodies it is assumed that there exists an unstressed state known as the natural state (3). When external loads are applied, the body deforms and reaches a different state of stress. When the external loads are removed, the elastic behavior requires that the body regain its natural state. Therefore, the energy supplied to the body by external means is fully recoverable; that is, no energy is dissipated during the deformation process. The constitutive equations derived from the internal energy considerations are called *Green elastic material models*. A *hyperelastic material* is defined as one that possesses a strain energy function, U_0. The subject of Green elastic models is discussed in various references (2,3,5–7); here we present some special cases, and an application of this concept.

Relation between U_0, Stress, and Strain

The internal strain energy density, U_0 (Fig. 6-4), that is, the strain energy per unit volume under deformation, can be defined in terms of strains. Hence

$$U_0 = U_0(\epsilon_{ij}) \tag{6-25}$$

Now we use the first law of thermodynamics, which states that "the work performed on a deforming body by external forces plus the heat, $\overline{Q}$, flowing into the system from outside equals the increase in the kinetic energy, T, plus the increase in internal strain energy, U" (2,3,8). Hence

$$\delta W_e + \delta \overline{Q} = \delta U + \delta T \tag{6-26}$$

Here δ denotes a change.

The law of conservation of kinetic energy, which states that "the work of all forces (internal W_i and external W_e) equals the increase of kinetic energy, δT of the system," gives

$$\delta W_e + \delta W_i = \delta T \tag{6-27}$$

From Eqs. (6-26) and (6-27), we have

$$\delta W_i = \delta T - \delta W_e = \delta T - \delta U - \delta T + \delta \overline{Q}$$

$$= \delta \overline{Q} - \delta U \tag{6-28a}$$

If there is no heat flow, $\delta \overline{Q} = 0$; hence

$$\delta W_i = -\delta U \tag{6-28b}$$

Now we write variation or change in the external work, W_e, in terms of the variation in the work by external body forces, F_i, and surface stresses, σ_i (Fig. 6-5):

$$\delta W_e = \int_V F_i \, \delta u_i \, dV + \int_S \sigma_{ij} n_i \, \delta u_i \, dS \tag{6-29}$$

where δ denotes variation; the u_i, $i = 1,2,3$, are the components of displacements; the n_j are the direction cosines of the outward normal to surface S; and

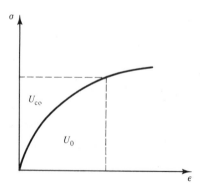

Figure 6-4 Strain energy density.

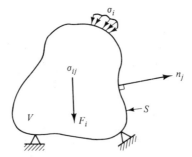

Figure 6-5 Body in equilibrium under external forces.

V denotes volume. It can be shown from Eq. (6-29) that

$$\delta W_e = \int_V \sigma_{ji}(\delta u_i)_{,j}\, dV \qquad (6\text{-}30)$$

The term $(\delta u_i)_{,j}$ denotes gradient of displacements

$$(\delta u_i)_{,j} = \delta\epsilon_{ij} + \delta\omega_{ij} \qquad (6\text{-}31)$$

where ϵ_{ij} is the strain tensor, which is symmetric, and ω_{ij} is the strain tensor, which is skew symmetric (see Chapter 3). Hence Eq. (6-30) gives

$$\delta W_e = \int_V \left(\sigma_{ji}\delta\epsilon_{ij} + \sigma_{ji}\delta\omega_{ij}\right) dV \qquad (6\text{-}32\text{a})$$

Since $\sigma_{ij} = \sigma_{ji}$, and ω_{ij} is skew symmetric, $\sigma_{ji}\delta\omega_{ij} = 0$. Hence

$$\delta W_e = \int_V \sigma_{ij}\delta\epsilon_{ij}\, dV \qquad (6\text{-}32\text{b})$$

If $\delta T = \delta\overline{Q} = 0$, Eq. (6-26) leads to

$$\delta W_e = \delta U$$

that is,

$$\delta U = \int_V \sigma_{ij}\delta\epsilon_{ij}\, dV \qquad (6\text{-}33\text{a})$$

or

$$\delta U = \int_V \delta U_0\, dV \qquad (6\text{-}33\text{b})$$

Hence

$$\delta U_0 = \sigma_{ij}\delta\epsilon_{ij} \qquad (6\text{-}34)$$

Now from Eq. (6-25), we have

$$\delta U_0 = \frac{\partial U_0}{\partial \epsilon_{ij}}\delta\epsilon_{ij} \qquad (6\text{-}35)$$

Comparison of Eqs. (6-34) and (6-35) leads to

$$\sigma_{ij} = \frac{\partial U_0}{\partial \epsilon_{ij}} \tag{6-36}$$

This relation is called the *Green elastic constitutive law*. By expressing U_0 in terms of components of strain or invariants of strain, we can derive Green elastic models of different orders.

We now express U_0 in terms of the invariants of strain as

$$U_0 = U_0(\bar{I}_{1\epsilon}, \bar{I}_{2\epsilon}, \bar{I}_{3\epsilon}) = U_0(I_1, I_2, I_3) \tag{6-37}$$

Simplified notation, I_1, I_2, I_3, is used here instead of $\bar{I}_{1\epsilon}$, $\bar{I}_{2\epsilon}$, and $\bar{I}_{3\epsilon}$ (Chapter 3). Then Eq. (6-36) gives

$$\sigma_{ij} = \frac{\partial U_0}{\partial I_1}\left(\frac{\partial I_1}{\partial \epsilon_{ij}}\right) + \frac{\partial U_0}{\partial I_2}\left(\frac{\partial I_2}{\partial \epsilon_{ij}}\right) + \frac{\partial U_0}{\partial I_3}\left(\frac{\partial I_3}{\partial \epsilon_{ij}}\right) \tag{6-38a}$$

Since

$$\frac{\partial I_1}{\partial \epsilon_{ij}} = \delta_{ij}$$

$$\frac{\partial I_2}{\partial \epsilon_{ij}} = \epsilon_{ij} \tag{6-38b}$$

$$\frac{\partial I_3}{\partial \epsilon_{ij}} = \epsilon_{im}\epsilon_{mj}$$

Equation (6-38a) becomes

$$\sigma_{ij} = \frac{\partial U_0}{\partial I_1}\delta_{ij} + \frac{\partial U_0}{\partial I_2}\epsilon_{ij} + \frac{\partial U_0}{\partial I_3}\epsilon_{im}\epsilon_{mj} \tag{6-38c}$$

We note that Eq. (6-38a) has the same form as that of the Cauchy elastic model, Eq. (6-6). The difference between the two is that the response functions ϕ_0, ϕ_1, and ϕ_2 in Eq. (6-6) are independent, while those in Eq. (6-38c) are dependent as follows:

$$\frac{\partial}{\partial I_2}\left(\frac{\partial U_0}{\partial I_1}\right) = \frac{\partial}{\partial I_1}\left(\frac{\partial U_0}{\partial I_2}\right) \tag{6-39a}$$

$$\frac{\partial}{\partial I_3}\left(\frac{\partial U_0}{\partial I_1}\right) = \frac{\partial}{\partial I_1}\left(\frac{\partial U_0}{\partial I_3}\right) \tag{6-39b}$$

$$\frac{\partial}{\partial I_2}\left(\frac{\partial U_0}{\partial I_3}\right) = \frac{\partial}{\partial I_3}\left(\frac{\partial U_0}{\partial I_2}\right) \tag{6-39c}$$

Thus the Green elastic model can be considered as a special case of the Cauchy elastic model.

The inverse relation to Eq. (6-36) can be derived as

$$\epsilon_{ij} = \frac{\partial U_{co}}{\partial \sigma_{ij}} (J_1, J_2, J_3) \tag{6-40a}$$

where U_{co} is the complementary energy density (Fig. 6-4), expressed in terms of the invariants of stress. Equation (6-40a) can be expressed as

$$\epsilon_{ij} = \frac{\partial U_{co}}{\partial J_1} \delta_{ij} + \frac{\partial U_{co}}{\partial J_2} \sigma_{ij} + \frac{\partial U_{co}}{\partial J_3} \sigma_{im}\sigma_{mj} \tag{6-40b}$$

First-Order Green Elastic Model

We can obtain models of various orders by retaining terms of appropriate orders of strain in Eq. (6-38c) and stress in Eq. (6-40b). For the first-order model from Eq. (6-38c), it is necessary to retain quadratic terms in strains in U_0 since after taking the first derivative, the stress–strain law should be of first order or linear.

If the body is initially unstrained, then the strain energy is due only to the new external forces. Under this assumption, the expression for obtaining the first-order model can be written as

$$U_0 = C_1 I_2 + C_2 I_1^2 \tag{6-41}$$

where C_1 and C_2 are material constants. Substitution of Eq. (6-41) in Eq. (6-38c) leads to

$$\sigma_{ij} = \frac{\partial \left(C_1 I_2 + C_2 I_1^2 \right)}{\partial I_1} \delta_{ij} + \frac{\partial \left(C_1 I_2 + C_2 I_1^2 \right)}{\partial I_2} \epsilon_{ij}$$

$$+ \frac{\partial \left(C_1 I_2 + C_2 I_1^2 \right)}{\partial I_3} \epsilon_{im}\epsilon_{mj}$$

$$= 2C_2 I_1 \delta_{ij} + C_1 \epsilon_{ij} \tag{6-42}$$

This expression is identical to Eq. (6-15a), which represents the first-order or linear Cauchy elastic model. It also is the same as the classical Hooke's law in the theory of elasticity. The constants C_1 and C_2 have a similar meaning as in Eq. (6-15a):

$$2C_2 = K - \frac{2G}{3}$$

or

$$C_2 = \frac{K}{2} - \frac{G}{3} = \frac{3K - 2G}{6} \tag{6-43a}$$

and

$$C_1 = 2G \tag{6-43b}$$

The expression for U_0 can now be expressed as

$$U_0 = 2G\left(I_2 - \frac{I_1^2}{6}\right) + \frac{K}{2}I_1^2$$

$$= 2GI_{2D} + \frac{K}{2}I_1^2 \tag{6-44}$$

Since for linear elastic materials

$$\frac{J_1}{3} = KI_1 \tag{6-45a}$$

$$\sqrt{J_{2D}} = 2G\sqrt{I_{2D}} \tag{6-45b}$$

the expression in Eq. (6-44) becomes

$$U_0 = \sqrt{J_{2D}}\sqrt{I_{2D}} + \frac{1}{2}\frac{J_1 I_1}{3} \tag{6-45c}$$

The first term in Eq. (6-44) represents the strain energy due to distortion or shear, and the second term represents strain energy due to volume changes.

It can also be shown that for linear elastic materials, the strain energy density U_0 and the complementary energy density U_{co} are equal; that is,

$$U_0 = U_{co} \tag{6-46a}$$

Therefore,

$$U_{co} = \sigma_{ij}\epsilon_{ij} - U_0 \tag{6-46b}$$

or

$$U_{co} = U_0 = \tfrac{1}{2}\sigma_{ij}\epsilon_{ij} \tag{6-46c}$$

This is a useful result.

Properties of Linear Elastic Parameters

The expression for U_0 in Eq. (6-41) can be expanded as

$$U_0 = C_1\left[\tfrac{1}{2}\left(\epsilon_{11}^2 + \epsilon_{22}^2 + \epsilon_{33}^2\right) + \epsilon_{12}^2 + \epsilon_{23}^2 + \epsilon_{13}^2\right]$$
$$+ C_2\left[\epsilon_{11}^2 + \epsilon_{22}^2 + \epsilon_{33}^2 + 2\epsilon_{11}\epsilon_{22} + 2\epsilon_{22}\epsilon_{33} + 2\epsilon_{11}\epsilon_{33}\right] \tag{6-47}$$

which can be written in quadratic form as

$$U_0 = \sum_{i=1}^{6}\sum_{j=1}^{6} C_{ij}\epsilon_i\epsilon_j \tag{6-48}$$

where ϵ_i and ϵ_j denote the six components of the strain tensor and C_{ij} in matrix notation is

$$
C_{ij} = \begin{pmatrix}
\frac{1}{2}C_1 + C_2 & C_2 & C_2 & 0 & 0 & 0 \\
C_2 & \frac{1}{2}C_1 + C_2 & C_2 & 0 & 0 & 0 \\
C_2 & C_2 & \frac{1}{2}C_1 + C_2 & 0 & 0 & 0 \\
0 & 0 & 0 & C_1 & 0 & 0 \\
0 & 0 & 0 & 0 & C_1 & 0 \\
0 & 0 & 0 & 0 & 0 & C_1
\end{pmatrix}
\qquad (6\text{-}49a)
$$

where $C_{ij} = C_{ji}$.

Since U_0 is a positive quantity, from the theory of quadratics, all minors of the diagonal elements of C_{ij} must be positive (7, 8). Hence

$$C_1 > 0$$

$$\frac{C_1}{4} + C_2 > 0 \qquad (6\text{-}49b)$$

$$\frac{C_1}{2} + 3C_2 > 0$$

In terms of familiar parameters G, K, E, and ν which are based on the theory of elasticity, the conditions in Eq. (6-49b) become

$$
\begin{aligned}
G &> 0 \\
K &> 0 \\
E &> 0 \\
-1 &< \nu < \tfrac{1}{2}
\end{aligned}
\qquad (6\text{-}49c)
$$

HIGHER-ORDER MODELS

In this section we consider both the Cauchy and Green elastic laws and discuss some of the useful higher-order models based on them. In this introductory treatment only the second-order models will be described in detail; the third- and higher-order models will be stated only briefly.

Second-Order Cauchy Elastic Model

Let us consider the Cauchy elastic law given by

$$\sigma_{ij} = \phi_0 \delta_{ij} + \phi_1 \epsilon_{ij} + \phi_2 \epsilon_{im}\epsilon_{mj} \qquad (6\text{-}6)$$

and its inverse form,

$$\epsilon_{ij} = \psi_0 \delta_{ij} + \psi_1 \sigma_{ij} + \psi_2 \sigma_{im}\sigma_{mj} \qquad (6\text{-}7)$$

The first term in these equations denotes the zero or scalar state of stress or strain; the second term denotes linear terms. In deriving the first-order models, we have included these two terms. Now, we consider the third term, given by $\epsilon_{im}\epsilon_{mj}$ and $\sigma_{im}\sigma_{mj}$.

We consider Eq. (6-6), and as stated before, write the response functions ϕ_0, ϕ_1, and ϕ_2 as polynomials in terms of the invariants of strain in such a way that Eq. (6-6) yields a second-order expression and includes all lower-order terms:

$$\phi_0 = a_1 I_1 + a_2 I_1^2 + a_3 I_2 \tag{6-50a}$$

$$\phi_1 = a_4 + a_5 I_1 \tag{6-50b}$$

$$\phi_2 = a_6$$

where a_i, $i = 1, 2, \ldots, 6$, are material parameters or constants. These constants must be determined experimentally. It may be noted that the invariants for ϕ_0, ϕ_1, and ϕ_2 in Eqs. (6-50) occur in decreasing order, starting with ϕ_0 as a quadratic and going to ϕ_2 as a constant. The units of a_i are the same as those of stress, σ_{ij}.

Substitution of Eqs. (6-50) in Eq. (6-6) gives

$$\sigma_{ij} = \left(a_1 I_1 + a_2 I_1^2 + a_3 I_2 \right)\delta_{ij} + \left(a_4 + a_5 I_1 \right)\epsilon_{ij} + a_6 \epsilon_{im}\epsilon_{mj} \tag{6-51a}$$

For a consistent polynomial expression, Eq. (6-51a) should specialize to linear elastic first-order law [Eq. (6-15a)]. Hence in the linear model only the first-order terms in Eq. (6-51a) are retained; thus

$$\sigma_{ij} = a_1 I_1 \delta_{ij} + a_4 \epsilon_{ij} \tag{6-51b}$$

Comparison of Eq. (6-51b) with Eq. (6-15a) gives

$$a_1 = K - \frac{2G}{3} \tag{6-51c}$$

$$a_4 = 2G \tag{6-51d}$$

Then Eq. (6-51a) becomes

$$\sigma_{ij} = \left[\left(K - \frac{2G}{3} \right)I_1 + a_2 I_1^2 + a_3 I_2 \right]\delta_{ij} + \left(2G + a_5 I_1 \right)\epsilon_{ij} + a_6 \epsilon_{im}\epsilon_{mj} \tag{6-52}$$

There appear six material parameters, K, G, a_2, a_3, a_5, and a_6 in Eq. (6-52). It is necessary to find them from appropriate (laboratory) tests. Each of these parameters or their combinations can usually be related to specific laboratory tests.

In order to understand the properties of the second-order law, we consider a few special stress and strain conditions.

Uniaxial strain state.   Here the strain tensor is given by

$$\epsilon_{ij} = \begin{pmatrix} \epsilon_{11} & 0 & 0 \\ 0 & 0 & 0 \\ 0 & 0 & 0 \end{pmatrix} \tag{6-53}$$

and its substitution in Eq. (6-52) gives

$$\sigma_{11} = \left(K - \frac{2G}{3} \right) \epsilon_{11} + a_2 \epsilon_{11}^2 + \frac{a_3}{2} \epsilon_{11}^2 + 2G\epsilon_{11} + a_5 \epsilon_{11}^2 + a_6 \epsilon_{11}^2$$

$$= \left(K + \frac{4G}{3} \right) \epsilon_{11} + \left(a_2 + \frac{a_3}{2} + a_5 + a_6 \right) \epsilon_{11}^2 \tag{6-54a}$$

$$\sigma_{22} = \left(K - \frac{2G}{3} \right) \epsilon_{11} + a_2 \epsilon_{11}^2 + \frac{a_3}{2} \epsilon_{11}^2 \tag{6-54b}$$

$$\sigma_{33} = \left(K - \frac{2G}{3} \right) \epsilon_{11} + a_2 \epsilon_{11}^2 + \frac{a_3}{2} \epsilon_{11}^2 \tag{6-54c}$$

$$\sigma_{12} = \sigma_{23} = \sigma_{13} = 0 \tag{6-54d}$$

Figure 6-6 shows the graphical representation of Eq. (6-54a) as a combination of the linear and nonlinear parts, the linear part being essentially the same as the classical (first-order) Hooke's law. It may be noted that the nonlinear portion of the response [Fig. 6-6(b)] includes the effect of the first-order term in strain, that is, ϵ_{11}.

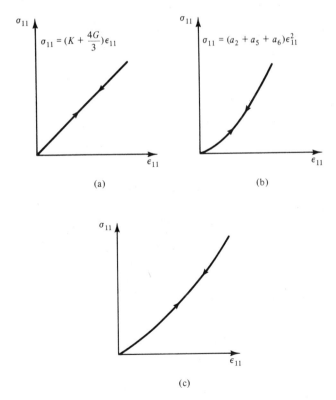

Figure 6-6 Symbolic stress–strain relationship for uniaxial strain state: (a) linear part; (b) nonlinear second-order part; (c) resultant stress–strain relationship.

The usual way to determine a_1 to a_6 would be to plot appropriate experimental results as in Fig. 6-6, and then find them by trial and error using least-squares and/or optimization techniques.

Simple shear strain state.   Now we consider the state of shearing deformation in a two-dimensional state where all other strain components vanish except ϵ_{12} (Fig. 6-7). We first analyze Eq. (6-6) for simple shear strain given by

$$\epsilon_{ij} = \begin{pmatrix} 0 & \epsilon_{12} & 0 \\ \epsilon_{12} & 0 & 0 \\ 0 & 0 & 0 \end{pmatrix} \tag{6-55a}$$

With this, Eq. (6-6) gives

$$\sigma_{ij} = \phi_0 \begin{pmatrix} 1 & 0 & 0 \\ 0 & 1 & 0 \\ 0 & 0 & 1 \end{pmatrix} + \phi_1 \begin{pmatrix} 0 & \epsilon_{12} & 0 \\ \epsilon_{12} & 0 & 0 \\ 0 & 0 & 0 \end{pmatrix} + \phi_2 \begin{pmatrix} \epsilon_{12}^2 & 0 & 0 \\ 0 & \epsilon_{12}^2 & 0 \\ 0 & 0 & 0 \end{pmatrix} \tag{6-55b}$$

Hence the stress components can be derived as

$$\sigma_{11} = \phi_0 + \phi_2 \epsilon_{12}^2$$
$$\sigma_{22} = \phi_0 + \phi_2 \epsilon_{12}^2$$
$$\sigma_{33} = \phi_0 \tag{6-56}$$
$$\sigma_{12} = \phi_1 \epsilon_{12}, \qquad \sigma_{23} = \sigma_{13} = 0$$

The deviatoric stress tensor can be expressed as

$$S_{ij} = \sigma_{ij} - \tfrac{1}{3}\sigma_{nn}\delta_{ij} \tag{6-57a}$$

Therefore,

$$
\begin{aligned}
S_{11} &= \sigma_{11} - \tfrac{1}{3}\sigma_{nn}\delta_{11} \\
&= \sigma_{11} - \tfrac{1}{3}(\sigma_{11} + \sigma_{22} + \sigma_{33}) \\
&= \sigma_{11} - \tfrac{1}{3}(2\sigma_{11} + \sigma_{33}) \\
&= \tfrac{1}{3}(\sigma_{11} - \sigma_{33}) \\
&= \tfrac{1}{3}\left(\phi_0 + \phi_2\epsilon_{12}^2\right) - \tfrac{1}{3}\phi_0 \\
&= \tfrac{1}{3}\phi_2\epsilon_{12}^2
\end{aligned} \tag{6-57b}
$$

$$S_{22} = \tfrac{1}{3}\phi_2\epsilon_{12}^2 \tag{6-57c}$$

$$S_{33} = \left(-\tfrac{2}{3}\phi_2\epsilon_{12}^2\right) \tag{6-57d}$$

Equations (6-56) imply that a simple shearing strain is accompanied by

No volume change

Figure 6-7 Simple shearing deformation.

nonuniform values of normal stresses, and Eq. (6-57b) shows that there exists a normal deviatoric stress on the planes of shearing.

It should be noted that for the first-order model, that is, Hooke's law, the foregoing states of normal stresses are not induced since ϕ_2 is equal to zero.

Now we consider the same derivation with respect to the special form Eq. (6-52) of Eq. (6-6). Substitution of strain components in Eq. (6-52) gives

$$\sigma_{11} = a_3 \epsilon_{12}^2 + a_6 \epsilon_{12}^2$$

$$= \sigma_{22} \tag{6-58a}$$

$$\sigma_{33} = a_3 \epsilon_{12}^2 \tag{6-58b}$$

$$\sigma_{12} = \left[\left(K - \frac{2G}{3}\right) I_1 + a_2 I_1^2 + a_3 \epsilon_{12}^2\right] \delta_{12}^0 + \left(2G + a_5^0 I_1\right)\epsilon_{12}$$

$$+ a_6\left(\epsilon_{11}^0 \epsilon_{12} + \epsilon_{12}\epsilon_{33}^0 + \epsilon_{12}\epsilon_{22}^0 + \epsilon_{13}^0 \epsilon_{32}\right)$$

$$= 2G\epsilon_{12} \tag{6-58c}$$

$$\sigma_{23} = \sigma_{13} = 0 \tag{6-58d}$$

Symbolic plots for $\sigma_{11}(\sigma_{22})$, σ_{33}, and σ_{12} are shown in Fig. 6-8.

It can be seen that the Cauchy elastic second-order law for simple shear strain state predicts a linear relation between shearing stress and shear strain [Eq. (6-58c)], and a second-order relation between normal stresses and shear strain [Eqs. (6-58a) and (6-58b)]. The former is the same as in the first-order model, that is, Hooke's law.

A state of simple shearing strain is accompanied by nonuniform states of normal stresses. Thus when we simulate a state of simple shear strain in a laboratory device for a second-order Cauchy elastic material, nonuniform

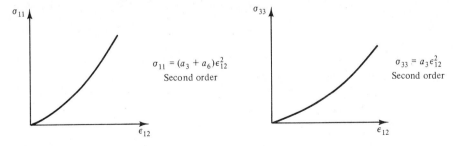

$\sigma_{11} = (a_3 + a_6)\epsilon_{12}^2$
Second order

$\sigma_{33} = a_3 \epsilon_{12}^2$
Second order

Figure 6-8 Symbolic response for simple shear strain state.

normal stresses are developed. This state also induces normal deviatoric stress on shearing planes of the following magnitudes:

$$S_{11} = S_{22} = \tfrac{1}{3} a_6 \epsilon_{12}^2 \qquad (6\text{-}58\text{e})$$

$$S_{33} = -\tfrac{2}{3} a_6 \epsilon_{12}^2 \qquad (6\text{-}58\text{f})$$

To find shear modulus G, it is necessary to perform a test that induces simple shear conditions in which the applied stress σ_{12} and the strain ϵ_{12} are known (measured). The values of a_3 and a_6 will have to be determined from measured responses (Fig. 6-8), expressing the relation between σ_{11} and ϵ_{12}, and between σ_{33} and ϵ_{12}. Then the known values of G, a_3, and a_6 will determine the special form of the second-order Cauchy elastic law [Eq. (6-52)], applicable for predicting the behavior of a medium subjected to simple shear strain.

Since the responses (Fig. 6-8) are nonlinear, and since the values of a_3 and a_6 are constant, for practical use it may become necessary to use weighted values of these constants, often defined at specific levels of stress or strain.

Pure shear stress state.   Consider the pure shear stress state given as

$$\sigma_{ij} = \begin{pmatrix} 0 & \sigma_{12} & 0 \\ \sigma_{12} & 0 & 0 \\ 0 & 0 & 0 \end{pmatrix} \qquad (6\text{-}59\text{a})$$

Substitution in Eq. (6-7) leads to

$$\epsilon_{ij} = \psi_0 \begin{pmatrix} 1 & 0 & 0 \\ 0 & 1 & 0 \\ 0 & 0 & 1 \end{pmatrix} + \psi_1 \begin{pmatrix} 0 & \sigma_{12} & 0 \\ \sigma_{12} & 0 & 0 \\ 0 & 0 & 0 \end{pmatrix} + \psi_2 \begin{pmatrix} \sigma_{12}^2 & 0 & 0 \\ 0 & \sigma_{12}^2 & 0 \\ 0 & 0 & 0 \end{pmatrix}$$

$$(6\text{-}59\text{b})$$

Hence

$$\epsilon_{11} = \psi_0 + \psi_2 \sigma_{12}^2$$

$$\epsilon_{22} = \psi_0 + \psi_2 \sigma_{12}^2$$

$$\epsilon_{33} = \psi_0 \qquad (6\text{-}59\text{c})$$

$$\epsilon_{12} = \psi_1 \sigma_{12}$$

$$\epsilon_{23} = \epsilon_{13} = 0$$

Therefore, the volumetric strain Δv is

$$\Delta v = \epsilon_{ii} = \epsilon_{11} + \epsilon_{22} + \epsilon_{33} = 3\psi_0 + 2\psi_2 \sigma_{12}^2 \qquad (6\text{-}59\text{d})$$

Also, the deviatoric strain tensor is

$$E_{ij} = \epsilon_{ij} - \tfrac{1}{3} \epsilon_{nn} \delta_{ij} \qquad (6\text{-}60\text{a})$$

Hence

$$E_{11} = \epsilon_{11} - \tfrac{1}{3}\epsilon_{nn}\delta_{11} = \epsilon_{11} - \tfrac{1}{3}(\epsilon_{11} + \epsilon_{22} + \epsilon_{33})$$
$$= \tfrac{2}{3}\epsilon_{11} - \tfrac{1}{3}\epsilon_{22} - \tfrac{1}{3}\epsilon_{33} = \tfrac{1}{3}(\epsilon_{11} - \epsilon_{33})$$
$$= \tfrac{1}{3}(\psi_0 + \psi_2\sigma_{12}^2) - \tfrac{1}{3}(\psi_0)$$
$$= \tfrac{1}{3}\psi_2\sigma_{12}^2$$
$$= E_{22} \tag{6-60b}$$

$$\left.\begin{aligned} E_{33} &= -\tfrac{2}{3}\psi_2\sigma_{12}^2 \\ E_{12} &= \psi_1\sigma_{12} \\ E_{23} &= 0 \\ E_{31} &= 0 \end{aligned}\right\} \tag{6-60c}$$

Equation (6-59d) indicates that a state of simple shear stress is accompanied by volume changes during deformation. Equation (6-60b) shows that there exist normal deviatoric strain with the volume changes. We note again that if the third term were not present in Eq. (6-7), that is, the model was first-order Cauchy elastic or Hooke's law, the volume change under shearing stress would not be predicted.

Comment. From the two foregoing states of shear strain and stress, it is evident that the first-order Cauchy elastic or the conventional Hooke's law cannot be used directly to simulate behavior of materials that experience volume changes under shear loading. This is particularly true for many geologic materials.

Hydrostatic stress state. For hydrostatic stress, state $\sigma_{11} = \sigma_{22} = \sigma_{33}$ and $\sigma_{12} = \sigma_{23} = \sigma_{13} = 0$. We have, from Eq. (6-7), volume change as

$$\epsilon_v = \epsilon_{11} + \epsilon_{22} + \epsilon_{33} = 3\psi_0 + 3\psi_1\sigma_{11} + 3\psi_2\sigma_{11}^2 \tag{6-61a}$$

and

$$\epsilon_{12} = \epsilon_{23} = \epsilon_{13} = 0 \tag{6-61b}$$

From the foregoing it is evident that a second-order, nonlinear Cauchy elastic model would predict volume change behavior under both shear and hydrostatic states of stress.

Third-Order Cauchy Elastic Model

It is possible to obtain third-, fourth-, and higher-order Cauchy elastic models based on Eqs. (6-6) and (6-7). In this introductory treatment, we do not plan to go into details of these models. However, brief details of the third-order model are given below.

The third-order Cauchy elastic law can be written by including all cubic terms or terms of the third order in strains. Hence

$$\sigma_{ij} = \left[\left(K - \tfrac{2}{3}G \right) I_1 + a_2 I_1^2 + a_3 I_2 + a_7 I_1^3 + a_8 I_1 I_2 + a_9 I_3 \right] \delta_{ij}$$
$$+ \left(2G + a_5 I_1 + a_{10} I_1^2 + a_{11} I_2 \right) \epsilon_{ij} + \left(a_6 + a_{12} I_1 \right) \epsilon_{im} \epsilon_{mj} \qquad (6\text{-}62)$$

We note that Eq. (6-52) will specialize to the first- and second-order laws if the second- and third-order terms in strains, respectively, were deleted from it.

The third-order law contains *twelve* material parameters to be determined from appropriate laboratory tests. As the order of the model increases, the number of parameters needed to define it also increases. Consequently, the task of their determination becomes more and more difficult.

Second-Order Green Elastic Model

We consider the general Green elastic law,

$$\sigma_{ij} = \frac{\partial U_0}{\partial I_1} \delta_{ij} + \frac{\partial U_0}{\partial I_2} \epsilon_{ij} + \frac{\partial U_0}{\partial I_3} \epsilon_{im} \epsilon_{mj} \qquad (6\text{-}38\text{c})$$

This relation can be used to develop second- and higher-order nonlinear models. For the second-order model, we retain terms up to the second power of strain in the stress–strain law. Then the strain energy density must be expressed as a cubic function in strains or invariants of strain. That is,

$$U_0 = b_0 + b_1 I_1 + b_2 I_1^2 + b_3 I_1^3 + b_4 I_1 I_2 + b_5 I_2 + b_6 I_3 \qquad (6\text{-}63\text{a})$$

where b_0, b_1, b_2, b_3, b_4, b_5, and b_6 are material constants. For a zero state of strain corresponding to a zero state of stress, $b_0 = b_1 = 0$. Then U_0 becomes

$$U_0 = b_2 I_1^2 + b_3 I_1^3 + b_4 I_1 I_2 + b_5 I_2 + b_6 I_3 \qquad (6\text{-}63\text{b})$$

or

$$U_0 = \left(b_2 I_1^2 + b_5 I_2 \right) + b_3 I_1^3 + b_4 I_1 I_2 + b_6 I_3 \qquad (6\text{-}63\text{c})$$

If the third-order terms are ignored, Eq. (6-63c) reduces to

$$U_0 = b_2 I_1^2 + b_5 I_2 \qquad (6\text{-}63\text{d})$$

which represents the *first-order* or *linear elastic* model [Eq. (6-41)]. The parameters b_2 and b_5 are given by

$$b_2 = \frac{\lambda}{2} = \frac{3K - 2G}{6} = \frac{E\nu}{2(1 + \nu)(1 - 2\nu)} \qquad (6\text{-}64\text{a})$$

$$b_5 = 2\mu = 2G = \frac{E}{1 + \nu} \qquad (6\text{-}64\text{b})$$

where λ and μ are Lamé's constants (1). Then

$$U_0 = \left(\frac{3K - 2G}{6} \right) I_1^2 + 2GI_2 + b_3 I_1^3 + b_4 I_1 I_2 + b_6 I_3 \qquad (6\text{-}65)$$

which contains a total of *five* parameters; note that the corresponding Cauchy model required *six* parameters.

Use of Eq. (6-65) with proper derivatives gives

$$\frac{\partial U_0}{\partial I_1} = \left(\frac{3K - 2G}{3} \right) I_1 + 3b_3 I_1^2 + b_4 I_2 \qquad (6\text{-}66a)$$

$$\frac{\partial U_0}{\partial I_2} = 2G + b_4 I_1 \qquad (6\text{-}66b)$$

$$\frac{\partial U_0}{\partial I_3} = b_6 \qquad (6\text{-}66c)$$

Substitution of Eqs. (6-66) into Eq. (6-38c) leads to

$$\sigma_{ij} = \left[\left(K - \frac{2G}{3} \right) I_1 + 3b_3 I_1^2 + b_4 I_2 \right] \delta_{ij} + (2G + b_4 I_1) \epsilon_{ij} + b_6 \epsilon_{im} \epsilon_{mj} \qquad (6\text{-}67)$$

It should be noted that Eq. (6-51a) for Cauchy elastic material has a similar form as Eq. (6-67), which describes Green elastic material. In fact, if $a_3 = a_5$, Eq. (6-51a) gives Green elastic material law. Hence Green elastic material can be considered as a special case of Cauchy material.

In order to understand the properties of the Green elastic law, we consider few special cases of stress and strain conditions.

Uniaxial strain state. The strain tensor for this case is defined in Eq. (6-53). Substitution of Eq. (6-53) into Eq. (6-67) gives

$$\sigma_{11} = \left(K - \frac{2G}{3} \right) \epsilon_{11} + 3b_3 \epsilon_{11}^2 + \frac{1}{2} b_4 \epsilon_{11}^2$$

$$+ (2G + b_4 \epsilon_{11}) \epsilon_{11} + b_6 \epsilon_{11}^2$$

$$= \left(K + \frac{4G}{3} \right) \epsilon_{11} + \left(3b_3 + \frac{3b_4}{2} + b_6 \right) \epsilon_{11}^2 \qquad (6\text{-}68a)$$

$$\sigma_{22} = \left(K - \frac{2G}{3} \right) \epsilon_{11} + \left(3b_3 + \frac{1}{2} b_4 \right) \epsilon_{11}^2 \qquad (6\text{-}68b)$$

$$\sigma_{33} = \left(K - \frac{2G}{3} \right) \epsilon_{11} + \left(3b_3 + \frac{1}{2} b_4 \right) \epsilon_{11}^2 \qquad (6\text{-}68c)$$

and

$$\sigma_{12} = \sigma_{23} = \sigma_{13} = 0 \qquad (6\text{-}68d)$$

Simple shear strain state.   The strain tensor for this case is defined in Eq. (6-55a). Substitution of this into Eq. (6-67) gives

$$\sigma_{11} = \tfrac{1}{2}b_4\epsilon_{12}^2 + b_6\epsilon_{12}^2 = \left(\tfrac{1}{2}b_4 + b_6\right)\epsilon_{12}^2 \qquad (6\text{-}69a)$$

$$\sigma_{22} = \left(\tfrac{1}{2}b_4 + b_6\right)\epsilon_{12}^2 \qquad (6\text{-}69b)$$

$$\sigma_{33} = \tfrac{1}{2}b_4\epsilon_{12}^2 \qquad (6\text{-}69c)$$

$$\sigma_{12} = 2G\epsilon_{12} \qquad (6\text{-}69d)$$

$$\sigma_{13} = \sigma_{23} = 0 \qquad (6\text{-}69e)$$

Equations (6-69a) to (6-69c) imply that a simple shearing strain is accompanied by nonuniform values of normal stresses.

The inverse relationship, that is, the strain–stress relationship, for the Green elastic material can be expressed in a similar form. This can be written as

$$\epsilon_{ij} = \left(2b_2'J_1 + 3b_3'J_1^2 + b_4'J_2\right)\delta_{ij} + \left(b_5' + b_4'J_1\right)\sigma_{ij} + b_6'\sigma_{im}\sigma_{mj} \qquad (6\text{-}70)$$

Here, b_2', b_3', b_4', b_5', and b_6' are the material parameters. This relationship can be specialized for special cases of stresses and strains.

Third-Order Green Elastic Model

$$U_0 = 2GI_2 + \left(\frac{K}{2} - \frac{G}{3}\right)I_1^2 + b_4I_1I_2 + b_3I_1^3 + b_6I_3$$

$$+ b_7I_1^4 + b_8I_2^2 + b_9I_1I_3 + b_{10}I_1^2I_2 \qquad (6\text{-}71a)$$

Here b_8 to b_{10} are additional constants to be determined. Use of Eq. (6-71a) in Eq. (6-38c) leads to

$$\sigma_{ij} = \left[\left(K - \tfrac{2}{3}G\right)I_1 + 3b_3I_1^2 + b_4I_2 + 4b_7I_1^3 + b_9I_3 + 2b_{10}I_1I_2\right]\delta_{ij}$$

$$+ \left(2G + b_4I_1 + 2b_8I_2 + b_{10}I_1^2\right)\epsilon_{ij} + \left(b_6 + b_9I_1\right)\epsilon_{im}\epsilon_{mj} \qquad (6\text{-}71b)$$

By expressing U_0 as a function of higher orders of the invariants of strain, we can obtain third-order, fourth-order, and so on Green elastic models. As the order increases, the number of parameters would increase. For example, a third-order model would require nine parameters. It becomes difficult to perform appropriate tests for determination of such a large number of parameters. Also, for many engineering purposes such higher-order models may not be needed. Such higher-order models will not be covered in this text.

DETERMINATION OF MATERIAL PARAMETERS

The material parameters that define the constitutive laws have to be determined from appropriate laboratory tests. In almost all the laboratory tests on geologic materials, the stress–strain response is measured from a known initial state of stress. Hence it is desirable to express the previously derived stress–strain relationships in incremental forms starting with an initial state of strain, ϵ_{ij}^0, corresponding to an initial state of stress, σ_{ij}^0. In the following section, the derivation of incremental stress–strain relationships and determination of material parameters are treated for the second-order Cauchy and Green elastic materials.

INCREMENTAL STRESS–STRAIN RELATIONSHIP: CAUCHY ELASTIC MATERIAL

The laboratory test data for the soil reported in Chapter 5 have been obtained from stress-controlled tests. Hence in this section we consider the inverse form of the Cauchy model [Eq. (6-7)], so that it can be readily used in determining parameters for the material described in Chapter 5. The inverse relationship for Cauchy material can be expressed as

$$\epsilon_{ij} = \left(a_1' J_1 + a_2' J_1^2 + a_3' J_2 \right) \delta_{ij} + \left(a_4' + a_5' J_1 \right) \sigma_{ij} + a_6' \sigma_{im} \sigma_{mj} \qquad (6\text{-}72)$$

where a_1', a_2', a_3', a_4', a_5', and a_6' are six material parameters to be determined. Substituting $J_1 = \sigma_{mm}$ and $J_2 = \sigma_{mn} \sigma_{nm}/2$ in Eq. (6-72), it is possible to express Eq. (6-72) as

$$\epsilon_{ij} = \left(a_1' \sigma_{mm} + a_2' \sigma_{mm}^2 + \tfrac{1}{2} a_3' \sigma_{mn} \sigma_{nm} \right) \delta_{ij} + \left(a_4' + a_5' \sigma_{mm} \right) \sigma_{ij} + a_6' \sigma_{im} \sigma_{mj}$$

$$(6\text{-}73)$$

As stated in the preceding section, the stress–strain response observed in laboratory tests is measured from a known initial state of stress, but the initial state of strain may be unknown. Therefore, to determine the material parameters, it is desirable to consider incremental form of stress–strain relations with an initial state of strain, ϵ_{ij}^0, corresponding to an initial state of stress σ_{ij}^0. Here, the increments are measured from the initial state, considering it as a fixed reference. If a stress increment (a change in stress) $\Delta \sigma_{ij}$ is added to the current state, then the strain state becomes $\epsilon_{ij} + \Delta \epsilon_{ij}$; here Δ denotes a change. By using Eq. (6-73), we can express the new state as

$$\begin{aligned}
\epsilon_{ij}^0 + \Delta \epsilon_{ij} =\ & \left[a_1' \left(\sigma_{mm}^0 + \Delta \sigma_{mm} \right) + a_2' \left(\sigma_{mm}^0 + \Delta \sigma_{mm} \right)^2 \right. \\
& \left. + \tfrac{1}{2} a_3' \left(\sigma_{mn}^0 + \Delta \sigma_{mn} \right) \left(\sigma_{nm}^0 + \Delta \sigma_{nm} \right) \right] \delta_{ij} \\
& + \left(a_4' + a_5' \sigma_{mm}^0 + a_5' \Delta \sigma_{mm} \right) \left(\sigma_{ij}^0 + \Delta \sigma_{ij} \right) \\
& + a_6' \left(\sigma_{im}^0 + \Delta \sigma_{im} \right) \left(\sigma_{mj}^0 + \Delta \sigma_{mj} \right)
\end{aligned} \qquad (6\text{-}74)$$

Here, the superscript 0 denotes the initial state at the start of an increment. By subtracting Eq. (6-73) from Eq. (6-74), the incremental form can be obtained as

$$\Delta\epsilon_{ij} = \left[a_1'\sigma_{mm} + 2a_2'\sigma_{mm}^0\Delta\sigma_{mm} + a_2'\Delta\sigma_{mm}^2 + a_3'\sigma_{mn}^0\Delta\sigma_{mn} + \tfrac{1}{2}a_3'\Delta\sigma_{mn}\Delta\sigma_{nm}\right]\delta_{ij}$$
$$+ \left[a_4'\Delta\sigma_{ij} + a_5'\sigma_{mm}^0\Delta\sigma_{ij} + a_5'\sigma_{ij}^0\Delta\sigma_{mm} + a_5'\Delta\sigma_{mm}\Delta\sigma_{ij}\right]$$
$$+ a_6'\left[2\sigma_{im}^0\Delta\sigma_{mj} + \Delta\sigma_{im}\Delta\sigma_{mj}\right] \tag{6-75}$$

By rearranging coefficients, we can write this as

$$\Delta\epsilon_{ij} = \left[a_1'\Delta\sigma_{mm} + a_2'\left(2\sigma_{mm}^0 + \Delta\sigma_{mm}\right)\Delta\sigma_{mm} + a_3'\left(\sigma_{mn}^0 + \frac{\Delta\sigma_{mn}}{2}\right)\Delta\sigma_{mn}\right]\delta_{ij}$$
$$+ \left[a_4'\Delta\sigma_{ij} + a_5'\left(\sigma_{mm}^0\Delta\sigma_{ij} + \sigma_{ij}^0\Delta\sigma_{mm} + \Delta\sigma_{mm}\Delta\sigma_{ij}\right)\right] + a_6'\left[2\sigma_{im}^0\Delta\sigma_{mj} + \Delta\sigma_{im}\Delta\sigma_{mj}\right] \tag{6-76}$$

If the initial state of stress is not hydrostatic, then under an increment of stress, the relation in Eq. (6-76) can exhibit what is termed as *stress-induced anisotropy*. It is a consequence of the nonlinear dependence of the strain increment on the total stress at the start of the increment. The material itself is isotropic and is defined by the six parameters a_i' $(i = 1, 2, \ldots, 6)$; however, during the incremental loading, the material can display different responses in different directions.

The relationship in Eq. (6-76) can be specialized to a few testing configurations in order to determine the material parameters. Let us consider an increment of stress state from an arbitrary initial hydrostatic condition. The stress tensor for the hydrostatic state takes the form

$$\sigma_{ij}^0 = \begin{pmatrix} \sigma & 0 & 0 \\ 0 & \sigma & 0 \\ 0 & 0 & \sigma \end{pmatrix} \tag{6-77a}$$

Since the states of stresses are referred to the principal axes, the incremental stress tensor can be expressed as

$$\Delta\sigma_{ij} = \begin{pmatrix} \Delta\sigma_1 & 0 & 0 \\ 0 & \Delta\sigma_2 & 0 \\ 0 & 0 & \Delta\sigma_3 \end{pmatrix} \tag{6-77b}$$

Substituting Eqs. (6-77) into Eq. (6-76), we obtain

$$\Delta\epsilon_{ij} = \left[a_1'\Delta\sigma_{mm} + a_2'(6\sigma + \Delta\sigma_{mm})\Delta\sigma_{mm} + a_3'\left(\sigma\Delta\sigma_{mm} + \tfrac{1}{2}\Delta\sigma_{mn}\Delta\sigma_{nm}\right)\right]\delta_{ij}$$
$$+ \left[a_4'\Delta\sigma_{ij} + a_5'\left(3\sigma\Delta\sigma_{ij} + \sigma\delta_{ij}\Delta\sigma_{mm} + \Delta\sigma_{mm}\Delta\sigma_{ij}\right)\right] + a_6'\left[2\sigma\Delta\sigma_{ij} + \Delta\sigma_{im}\Delta\sigma_{mj}\right] \tag{6-78}$$

In order to determine the material parameters in Eq. (6-78), let us consider some special cases of Eq. (6-77b).

Case 1: Hydrostatic state.   Here the incremental stress tensor [Eq. (6-77b)] becomes

$$\Delta\sigma_{ij} = \begin{pmatrix} \Delta\sigma_1 & 0 & 0 \\ 0 & \Delta\sigma_1 & 0 \\ 0 & 0 & \Delta\sigma_1 \end{pmatrix} \tag{6-79a}$$

For this case, Eq. (6-78) reduces to

$$\Delta\epsilon_1 = \left[(3a_1' + a_4') + (18a_2' + 3a_3' + 6a_5' + 2a_6')\sigma \right] \Delta\sigma_1$$
$$+ \left[9a_2' + \tfrac{3}{2}a_3' + 3a_5' + a_6' \right] \Delta\sigma_1^2 \tag{6-79b}$$

Case 2: Conventional triaxial compression (CTC) state.   For the CTC state of stress [Eq. (6-77b)],

$$\Delta\sigma_{ij} = \begin{pmatrix} \Delta\sigma_1 & 0 & 0 \\ 0 & 0 & 0 \\ 0 & 0 & 0 \end{pmatrix} \tag{6-80a}$$

For this case Eq. (6-78) reduces to

$$\Delta\epsilon_1 = \left[(a_1' + a_4') + (6a_2' + a_3' + 4a_5' + 2a_6')\sigma \right] \Delta\sigma_1 + \left[a_2' + \tfrac{1}{2}a_3' + a_5' + a_6' \right] \Delta\sigma_1^2 \tag{6-80b}$$

$$\Delta\epsilon_2 = \left[a_1' + (6a_2' + a_3' + a_5')\sigma \right] \Delta\sigma_1 + \left(a_2' + \tfrac{1}{2}a_3' \right) \Delta\sigma_1^2 \tag{6-80c}$$

Case 3: Conventional triaxial extension (CTE) state.   The incremental stress tensor [Eq. (6-77b)] becomes

$$\Delta\sigma_{ij} = \begin{pmatrix} 0 & 0 & 0 \\ 0 & \Delta\sigma_2 & 0 \\ 0 & 0 & \Delta\sigma_2 \end{pmatrix} \tag{6-81a}$$

For this case, Eq. (6-78) reduces to

$$\Delta\epsilon_1 = \left[2a_1' + (12a_2' + 2a_3' + 2a_5')\sigma \right] \Delta\sigma_2 + (4a_2' + a_3') \Delta\sigma_2^2 \tag{6-81b}$$

$$\Delta\epsilon_2 = \left[(2a_1' + a_4') + (12a_2' + 2a_3' + 5a_5' + 2a_6')\sigma \right] \Delta\sigma_2$$
$$+ \left[4a_2' + a_3' + 2a_5' + a_6' \right] \Delta\sigma_2^2 \tag{6-81c}$$

Case 4: Reduced triaxial extension (RTE) state.   The incremental stress tensor [Eq. (6-77b)] becomes

$$\Delta\sigma_{ij} = \begin{pmatrix} -\Delta\sigma_1 & 0 & 0 \\ 0 & 0 & 0 \\ 0 & 0 & 0 \end{pmatrix} \tag{6-82a}$$

For this case, Eq. (6-78) reduces to

$$\Delta\epsilon_1 = -\left[(a_1' + a_4') + (6a_2' + a_3' + 4a_5' + 2a_6')\sigma \right] \Delta\sigma_1$$
$$+ \left[a_2' + \tfrac{1}{2}a_3' + a_5' + a_6' \right] \Delta\sigma_1^2 \tag{6-82b}$$

$$\Delta\epsilon_2 = -\left[a_1' + (6a_2' + a_3' + a_5')\sigma \right] \Delta\sigma_1 + \left(a_2' + \tfrac{1}{2}a_3' \right) \Delta\sigma_1^2 \tag{6-82c}$$

Case 5: Triaxial compression (TC) state. The incremental stress tensor [Eq. (6-77b)] becomes

$$\Delta\sigma_{ij} = \begin{pmatrix} \Delta\sigma_1 & 0 & 0 \\ 0 & -\frac{1}{2}\Delta\sigma_1 & 0 \\ 0 & 0 & -\frac{1}{2}\Delta\sigma_1 \end{pmatrix} \tag{6-83a}$$

For this case, Eq. (6-78) reduces to

$$\Delta\epsilon_1 = \left[a_4' + (3a_5' + 2a_6')\sigma \right] \Delta\sigma_1 + \left[\frac{3}{4}a_3' + a_6' \right] \Delta\sigma_1^2 \tag{6-83b}$$

$$\Delta\epsilon_2 = -\left[\frac{1}{2}a_4' + \left(\frac{3}{2}a_5' + a_6' \right)\sigma \right] \Delta\sigma_1 + \left[\frac{3}{4}a_3' + \frac{1}{4}a_6' \right] \Delta\sigma_1^2 \tag{6-83c}$$

Case 6: Triaxial extension (TE) state. The incremental stress tensor [Eq. (6-77b)] becomes

$$\Delta\sigma_{ij} = \begin{pmatrix} -\Delta\sigma_1 & 0 & 0 \\ 0 & \frac{1}{2}\Delta\sigma_1 & 0 \\ 0 & 0 & \frac{1}{2}\Delta\sigma_1 \end{pmatrix} \tag{6-84a}$$

This can be substituted in Eq. (6-78) to obtain

$$\Delta\epsilon_1 = -\left[a_4' + (3a_5' + 2a_6')\sigma \right] \Delta\sigma_1 + \left[\frac{3}{4}a_3' + a_6' \right] \Delta\sigma_1^2 \tag{6-84b}$$

$$\Delta\epsilon_2 = \left[\frac{1}{2}a_4' + \left(\frac{3}{2}a_5' + a_6' \right)\sigma \right] \Delta\sigma_1 + \left[\frac{3}{4}a_3' + \frac{1}{4}a_6' \right] \Delta\sigma_1^2 \tag{6-84c}$$

Case 7: Simple shear (SS) state of stress. In terms of the principal stresses, the stress tensor for the SS state can be expressed as

$$\Delta\sigma_{ij} = \begin{pmatrix} \Delta\sigma_1 & 0 & 0 \\ 0 & 0 & 0 \\ 0 & 0 & -\Delta\sigma_1 \end{pmatrix} \tag{6-85a}$$

Therefore, Eq. (6-78) takes the following form:

$$\Delta\epsilon_1 = \left[a_4' + (3a_5' + 2a_6')\sigma \right] \Delta\sigma_1 + \left[a_3' + a_6' \right] \Delta\sigma_1^2 \tag{6-85b}$$

$$\Delta\epsilon_2 = a_3' \Delta\sigma_1^2 \tag{6-85c}$$

$$\Delta\epsilon_3 = -\left[a_4' + (3a_5' + 2a_6')\sigma \right] \Delta\sigma_1 + \left[a_3' + a_6' \right] \Delta\sigma_1^2 \tag{6-85d}$$

PROCEDURE FOR EVALUATION OF PARAMETERS

Equations (6-79) to (6-85) give relationships between stress and strain under different testing conditions. Hence the material parameters can be determined by using some or all of Eqs. (6-79) to (6-85) with data from corresponding testing configurations. For the second-order Cauchy material under consideration, there are six independent material parameters to be determined.

In order to determine the six parameters, we need at least six independent equations with the parameters as unknowns. By inspection of Eq. (6-79b), it can be seen that six data points from a hydrostatic compression test would give six simultaneous equations, and hence the six material parameters can be obtained by solving them. This can also be done with the CTC test data by using Eq. (6-80b). When the material parameters are determined from only one type of test, the model may not be able to reproduce behavior of a different kind of test. It should be noted that if the material is ideally a second-order Cauchy elastic type, then the parameters will not depend on the type of test.

There are unavoidable errors in testing procedures, and hence there may occur variations in the parameters if they are determined from different tests. Hence if we could use the data obtained from tests conducted under different stress paths, the errors can be rounded over the whole range of the stress space.

As an example, let us consider the case of the hydrostatic compression (HC) state. The relationship between the strain and the stress is given in Eq. (6-79b). For a given value of initial hydrostatic state, Eq. (6-79b) can be written as

$$(\Delta\epsilon_1)_i = [A](\Delta\sigma_1)_i + [B](\Delta\sigma_1)_i^2 \qquad (i = 1, 2, \ldots, n) \qquad (6\text{-}86)$$

Here A and B are functions of a_1' to a_6', and subscript i denotes the data point number. If more than two data points are selected, there will be more than two equations to determine A and B. That is, there will be an overdetermined set of equations in A and B. In order to obtain appropriate values of A and B, the overdetermined set of simultaneous equations can be solved by using a numerical method such as the *least-squares technique* (9); this is described in the following section. For n number of data points, Eq. (6-79b) can be written as

$$
\begin{aligned}
(\Delta\epsilon_1)_1 &= A(\Delta\sigma_1)_1 + B(\Delta\sigma_1)_1^2 \\
(\Delta\epsilon_1)_2 &= A(\Delta\sigma_1)_2 + B(\Delta\sigma_1)_2^2 \\
&\cdots\cdots\cdots\cdots\cdots\cdots\cdots \\
(\Delta\epsilon_1)_n &= A(\Delta\sigma_1)_n + B(\Delta\sigma_1)_n^2
\end{aligned}
\qquad (6\text{-}87a)
$$

In matrix form, Eq. (6-87a) can be expressed as

$$
\begin{Bmatrix} (\Delta\epsilon_1)_1 \\ (\Delta\epsilon_1)_2 \\ \cdots \\ (\Delta\epsilon_1)_n \end{Bmatrix}
=
\begin{bmatrix} (\Delta\sigma_1)_1 & (\Delta\sigma_1)_1^2 \\ (\Delta\sigma_1)_2 & (\Delta\sigma_1)_2^2 \\ \cdots & \cdots \\ (\Delta\sigma_1)_n & (\Delta\sigma_1)_n^2 \end{bmatrix}
\begin{Bmatrix} A \\ \\ B \end{Bmatrix}
\qquad (6\text{-}87b)
$$

or

$$
\begin{array}{ccc}
\{\Delta\epsilon_1\} & = & [\mathbf{S}] \quad \langle\mathbf{X}\rangle \\
(n \times 1) & & (n \times 2)\,(2 \times 1)
\end{array}
\qquad (6\text{-}87c)
$$

where n is the number of points chosen on a given stress–strain curve. This can be solved by using the least-squares method as (9)

$$[S]^T \quad [W] \quad [S] \quad \{X\} = [S]^T \quad [W] \quad \{\Delta\epsilon_1\} \qquad (6\text{-}88)$$
$$(2 \times n) \; (n \times n) \; (n \times 2) \; (2 \times 1) \quad (2 \times n) \; (n \times n) \; (n \times 1)$$

Here $[W]$ is a diagonal matrix having the weight given for each data point. When equal weights are given to all data points, $[W]$ is a unit diagonal matrix. Equation (6-88) can be written as

$$[D] \quad \{X\} = \{q\} \qquad (6\text{-}89)$$
$$(2 \times 2) \; (2 \times 1) \quad (2 \times 1)$$

By inverting Eq. (6-89), the values of A and B can be found. This will give two equations containing the material constants and known values of A and B. This procedure can be repeated for a number of tests. Therefore, one can obtain as many sets of equations as the number of tests. These equations can also be overdetermined, and can be solved for $a_1', a_2', \ldots, a_6'$, by using the least-squares technique again.

Further details and a computer routine for the least-squares method are given in Appendix 3.

Example 6-1

In this example the material parameters for the inverse second-order Cauchy model are determined. The experimental data for soil given in Chapter 5 are used. We consider a few testing configurations, involving initial density of 2.0 g/cm^3 or close to it.

It may be noted that the higher order elastic models are essentially relevant for (monotonic) loading; hence the envelope of leading is used, and the unloading–reloading curves are not considered. Also, for materials exhibiting anisotropy, say in two extension strains in the CTC tests (Fig. 5-15), an average value is used; this is because these models are valid for isotropic materials.

Hydrostatic test: The data shown in Fig. 5-14(b) are used. The corresponding density is 1.82 g/cm^3, and the initial confining pressure σ_0 is 2.0 psi (13.8 kPa). Substituting the value of σ_0 for σ in Eq. (6-79b), the following expression can be obtained:

$$\Delta\epsilon_1 = A\,\Delta\sigma_1 + B\,\Delta\sigma_1^2 \qquad (6\text{-}90)$$

where

$$A = 3a_1' + 36a_2' + 6a_3' + a_4' + 12a_5' + 4a_6'$$

$$B = 9a_2' + \tfrac{3}{2}a_3' + 3a_5' + a_6'$$

By using the least-squares fitting method described previously, we obtain values of A and B; hence

$$A = 3a_1' + 36a_2' + 6a_3' + a_4' + 12a_5' + 4a_6' = 0.004 \qquad (6\text{-}91a)$$

$$B = 9a_2' + \tfrac{3}{2}a_3' + 3a_5' + a_6' = -0.0000467 \qquad (6\text{-}91b)$$

Triaxial compression test: Experimental data shown in Fig. 5-19(a) are used; the corresponding density is 2.04 g/cm^3, and the initial confining pressure is 10 psi (69 kPa). Substituting the value of the initial pressure in Eq. (6-83b), we have

$$\Delta\epsilon_1 = A\,\Delta\sigma_1 + B\,\Delta\sigma_1^2 \tag{6-92}$$

where

$$A = a_4' + 30a_5' + 20a_6'$$
$$B = \tfrac{3}{4}a_3' + a_6'$$

Use of the experimental data gives the following values for A and B:

$$A = a_4' + 30a_5' + 20a_6' = -0.00405 \tag{6-93a}$$
$$B = \tfrac{3}{4}a_3' + a_6' = 0.0011 \tag{6-93b}$$

Using Eq. (6-83c) for $\sigma = 10$, we obtain

$$\Delta\epsilon_2 = A\,\Delta\sigma_1 + B\,\Delta\sigma_1^2 \tag{6-94}$$

where

$$A = -\tfrac{1}{2}a_4' - 15a_5' - 10a_6'$$
$$B = \tfrac{3}{4}a_3' + \tfrac{1}{4}a_6'$$

Use of the experimental data in Fig. 5-19(a) gives

$$A = -\tfrac{1}{2}a_4' - 15a_5' - 10a_6' = 0.00318 \tag{6-95a}$$
$$B = \tfrac{3}{4}a_3' + \tfrac{1}{4}a_6' = -0.000603 \tag{6-95b}$$

Simple shear test: Figure 5-21(b) shows the data obtained from a SS test at an initial density of 2.05 g/cm^3. The initial confinement is 10 psi (69 kPa). By using Eq. (6-85b) for $\sigma = 10$ psi, we obtain

$$\Delta\epsilon_1 = A\,\Delta\sigma_1 + B\,\Delta\sigma_1^2 \tag{6-96}$$

where

$$A = a_4' + 30a_5' + 20a_6'$$
$$B = a_3' + a_6'$$

By using the least-squares fitting method, we have

$$A = a_4' + 30a_5' + 20a_6' = -0.00423 \tag{6-97a}$$
$$B = a_3' + a_6' = 0.00138 \tag{6-97b}$$

Use of Eq. (6-85d) for $\sigma = 10$ psi (69 kPa) gives

$$\Delta\epsilon_3 = A\,\Delta\sigma_1 + B\,\Delta\sigma_1^2 \tag{6-98}$$

where

$$A = -a_4' - 30a_5' - 20a_6'$$
$$B = a_3' + a_6'$$

Use of the experimental data in Fig. 5-21(b) gives the values of A and B as

$$A = -a_4' - 30a_5' - 20a_6' = 0.0054 \tag{6-99a}$$
$$B = a_3' + a_4' = -0.00138 \tag{6-99b}$$

Conventional triaxial extension: Here the data shown in Fig. 5-17 with the initial density of 1.96 g/cm^3 are used. The initial confining pressure is 10 psi (69 kPa). Substituting this value of σ_0 in Eq. (6-81b), the following expression for $\Delta\epsilon_1$ can be obtained:

$$\Delta\epsilon_1 = A\,\Delta\sigma_2 + B\,\Delta\sigma_2^2 \tag{6-100}$$

where

$$A = 2a_1' + 120a_2' + 20a_3' + 20a_5'$$

$$B = 4a_2' + a_3'$$

By using the least-squares method, values of A and B can be found. In this case,

$$A = 2a_1' + 120a_2' + 20a_3' + 20a_5' = -0.000544 \tag{6-101a}$$

$$B = 4a_2' + a_3' = -0.000131 \tag{6-101b}$$

Upon substitution of the value of σ_0, the expression given in Eq. (6-81c) for $\Delta\epsilon_2$ reduces to the following form:

$$\Delta\epsilon_2 = A\,\Delta\sigma_2 + B\,\Delta\sigma_2^2 \tag{6-102}$$

where

$$A = 2a_1' + a_4' + 120a_2' + 20a_3' + 50a_5' + 20a_6'$$

$$B = 4a_2' + a_3' + 2a_5' + a_6'$$

Analysis of experimental data gives

$$A = 2a_1' + a_4' + 120a_2' + 20a_3' + 50a_5' + 20a_6' = 0.00268 \tag{6-103a}$$

$$B = 4a_2' + a_3' + 2a_5' + a_6' = 0.0000557 \tag{6-103b}$$

A summary of all the computations above is given in Table 6-1.

TABLE 6-1   Summary of Computations for Example 6-1

Test	Equation	Coefficients of Equations						Value of Constant
		a_1'	a_2'	a_3'	a_4'	a_5'	a_6'	
HC	(6-91a)	3	36	6	1	12	4	0.4×10^{-2}
HC	(6-91b)	0	9	1.5	0	3	1	-0.467×10^{-4}
TC	(6-93a)	0	0	0	1	30	20	-0.405×10^{-2}
TC	(6-93b)	0	0	0.75	0	0	1	0.11×10^{-2}
TC	(6-95a)	0	0	0	-0.5	-15	-10	0.318×10^{-2}
TC	(6-95b)	0	0	0.75	0	0	0.25	-0.603×10^{-2}
SS	(6-97a)	0	0	0	1	30	20	-0.423×10^{-2}
SS	(6-97b)	0	0	1	0	0	1	0.138×10^{-2}
SS	(6-99a)	0	0	0	-1	-30	-20	0.540×10^{-2}
SS	(6-99b)	0	0	1	0	0	1	-0.138×10^{-2}
CTE	(6-101a)	2	120	20	0	20	0	-0.544×10^{-3}
CTE	(6-101b)	0	4	1	0	0	0	-0.131×10^{-2}
CTE	(6-103a)	2	120	20	1	50	20	0.268×10^{-2}
CTE	(6-103b)	0	4	1	0	2	1	0.557×10^{-4}

By using the least-squares fitting method, the equations given in Table 6-1 can be solved to obtain the values of a'_1, a'_2, a'_3, a'_4, a'_5, and a'_6 as follows:

$$a'_1 = 0.00628 \ (\text{psi})^{-1}$$
$$a'_2 = 0.00339 \ (\text{psi})^{-2}$$
$$a'_3 = -0.00161 \ (\text{psi})^{-2}$$
$$a'_4 = -0.0140 \ (\text{psi})^{-1} \qquad (6\text{-}104)$$
$$a'_5 = -0.000907 \ (\text{psi})^{-2}$$
$$a'_6 = 0.00188 \ (\text{psi})^{-2}$$

Comment. In order to study the accuracy of the second-order Cauchy model just established, the observed deformation behavior is predicted by using the model with the computed parameters [Eq. (6-104)] for a few stress paths as shown in Figs. 6-9 and 6-10. These figures show that the predictions

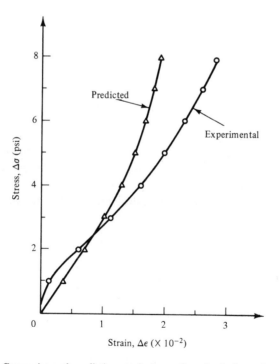

Figure 6-9 Comparison of predictions and observations for hydrostatic compression test using the Cauchy elastic model: $\sigma_0 = 2.75$ psi [corresponds to zero state of strain in Fig. 5-14(b)].

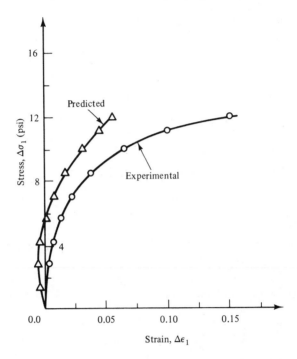

Figure 6-10 Comparison of predictions and observations for triaxial compression test using the Cauchy elastic model [Fig. 5-19(a)]: $\sigma_0 = 10$ psi.

seem to match the observed data satisfactorily only in an approximate sense. This can imply that the material does not behave as a second-order Cauchy model. The predictions may have been improved if the initial slopes of stress–strain curves were enforced as a condition in the determination of parameters. However, in the foregoing example, the initial slopes were not considered, since it was meant only for elaborating the numerical procedure for evaluating material parameters. Subsequently, in Example 6-2, the procedure for incorporating initial slopes is described.

INCREMENTAL STRESS–STRAIN RELATIONSHIP: GREEN ELASTIC MATERIAL

In this section the inverse relationship, that is, the strain–stress relationship in Eq. (6-70) for second-order Green material, will be considered first. Then the incremental form of the stress–strain relationship, Eq. (6-67), will be obtained.

By using a procedure similar to that described in the preceding section [Eqs. (6-73) to (6-76)], the incremental strain–stress relationship for the Green elastic material can be obtained as

$$\Delta\epsilon_{ij} = \left[\left(2b_2' + 6b_3'\sigma_{mm}^0\right)\delta_{ij} + b_4'\sigma_{ij}^0\right]\Delta\sigma_{mm} + \left(b_5' + b_4'\sigma_{mm}^0\right)\Delta\sigma_{ij}$$
$$+ b_4'\sigma_{mn}^0 \Delta\sigma_{mn}\epsilon_{ij} + b_6'\left(\sigma_{im}^0 \Delta\sigma_{mj} + \sigma_{mj}^0\Delta\sigma_{im}\right)$$
$$+ \left[3b_3'(\Delta\sigma_{mm})^2 + \tfrac{1}{2}b_4'\Delta\sigma_{mn}\Delta\sigma_{nm}\right]\delta_{ij}$$
$$+ b_4'\Delta\sigma_{mm}\Delta\sigma_{ij} + b_6'\Delta\sigma_{im}\Delta\sigma_{mj} \tag{6-105}$$

Here, the superscript 0 denotes initial conditions before application of the increment. Since most of the laboratory tests are carried out from an initial hydrostatic state of stress, let us consider an arbitrary initial hydrostatic condition [Eq. (6-77a)]. Since all the states of stresses are referred to the principal axes, the incremental stress tensor during testing takes the general form given in Eq. (6-77b). Substituting Eq. (6-77b) into Eq. (6-105), we obtain

$$\Delta\epsilon_{ij} = \left[6b_2' + (54b_3' + 3b_4')\sigma\right]\Delta p\delta_{ij} + (b_5' + 3b_4')\Delta\sigma_{ij}$$
$$+ \left(3b_4'\sigma\Delta p\delta_{ij} + 2b_6'\sigma\Delta\sigma_{ij}\right)$$
$$+ \left[27b_3'(\Delta p)^2 + \tfrac{1}{2}b_4'(3\Delta p)^2\right.$$
$$\left. - \tfrac{1}{2}b_4'(2\,\Delta\sigma_1\,\Delta\sigma_2 + 2\,\Delta\sigma_1\,\Delta\sigma_3 + 2\,\Delta\sigma_2\,\Delta\sigma_3)\right]\delta_{ij}$$
$$+ 3b_4'\Delta p\,\Delta\sigma_{ij} + b_6'\Delta\sigma_{im}\,\Delta\sigma_{mj} \tag{6-106}$$

where $\Delta p = \tfrac{1}{3}(\Delta\sigma_1 + \Delta\sigma_2 + \Delta\sigma_3)$. In a manner similar to one described for the Cauchy material, Eq. (6-106) can be specialized for selected stress paths. The data utilized in Example 6-1 can be used to evaluate parameters in Eq. (6-106) for the second-order Green materials.

Example 6-2

In this example, material parameters for the inverse second-order Green model are determined for the experimental data given in Chapter 5. Here the data used are the same as those used in Example 6-1 for the Cauchy model. In this example, the initial slopes of observed stress–strain curves are incorporated as shown below.

Hydrostatic test: By simplifying Eq. (6-106), following expression can be obtained.

$$\Delta\epsilon = A\,\Delta\sigma + B\,\Delta\sigma^2 \tag{6-107}$$

where

$$A = (6b_2' + b_5') + (54b_3' + 9b_4' + 2b_6')\sigma$$
$$B = 27b_3' + \tfrac{9}{2}b_4' + b_6'$$

The slope of the stress–strain curve can be related to parameter A as

$$\left(\frac{\Delta\epsilon}{\Delta\sigma}\right)_{\Delta\sigma\to 0} = A = \frac{1}{\text{slope}} \tag{6-108a}$$

Once the value of A is determined from Eq. (6-108a), the value of B can be found from

Eq. (6-107) by using a number of data points with a least-squares method. For this test the following values of A and B have been computed; the initial slope was computed at $\sigma = 2.75$ psi.

$$A = (6b_2' + b_5') + (54b_3' + 9b_4' + 2b_6')2.75 = 0.00520 \qquad (6\text{-}108b)$$

$$B = 0.00013 \qquad (6\text{-}108c)$$

Triaxial compression test: $\sigma_0 = 10$ psi [Fig. 5-19(a)]. For this test, Eq. (6-106) can be simplified as

$$\Delta\epsilon_1 = A\,\Delta\sigma_1 + B\,\Delta\sigma_1^2 \qquad (6\text{-}109)$$

where

$$A = b_5' + 30b_4' + 20b_6'$$

$$B = b_6' + \tfrac{3}{4}b_4'$$

The initial slope of the stress–strain curve can be related as

$$\left(\frac{\Delta\epsilon_1}{\Delta\sigma_1}\right)_{\Delta\sigma_1 \to 0} = A = 0.00166 \qquad (6\text{-}110a)$$

Substituting Eq. (6-110a) into Eq. (6-109), and considering a number of values of $\Delta\sigma$, the value of B was found as

$$B = 0.000381 \qquad (6\text{-}110b)$$

For the same test, $\Delta\epsilon_2$ can be expressed as

$$\Delta\epsilon_2 = A\,\Delta\sigma_1 + B\,\Delta\sigma_1^2 \qquad (6\text{-}111)$$

where

$$A = -\left(\frac{b_5'}{2} + 15b_4' + 10b_6'\right)$$

$$B = \tfrac{1}{4}(b_6' + 3b_4')$$

By incorporating the initial slope in a manner similar to that described previously, the values of A and B have been found as

$$A = 0.00083 \qquad (6\text{-}112a)$$

$$B = -0.0000866 \qquad (6\text{-}112b)$$

Simple shear test: $\sigma_0 = 10$ psi [Fig. 5-21(b)]. For a simple shear test, Eq. (6-106) can be simplified as

$$\Delta\epsilon_1 = A\,\Delta\sigma_1 + B\,\Delta\sigma_1^2 \qquad (6\text{-}113)$$

where

$$A = b_5' + 30b_4' + 20b_6'$$

$$B = b_4' + b_6'$$

By incorporating the initial slope, the value of A can be computed as

$$A = 0.00170 \tag{6-114a}$$

The value of B can be found from Eq. (6-113) by substituting the value of A for a number of $\Delta\sigma$ values, and by using a least-squares method, as

$$B = 0.000381 \tag{6-114b}$$

Now, by combining Eqs. (6-108), (6-110), (6-112), and (6-114), and using a least-squares method, the following values of b'_i were found:

$$b'_2 = 0.133 \times 10^{-2} \, (\text{psi})^{-1}$$

$$b'_3 = 0.223 \times 10^{-4} \, (\text{psi})^{-2}$$

$$b'_4 = -0.306 \times 10^{-2} \, (\text{psi})^{-2} \tag{6-115}$$

$$b'_5 = -0.209 \times 10^{-2} \, (\text{psi})^{-1}$$

$$b'_6 = 0.647 \times 10^{-3} \, (\text{psi})^{-2}$$

Figures 6-11 to 6-13 show a comparison of measured and predicted behavior for three stress paths. The predicted behavior for the foregoing stress paths is in fairly good agreement with measured values.

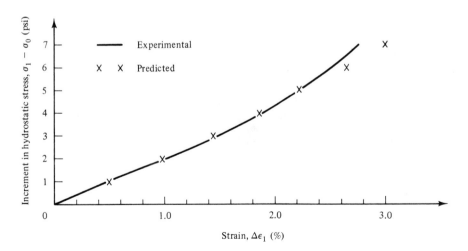

Figure 6-11 Comparison of predictions and observations for hydrostatic compression test using the Green elastic model: $\sigma_0 = 2.75$ psi [corresponds to zero state of strain, Fig. 5-14(b)].

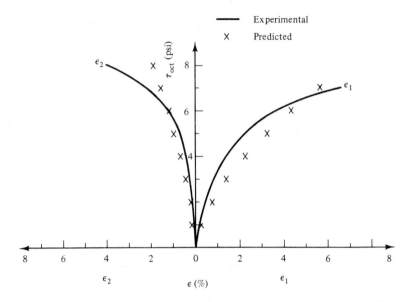

Figure 6-12 Comparison of predictions and observations for triaxial compression test using the Green elastic model [Fig. 5-19(a)]: $\sigma_0 = 10$ psi.

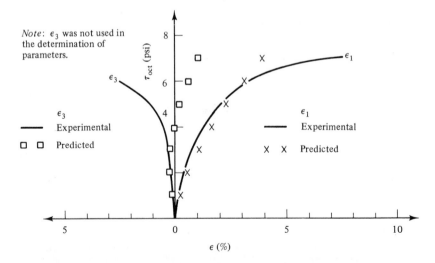

Figure 6-13 Comparison of predictions and observations for simple shear test using the Green elastic model [Fig. 5-21(b)]: $\sigma_0 = 10$ psi.

Example 6-3

In this example we consider the stress–strain (not strain–stress) relationship for second-order Green elastic material, as given in Eq. (6-67). We start from an initial state of strain, ϵ_{ij}^0, and stress, σ_{ij}^0. Then we add an increment of load causing an increment of stress, $\Delta\sigma_{ij}$, and an increment of strain, $\Delta\epsilon_{ij}$. Then Eq. (6-67) becomes

$$\sigma_{ij}^0 + \Delta\sigma_{ij} = \left[\left(K - \frac{2G}{3} \right) \left(\epsilon_{nn}^0 + \Delta\epsilon_{nn} \right) + 3b_3 \left(\epsilon_{nn}^0 + \Delta\epsilon_{nn} \right)^2 \right.$$
$$+ \tfrac{1}{2} b_4 \left(\epsilon_{nm}^0 + \Delta\epsilon_{nm} \right) \left(\epsilon_{nm}^0 + \Delta\epsilon_{nm} \right) \left. \right] \delta_{ij}$$
$$+ \left[2G + b_4 \left(\epsilon_{nn}^0 + \Delta\epsilon_{nn} \right) \right] \left(\epsilon_{ij}^0 + \Delta\epsilon_{ij} \right) + b_6 \left(\epsilon_{ik}^0 + \Delta\epsilon_{ik} \right) \left(\epsilon_{kj}^0 + \Delta\epsilon_{kj} \right)$$

$$(6\text{-}116)$$

If Eq. (6-67) is subtracted from Eq. (6-116), we obtain

$$\Delta\sigma_{ij} = \left[\left(K - \frac{2G}{3} \right) + 6b_3 \epsilon_{nn}^0 + b_4 \epsilon_{ij}^0 \right] \Delta\epsilon_{nn}$$
$$+ \left(2G + b\epsilon_{nn}^0 \right) \Delta\epsilon_{ij} + b_4 \epsilon_{nm}^0 \Delta\epsilon_{nm} \delta_{ij} + b_6 \left(\epsilon_{ik}^0 \Delta\epsilon_{kj} + \epsilon_{kj}^0 \Delta\epsilon_{ik} \right)$$
$$+ \left[3b_3 \left(\Delta\epsilon_{nn} \right)^2 + \tfrac{1}{2} b_4 \Delta\epsilon_{nm} \Delta\epsilon_{nm} \right] \delta_{ij} + b_4 \Delta\epsilon_{nn} \Delta\epsilon_{ij} + b_6 \Delta\epsilon_{ik} \Delta\epsilon_{kj}$$

$$(6\text{-}117)$$

In matrix notation, the relationship above can be expressed as

$$\{\Delta\sigma\} = [\mathbf{C}_L]\{\Delta\epsilon\} + [\mathbf{C}_{NL}]\{\Delta\epsilon^2\} \qquad (6\text{-}118)$$

where $[\mathbf{C}_L]$ corresponds to the linear part and $[\mathbf{C}_{NL}]$ to the nonlinear part. If the higher-order terms are neglected, the relationship above reduces to

$$\Delta\sigma_{ij} = C_{ijkl}^L \Delta\epsilon_{kl} \qquad (6\text{-}119a)$$

or

$$\{\Delta\sigma\} = [\mathbf{C}_L]\{\Delta\epsilon_{kl}\} \qquad (6\text{-}119b)$$

where

$$C_{ijkl}^L = \left[\left(K - \frac{2G}{3} \right) + 6b_3 \epsilon_{nn}^0 \right] \delta_{ij}\delta_{kl} + \left(2G + b_4 \epsilon_{nn}^0 \right) \delta_{ik}\delta_{jl}$$
$$+ b_4 \left(\epsilon_{ij}^0 \delta_{kl} + \epsilon_{kl}^0 \delta_{ij} \right) + b_6 \left(\epsilon_{ik}^0 \delta_{jl} + \epsilon_{ij}^0 \delta_{ik} \right) \qquad (6\text{-}120)$$

It can be noted that if the initial state of strain is zero, Eq. (6-120) reduces to the incremental Hooke's law. If the initial state of stress is neither hydrostatic nor nonzero, the material tensor in Eq. (6-120) becomes anisotropic, which is referred to as *stress-induced anisotropy*. Equation (6-120) has been employed by Chang et al. (5) for modeling the behavior of Ottawa sand.

The work of Chang et al. (5) has been one of the initial studies of hyperelastic (Green elastic) model for simulation of the behavior of Ottawa sand. The following material parameters for the Ottawa sand have been reported by Chang et al. (5) using a curve-fitting approach:

$$K - \frac{2G}{3} = 4.6 \times 10^3 \text{ psi } \left(31.69 \times 10^3 \text{ kN/m}^2 \right)$$

$$G = 4.6 \times 10^3 \text{ psi } \left(31.69 \times 10^3 \text{ kN/m}^2 \right)$$

$$b_3 = -1.2 \times 10^6 \text{ psi } \left(-8.27 \times 10^6 \text{ kN/m}^2 \right)$$

$$b_4 = -0.2 \times 10^6 \text{ psi } \left(-1.378 \times 10^6 \text{ kN/m}^2 \right)$$

$$b_6 = 1.2 \times 10^6 \text{ psi } \left(-1.378 \times 10^6 \text{ kN/m}^2 \right)$$

Figures 6-14 to 6-16 show results of comparisons between predictions and laboratory measurements for hydrostatic, simple shear, and one-dimensional compression tests, respectively, as reported by Chang et al. (5). These test data have been obtained by using a truly triaxial or multiaxial testing device. It should be noted that the constants above simulate the material behavior satisfactorily only over limited ranges of the behavior.

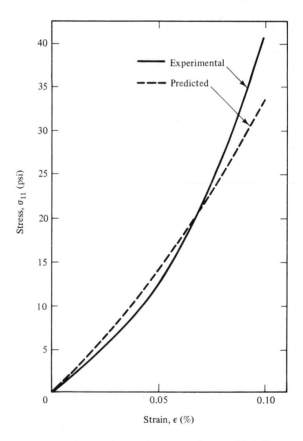

Figure 6-14 Hydrostatic compression test (Ref. 5).

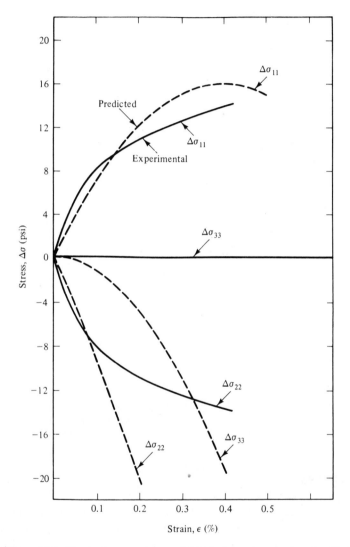

Figure 6-15 Simple shear test from an initial hydrostatic state: $b_0 = 7 \times 10^{-4}$ (Ref. 5).

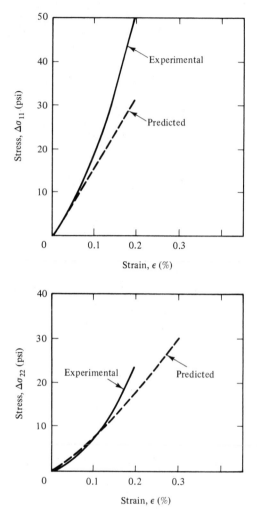

Figure 6-16 One-dimensional compression test: $b_0 = 2.4 \times 10^{-4}$ (Ref. 5).

APPLICATIONS

Applications of the linear elastic (first-order Green or Cauchy) models for solving boundary value problems are common in the literature, and hence will not be considered herein. In fact, it is the simplest model one can use to describe the material behavior, although in many instances it does not appropriately describe the real behavior.

Second-order hyperelastic models described previously in this chapter have been used in conjunction with the finite element method for solving some boundary value problems (5, 10). The second-order Cauchy elastic model given in Eq. (6-52), in general, will lead to a nonsymmetric constitutive matrix, while the Green elastic model leads to a symmetric constitutive matrix. Therefore, from a viewpoint of computational effort, implementation of second-order Green elastic model can be easier than the second-order Cauchy elastic model. The incremental stress–strain relationship for second-order hyperelastic models are given in Eqs. (6-76) and (6-117), and can easily be used with a nonlinear finite element formulation.

Chang et al. (5) have implemented the second-order Green elastic model in an incremental finite element procedure which uses triangular elements. As an example, they solved a problem relating to plane strain pure shear test behavior. The conditions used in their example were

$$\epsilon_{11}^0 = \epsilon_{22}^0 = \epsilon_{33}^0 = b \ (\text{constant})$$

$$\epsilon_{ij}^0 = 0 \qquad \text{for } i \neq j$$

$$\Delta\epsilon_{11} = +\epsilon$$

$$\Delta\epsilon_{22} = -\epsilon$$

$$\Delta\epsilon_{33} = 0$$

$$\Delta\epsilon_{ij} = 0 \qquad \text{for } i \neq j$$

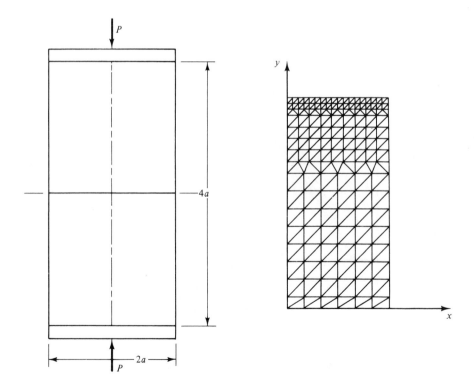

Figure 6-17 Plane strain test geometry and finite element mesh (Ref. 5).

The finite element mesh used in the analysis is shown in Fig. 6-17. The material parameters used in the analysis were given in Example 6-3. A comparison of finite element results with the closed-form solution is shown in Fig. 6-18. The results obtained with 32 increments seem to compare well with the exact solution.

A third-order hyperelastic model has been used by Ko and Mason (10) to characterize the behavior of a sand. They utilized a Green elastic model in this study. The parameters for the model were evaluated from laboratory data obtained by using a truly triaxial device. The accuracy of the model has been verified by predicting the response under different stress paths. They have then used the constitutive model in conjunction with a finite element procedure to solve a laboratory footing problem. The finite element predictions are compared with experimental results as shown in Fig. 6-19. Overall, the third-order Green elastic model was demonstrated to be adequate to describe the behavior

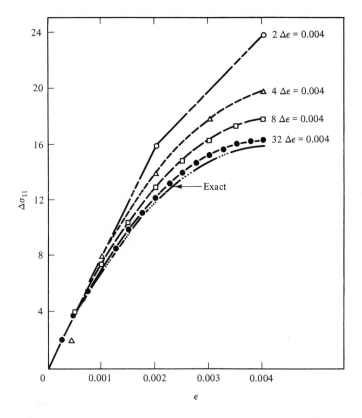

Figure 6-18 Comparison of incremental analysis of plane strain pure shear test with exact solution (Ref. 5).

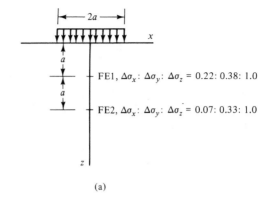

(a)

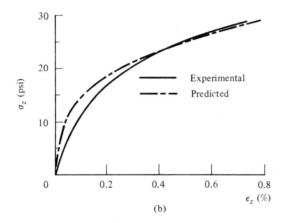

(b)

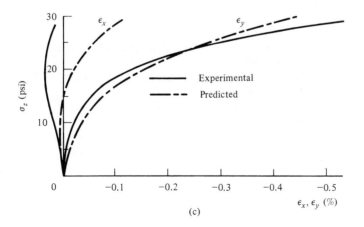

(c)

Figure 6-19 Verification of model prediction (Ref. 10).

of the sand. Saleeb and Chen (11) have presented application of hyperelastic (Green) models for simulating behavior of some soils.

It thus appears that the higher-order hyperelastic models can be used successfully for describing the monotonic loading behavior of granular soils. However, there are certain limitations associated with hyperelastic models. The second-order model, because of its mathematical form, can exhibit material instability beyond a certain level of strain; the magnitude of this critical strain depends on the material parameters of the model. Chang et al. (5) have shown how this instability problem was handled in the computations. In the event that certain elements are affected by instability in a localized region, the errors induced in the entire problem are restricted to the immediate vicinity of those elements.

The hyperelastic models are based on concepts of nonlinear elasticity, and hence they may not be suitable to model the general behavior of geologic media which exhibit inelastic deformations. However, they can be used to model the stress–strain behavior of such materials subject to only monotonically increasing loads. That is, these models may not be suitable to describe the behavior of geologic media subjected to loading, unloading, and reloading.

PROBLEMS

6-1. Express the stress–strain relationship for the second-order Green elastic law for an isotropic medium. The strain energy density U_0 has the form

$$U_0 = U_0(I_2, I_3)$$

6-2. Using $\sigma_{ij} = \partial U_0/\partial \epsilon_{ij}$, obtain a derivation of the relationship

$$\epsilon_{ij} = \frac{\partial U_{co}}{\partial \sigma_{ij}} \tag{6-40a}$$

6-3. (a) Choose values for K, G, a_2, a_3, a_5, and a_6, and then plot the expressions in Eqs. (6-54a) and (6-54b).
   (b) Choose values for a_6, and plot S_{11} versus ϵ_{12}^2, and S_{11} versus ϵ_{12}, and comment on the determination of a_6 from such data.

6-4. Derive second-order Cauchy elastic [Eq. (6-52)] and second-order Green elastic [Eq. (6-67)] models for (a) hydrostatic and (b) cylindrical triaxial strain states. Plot (symbolic) curves to express the stress–strain relations in parts (a) and (b). Comment on the softening or hardening deviation from linearity, if any.

6-5. Derive the second-order Green elastic model for the inverse relation expressing ϵ_{ij} as

$$\epsilon_{ij} = \frac{\partial U_{co}}{\partial J_1} \delta_{ij} + \frac{\partial U_{co}}{\partial J_2} \sigma_{ij} + \frac{\partial U_{co}}{\partial J_3} \sigma_{ik}\sigma_{kj}$$

for (a) a uniaxial state of stress, and (b) hydrostatic states of stress. Plot the strain–stress relations. Comment on the deviatoric and volumetric behavior induced for these states of stress.

6-6. If the (initial) state of strain and hence stress were nonzero in Eq. (6-63), discuss and derive the second-order Green elastic law in the incremental form.

6-7. Derive the stress–strain relation for the second-order Green and Cauchy elastic models for (a) one-dimensional compression, and (b) plane strain idealization. [*Hint*: Use Eqs. (6-52) and (6-67). Derive "incremental" forms.]

6-8. **(a)** Show that the second-order Cauchy and Green elastic models predict isotropic material behavior.

(b) Show that the incremental form of the second-order Cauchy model can predict stress-induced anisotropy.

[*Hint*: Consider initial hydrostatic state of (principal) stress with hydrostatic or triaxial (principal) stress increments and show that material is isotropic; however, if the initial state is anisotropic or nonhydrostatic, the material will show stress-induced anisotropy under increments of stress. Find relations between $\Delta\epsilon_{11}$ and $\Delta\sigma_{11}$ and $\Delta\epsilon_{22}$ and $\Delta\sigma_{22}$ or $\Delta\epsilon_{33}$ and $\Delta\sigma_{33}$ and show that they indicate anisotropy.]

REFERENCES

1. Timoshenko, S. P., and Goodier, J. N., *Theory of Elasticity*, McGraw-Hill Book Company, New York, 1970.

2. Fung, Y. C., *Foundations of Solid Mechanics*, Prentice-Hall, Inc., Englewood Cliffs, N.J., 1965.

3. Eringen, A. C., *Nonlinear Theory of Continuous Media*, McGraw-Hill Book Company, New York, 1962.

4. Franklin, J. N., *Matrix Theory*, Prentice-Hall, Inc., Englewood Cliffs, N.J., 1968.

5. Chang, T. Y., Ko, H. Y., Scott, R. F., and Westman, R. A., "An Integrated Approach to the Stress Analysis of Granular Materials," Report, California Institute of Technology, Pasadena, Calif., 1967.

6. Desai, C. S., "Overview, Trends and Projections: Theory and Application of the FEM in Geotechnical Engineering," State-of-the-Art Paper, *Proc. Symp. Appl. FEM Geotech. Eng.*, Vicksburg, Miss., 1972.

7. Rohani, B., "Mechanical Constitutive Models for Engineering Materials," *Report S-77-19*, U.S. Army Corps of Engineers Waterways Exp. Stn., Vicksburg, Miss., 1977.

8. Langhaar, H. L., *Energy Methods in Applied Mechanics*, John Wiley & Sons, Inc., New York, 1962.

9. Carnahan, B., Luther, H. A., and Wilkes, J. O., *Applied Numerical Methods*, John Wiley & Sons, Inc., New York, 1969.

10. Ko, H. Y., and Mason, R. M., "Nonlinear Characterization and Analysis of Sand,"

Proc. 2nd Int. Conf. Numer. Methods Geomech., C. S. Desai (Ed.), Blacksburg, Va., ASCE, June 1976.

11. Saleeb, A. F., and Chen, W. F., "Nonlinear Hyperelastic (Green) Constitutive Models for Soils: Predictions and Comparisons," *Proc. Workshop on Limit Equilibrium, Plasticity and Generalized Stress–Strain in Geotech. Eng.*, McGill Univ., Montreal, May 1980, ASCE, New York, 1981, pp. 265–285.

7

HYPOELASTICITY

In Chapter 6 we presented models based on higher-order elasticity, in which the state of stress depends on the current state of deformation (strain), and the behavior is not dependent on the stress path followed. For many materials, including geologic materials, this may not be realistic, since the behavior may depend on the stress path followed. The total deformation of such materials can be decomposed into a recoverable part and an irrecoverable part.

The concept, called *hypoelasticity*, constitutes a generalized incremental law in which the behavior can be simulated from increment to increment rather than for the entire load or stress at a time. In hypoelasticity, the increment of stress is expressed as a function of stress and increment of strain. Hypoelasticity was first introduced by Truesdell (1) and its details can be found in Refs. 1 and 2.

Literally, "hypo" means "in a lower sense" or "to a lower degree." Hence *hypoelastic* can imply a material that is elastic to a lower or incremental sense; in contrast, *hyperelastic* means elastic to a higher degree. As will be seen subsequently, a hypoelastic material can be interpreted to be capable of allowing for inelastic or plastic behavior.

The hypoelastic concept can provide simulation of constitutive behavior in a smooth manner, and hence can be used for hardening or softening materials. Its incremental nature allows for path dependence. Hence hypoelastic models can be appropriate for some geologic materials.

It is not intended to give in this text all details of hypoelastic models; they are available in various publications (1–4). Here we have considered some of the aspects in a simplified way with special attention to their potential application.

GENERAL EQUATIONS

We can express a general form of constitutive laws as (see Chapter 5)

$$f(\sigma, \dot{\sigma}, \epsilon, \dot{\epsilon}) = 0 \tag{7-1}$$

where σ is the stress, $\dot{\sigma}$ the rate of stress, ϵ the (infinitesimal) strain, and $\dot{\epsilon}$ the rate of strain.

Although expression (7-1) involves a time derivative (denoted by an overdot), it can represent a time-independent incremental relationship if the equation is homogeneous in time. That is, if time appears to the same dimension in all the terms of Eq. (7-1), time may be dropped from the equation, and the overdot or the rate can mean simply an incremental quantity.

A special form of Eq. (7-1) can be expressed as

$$\dot{\sigma}_{ij} = f_{ij}(\sigma_{kl}, \dot{\epsilon}_{kl}) \tag{7-2}$$

The relationship in Eq. (7-2) is a general expression that can include anisotropic behavior, and it can be modified to include temperature dependence (1). A constitutive law should possess dimensional invariance, that is, all of its terms should have the same dimensions and in the hypoelastic law, the response functions in C_{ijkl} are expressed only in terms of stresses and are not functions of time (1). Hence Eq. (7-2) can be written as

$$\dot{\sigma}_{ij} = C_{ijkl}\dot{\epsilon}_{kl} \tag{7-3}$$

where C_{ijkl} is a function of only the stress tensor. Since the relationship (7-3) is homogeneous in time and we are interested only in time-independent behavior in this chapter, the overdot means simply an incremental quantity. For an *isotropic* hypoelastic material, Eq. (7-3) can be expressed as (1–9)

$$\begin{aligned}
\dot{\sigma}_{ij} =\ & \alpha_0 \dot{\epsilon}_{kk}\delta_{ij} + \alpha_1 \dot{\epsilon}_{ij} + \alpha_2 \dot{\epsilon}_{kk}\sigma_{ij} \\
& + \alpha_3 \sigma_{mn}\dot{\epsilon}_{nm}\delta_{ij} + \alpha_4\left(\sigma_{im}\dot{\epsilon}_{mj} + \dot{\epsilon}_{im}\sigma_{mj}\right) \\
& + \alpha_5 \dot{\epsilon}_{kk}\sigma_{im}\sigma_{mj} + \alpha_6 \sigma_{mn}\dot{\epsilon}_{nm}\sigma_{ij} + \alpha_7 \sigma_{mn}\sigma_{nk}\dot{\epsilon}_{km}\delta_{ij} \\
& + \alpha_8\left(\sigma_{im}\sigma_{mk}\dot{\epsilon}_{kj} + \dot{\epsilon}_{im}\sigma_{mk}\sigma_{kj}\right) + \alpha_9 \sigma_{mn}\dot{\epsilon}_{nm}\sigma_{ik}\sigma_{kj} \\
& + \alpha_{10}\sigma_{mn}\sigma_{nk}\dot{\epsilon}_{km}\sigma_{ij} + \alpha_{11}\sigma_{mn}\sigma_{nk}\dot{\epsilon}_{km}\sigma_{ir}\sigma_{rj}
\end{aligned} \tag{7-4a}$$

or

$$\begin{aligned}
d\sigma_{ij} =\ & \alpha_0 d\epsilon_{kk}\,\delta_{ij} + \alpha_1 d\epsilon_{ij} + \alpha_2 d\epsilon_{kk}\,\sigma_{ij} + \alpha_3 \sigma_{mn}d\epsilon_{nm}\,\delta_{ij} \\
& + \alpha_4\left(\sigma_{im}d\epsilon_{mj} + d\epsilon_{im}\sigma_{mj}\right) + \alpha_5 d\epsilon_{kk}\sigma_{im}\sigma_{mj} \\
& + \alpha_6 \sigma_{mn}d\epsilon_{nm}\sigma_{ij} + \alpha_7 \sigma_{mn}\sigma_{nk}d\epsilon_{km}\,\delta_{ij} \\
& + \alpha_8\left(\sigma_{im}\sigma_{mk}d\epsilon_{kj} + d\epsilon_{im}\sigma_{mk}\sigma_{kj}\right) + \alpha_9 \sigma_{mn}d\epsilon_{nm}\sigma_{ik}\sigma_{kj} \\
& + \alpha_{10}\sigma_{mn}\sigma_{nk}d\epsilon_{km}\sigma_{ij} + \alpha_{11}\sigma_{mn}\sigma_{nk}d\epsilon_{km}\sigma_{ir}\sigma_{rj}
\end{aligned} \tag{7-4b}$$

where $\alpha_0, \alpha_1, \ldots, \alpha_{11}$ are functions of invariants of stress, and $d\sigma_{ij}$ and $d\epsilon_{ij}$ are the incremental stress and incremental strain tensors, respectively. Definition of the general hypoelastic model in Eq. (7-4) requires 12 response functions, α_i. Hypoelastic models have a greater generality, and in fact, the theory of incremental elasticity can be considered a special case of hypoelasticity. Also, as discussed in Chapter 11, hypoelasticity and plasticity can have common features and a common basis.

In elasticity, stresses reached from a neutral state should be compatible with elastic strains; neutral state implies a state of zero stress and strain. On the other hand, in hypoelasticity any state of stress that satisfies equilibrium conditions is admissible. The general relation [Eq. (7-4)] can be used to obtain its special forms of various *orders* or *grades*. Since the word "grade" is commonly used in literature, we shall use that term herein.

MODELS OF VARIOUS ORDERS

Grade-Zero Hypoelastic Model

If the law in Eq. (7-4b) is assumed to be independent of stress, the material is called *grade-zero hypoelastic*. For this case,

$$\alpha_2 = \alpha_3 = \cdots = \alpha_{11} = 0 \qquad (7-5)$$

Hence Eq. (7-4b) becomes

$$d\sigma_{ij} = \alpha_0 d\epsilon_{kk}\delta_{ij} + \alpha_1 d\epsilon_{ij} \qquad (7-6)$$

where $\alpha_0 = \lambda$ and $\alpha_1 = \mu$, and λ and μ are Lamé's constants (see Chapter 6). With this, Eq. (7-6) is the incremental form of the generalized Hooke's law, which can be written as

$$d\sigma_{ij} = \left(K - \frac{2G}{3} \right) d\epsilon_{kk}\delta_{ij} + 2Gd\epsilon_{ij} \qquad (7-7)$$

This model requires only two parameters, K and G, and is valid for linear and isotropic media. It has provided a basis for a number of approximate but simplified models for the characterization of many materials. We shall discuss them further in Chapter 8.

Grade-One or Linear Hypoelastic Model

If the right-hand side of Eq. (7-4b) is assumed to be a linear function of stress, the material is called a *grade-one hypoelastic material*. For this case we define

$$
\begin{aligned}
\alpha_0 &= a_0 + a_2 J_1 = a_0 + a_2\sigma_{nn} \\
\alpha_1 &= a_1 + a_3 J_1 = a_1 + a_3\sigma_{nn} \\
\alpha_2 &= a_4 \\
\alpha_3 &= a_6 \\
\alpha_4 &= a_5 \\
\alpha_5 &= \alpha_6 = \cdots = \alpha_{11} = 0
\end{aligned} \qquad (7-8)
$$

where $a_0, a_1, \ldots, a_6$ are material constants pertaining to grade-one hypoelastic model. Then the grade-one hypoelastic model can be expressed as

$$d\sigma_{ij} = a_0 d\epsilon_{kk}\,\delta_{ij} + a_1 d\epsilon_{ij} + a_2\sigma_{nn} d\epsilon_{kk}\,\delta_{ij} + a_3\sigma_{nn} d\epsilon_{ij} + a_4 d\epsilon_{kk}\sigma_{ij}$$

$$+ a_5\left(\sigma_{im} d\epsilon_{mj} + d\epsilon_{im}\sigma_{mj}\right) + a_6\sigma_{mn} d\epsilon_{nm}\,\delta_{ij} \tag{7-9}$$

and the constitutive tensor [Eq. (7-3)] is expressed as (5)

$$C_{ijkl} = \left(a_0 + a_2\sigma_{pp}\right)\delta_{ij}\delta_{kl} + \tfrac{1}{2}\left(a_1 + a_3\sigma_{pp}\right)\left(\delta_{ik}\delta_{jl} - \delta_{jk}\delta_{il}\right)$$

$$+ a_4\sigma_{ij}\delta_{kl} + \tfrac{1}{2}a_5\left(\sigma_{jk}\delta_{li} + \sigma_{jl}\delta_{ki} + \sigma_{ik}\delta_{lj} + \sigma_{il}\delta_{kj}\right) + a_6\sigma_{kl}\delta_{ij} \tag{7-10}$$

Equation (7-9) represents the differential equation in the incremental form for grade-one hypoelastic model. The material parameters a_0 to a_6 can be determined by integrating specialized forms of Eq. (7-9) for a specific test along a given stress path. It is obvious that determination of the seven parameters a_0 to a_6 would require data from a number of appropriate (laboratory) tests. Since it is difficult to define unique number of tests to determine these parameters, the usual procedure is to obtain as many tests as possible. Then we use curve fitting and other optimization techniques to find the *best* values of the parameters as described in Chapter 6. We now consider a few common tests and show how the integration along the specific stress paths is achieved.

Hydrostatic state of stress. For hydrostatic state of stress with isotropic medium, we have

$$\sigma_{11} = \sigma_{22} = \sigma_{33} = \sigma$$
$$\sigma_{12} = \sigma_{23} = \sigma_{13} = 0$$
$$\epsilon_{11} = \epsilon_{22} = \epsilon_{33} = \epsilon$$
$$\epsilon_{12} = \epsilon_{23} = \epsilon_{13} = 0$$

These values can be considered as boundary conditions for the specialized form of the differential equation [Eq. (7-9)], given by

$$d\sigma_{ii} = 3a_0 d\epsilon_{nn} + a_1 d\epsilon_{ii} + 3a_2\left(\sigma_{nn}\right) d\epsilon_{kk}$$
$$+ a_3\sigma_{nn} d\epsilon_{ii} + a_4\sigma_{ii} d\epsilon_{kk}$$
$$+ a_5\left(\sigma_{ik} d\epsilon_{ik} + \sigma_{ik} d\epsilon_{ik}\right) + a_6\left(\sigma_{kl} d\epsilon_{kl}\right)\delta_{ii}$$

or

$$3d\sigma = 3 \times 3a_0 d\epsilon + 3a_1 d\epsilon + 3 \times 3 \times 3a_2\sigma d\epsilon$$
$$+ 3 \times 3a_3\sigma d\epsilon + 3 \times 3a_4\sigma d\epsilon$$
$$+ 3 \times 2a_5\left(\sigma d\epsilon\right) + 3 \times 3a_6 d\epsilon$$

or

$$d\sigma = \left(3a_0 + a_1\right) d\epsilon + \left(9a_2 + 3a_3 + 3a_4 + 2a_5 + 3a_6\right)\sigma d\epsilon$$

or

$$d\sigma = c_1 d\epsilon + c_2\sigma d\epsilon$$
$$= \left(c_1 + c_2\sigma\right) d\epsilon \tag{7-11a}$$

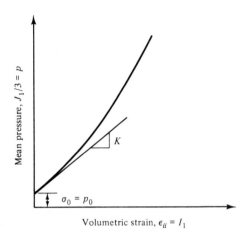

Figure 7-1 Symbolic mean pressure versus volumetric strain response.

where $c_1 = 3a_0 + a_1$ and $c_2 = (9a_2 + 3a_3 + 3a_4 + 2a_5 + 3a_6)$. Equation (7-11a) can be written as

$$\frac{d\sigma}{c_1 + c_2\sigma} = d\epsilon \qquad (7\text{-}11b)$$

Now we can integrate Eq. (7-11b) as

$$\int_{\sigma_0}^{\sigma} \frac{d\sigma}{c_1 + c_2\sigma} = \int_0^{\epsilon} d\epsilon \qquad (7\text{-}12a)$$

(here the initial strain is assumed to be at zero reference state), to yield

$$\sigma = \left(\frac{c_1}{c_2} + \sigma_0\right)e^{c_2\epsilon} - \frac{c_1}{c_2} \qquad (7\text{-}12b)$$

where σ_0 is the initial hydrostatic stress. The inverse relation can be obtained as

$$\epsilon = \frac{1}{c_2} \ln\left(\frac{c_1 + c_2\sigma}{c_1 + c_2\sigma_0}\right) \qquad (7\text{-}12c)$$

The parameters c_1 and c_2 in Eq. (7-12) can be obtained by curve fitting on stress–strain data from hydrostatic (compression) tests (Fig. 7-1), that is, (mean) pressure–volumetric strain curves.

Conventional triaxial compression. In this test configuration, we use cylindrical specimen of (geologic) materials. The cell pressure or the confining stress $\sigma_{33} = \sigma_{22}$ is held constant and the axial stress σ_{11} is increased. The stress tensor is given by

$$\sigma_{ij} = \begin{pmatrix} \sigma_{11} & 0 & 0 \\ 0 & \sigma_{33} & 0 \\ 0 & 0 & \sigma_{33} \end{pmatrix} \qquad (7\text{-}13)$$

Figure 7-2 shows symbolic curves from CTC tests. Since the confining or hydrostatic stress is kept constant, we have $d\sigma_{22} = d\sigma_{33} = 0$. With these conditions, Eq. (7-9) specializes to

$$
\begin{aligned}
d\sigma_{11} = & \; a_0(d\epsilon_{11} + d\epsilon_{22} + d\epsilon_{33}) + a_1 d\epsilon_{11} + a_2(\sigma_{11} + 2\sigma_{33})(d\epsilon_{11} \\
& + d\epsilon_{22} + d\epsilon_{33}) + a_3(\sigma_{11} + 2\sigma_{33})\, d\epsilon_{11} \\
& + a_4\sigma_{11}(d\epsilon_{11} + d\epsilon_{22} + d\epsilon_{33}) \\
& + a_5\left(\sigma_{11} d\epsilon_{11} + \sigma_{12}\cancel{d\epsilon_{12}}^{0} + \sigma_{33}\cancel{d\epsilon_{13}}^{0} + \sigma_{11} d\epsilon_{11} \right. \\
& \left. + \sigma_{12}\cancel{d\epsilon_{12}}^{0} + \sigma_{13}\cancel{d\epsilon_{13}}^{0}\right) \\
& + a_6(\sigma_{11} d\epsilon_{11} + \sigma_{22} d\epsilon_{22} + \sigma_{33} d\epsilon_{33}) \\
= & \left[a_0 + a_1 + a_2(\sigma_{11} + 2\sigma_{33}) + a_3(\sigma_{11} + 2\sigma_{33}) \right. \\
& \left. + a_4\sigma_{11} + 2a_5\sigma_{11} + a_6\sigma_{11}\right] d\epsilon_{11} \\
& + 2\left[a_0 + a_2(\sigma_{11} + 2\sigma_{33}) + a_4\sigma_{11} + a_6\sigma_{33}\right] d\epsilon_{33} \qquad \text{(7-14a)}
\end{aligned}
$$

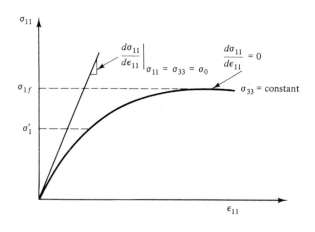

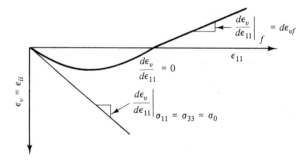

Figure 7-2 Symbolic response from CTC test.

and

$$d\sigma_{33} = a_0(d\epsilon_{11} + 2d\epsilon_{33}) + a_1 d\epsilon_{33}$$
$$+ a_2(\sigma_{11} + 2\sigma_{33})(d\epsilon_{11} + 2 d\epsilon_{33}) + a_3(\sigma_{11} + 2\sigma_{33}) d\epsilon_{33}$$
$$+ a_4\sigma_{33}(d\epsilon_{11} + 2 d\epsilon_{33}) + a_5(\sigma_{22} d\epsilon_{22} + \sigma_{22} d\epsilon_{22})$$
$$+ a_6(\sigma_{11} d\epsilon_{11} + 2\sigma_{33} d\epsilon_{33})$$

or

$$d\sigma_{33} = \left[a_0 + a_2(\sigma_{11} + 2\sigma_{33}) + a_4\sigma_{33} + a_6\sigma_{11} \right] d\epsilon_{11}$$
$$+ \left[2a_0 + a_1 + 2a_2(\sigma_{11} + 2\sigma_{33}) + a_3(\sigma_{11} + 2\sigma_{33}) \right.$$
$$\left. + 2a_4\sigma_{33} + 2a_5\sigma_{33} + 2a_6\sigma_{33} \right] d\epsilon_{33} \qquad (7\text{-}14\text{b})$$

Because $d\sigma_{33} = 0$ in the CTC test, Eq. (7-14b) gives

$$\frac{d\epsilon_{33}}{d\epsilon_{11}} = -\frac{a_0 + a_2(\sigma_{11} + 2\sigma_{33}) + a_4\sigma_{33} + a_6\sigma_{11}}{2a_0 + a_1 + 2a_2(\sigma_{11} + 2\sigma_{33}) + a_3(\sigma_{11} + 2\sigma_{33}) + 2a_4\sigma_{33} + 2a_5\sigma_{33} + 2a_6\sigma_{33}}$$

$$= \frac{A_1}{B_1} \qquad (7\text{-}15)$$

If we substitute for $d\epsilon_{33}$ from Eq. (7-15) in Eq. (7-14a), the ratio of the increment of axial stress to that of axial strain is

$$\frac{d\sigma_{11}}{d\epsilon_{11}} = \frac{\beta_1}{\beta_2} \qquad (7\text{-}16\text{a})$$

where

$$\beta_1 = \left[(a_0 + a_1) + (a_2 + a_3 + a_4 + 2a_5 + a_6)\sigma_{11} \right.$$
$$\left. + (2a_2 + 2a_3)\sigma_{33} \right]\beta_2 - 2\left[a_0 + (a_2 + a_4)\sigma_{11} \right.$$
$$\left. + (2a_2 + a_6)\sigma_{33} \right]\left[a_0 + (a_2 + a_6)\sigma_{11} + (2a_2 + a_4)\sigma_{33} \right] \qquad (7\text{-}16\text{b})$$
$$\beta_2 = (2a_0 + a_1) + (2a_2 + a_3)\sigma_{11} + 2(2a_2 + a_3 + a_4 + a_5 + a_6)\sigma_{33}$$
$$\qquad (7\text{-}16\text{c})$$

In the CTC tests, we often measure volumetric strain $d\epsilon_v = d\epsilon_{ii} = (d\epsilon_{11} + d\epsilon_{22} + d\epsilon_{33})$ and plot it with respect to the axial strain, $d\epsilon_{11}$. In order to find the ratio of $d\epsilon_{ii}/d\epsilon_{11}$, we use Eq. (7-15) as

$$\frac{2 d\epsilon_{33}}{d\epsilon_{11}} + 1 = \frac{2A_1}{B_1} + 1 \qquad (7\text{-}17\text{a})$$

which leads to

$$\frac{d\epsilon_v}{d\epsilon_{11}} = \frac{d\epsilon_{ii}}{d\epsilon_{11}} = \frac{a_1 + (a_3 - 2a_6)\sigma_{11} + 2(a_3 + a_5 + a_6)\sigma_{33}}{2a_0 + a_1 + (2a_2 + a_3)\sigma_{11} + 2(2a_2 + a_3 + a_4 + a_5 + a_6)\sigma_{33}}$$
$$\qquad (7\text{-}17\text{b})$$

Equations (7-16a) and (7-17b) give two differential equations that can be integrated. It can be seen that they are complicated, and closed-form integration can be quite difficult. Hence a simpler procedure would be to fit given data of σ_{11} versus ϵ_{11} and ϵ_v versus ϵ_{11} for a selected number of points on the curves (Fig. 7-2). One such point would be the *initial* hydrostatic stress $\sigma_{11} = \sigma_{33} = \sigma_0$. Then the initial slopes of curves in Fig. 7-2 according to Eqs. (7-16a) and (7-17b) are given by (Fig. 7-2)

$$\left.\frac{d\sigma_{11}}{d\epsilon_{11}}\right|_{\sigma_{11}=\sigma_{33}=\sigma_0} = \frac{d(\sigma_{11}-\sigma_{33})}{d\epsilon_{11}} = \frac{A_2}{A_3} \qquad (7\text{-}18)$$

$$\left.\frac{d\epsilon_v}{d\epsilon_{11}}\right|_{\sigma_{11}=\sigma_{33}=\sigma_0} = \frac{A_4}{A_3} \qquad (7\text{-}19)$$

where

$$A_2 = \left[(a_0 + a_1) + (3a_2 + 3a_3 + a_4 + 2a_5 + a_6)\sigma_0\right]A_3$$
$$\quad - 2\left[a_0 + (3a_2 + a_4 + a_6)\sigma_0\right]\left[a_0 + (3a_2 + a_4 + a_6)\sigma_0\right]$$
$$A_3 = 2a_0 + a_1 + (6a_2 + 3a_3 + 2a_4 + 2a_5 + 2a_6)\sigma_0$$
$$A_4 = a_0 + (3a_2 + a_4 + a_6)\sigma_0$$

Coon and Evans (6), Vagneron et al. (8), and Mysore (10) have presented and used relations of the same form as in Eqs. (7-16) and (7-17) for CTC tests. In view of the complexity in these equations, an approximate procedure can be used to derive the material parameters. Here we can use curve fitting only for a limited number of points, say for the initial slopes and failure condition. For the former, let us express the ratio of slopes in Eqs. (7-16a) and (7-17b) as

$$\left.\frac{d\sigma_{11}}{d\epsilon_{11}} \bigg/ \frac{d\epsilon_v}{d\epsilon_{11}}\right|_{\sigma_{11}=\sigma_{33}=\sigma_0} = (3a_0 + a_1) + (9a_2 + 3a_3 + 3a_4 + 2a_5 + 3a_6)\sigma_0$$
$$= c_1 + c_2\sigma_0 \qquad (7\text{-}20)$$

The parameters c_1 and c_2 are similar to those in Eq. (7-11a) for hydrostatic compression test. Because of this dependence of the two initial slopes, we can use only one of them as a basis for determining the parameters.

At failure, the slope in Eq. (7-16a) can be assumed to vanish; hence from Eq. (7-16a) we have

$$\left[(a_0 + a_1) + (a_2 + a_3 + a_4 + 2a_1 + a_6)\sigma_{1f} + (2a_2 + 2a_3)\sigma_{33}\right]$$
$$\times \left[(2a_0 + a_1) + (2a_2 + a_3)\sigma_{1f} + 2(2a_2 + a_3 + a_4 + a_5 + a_6)\sigma_{33}\right]$$
$$- 2\left[a_0 + (a_2 + a_4)\sigma_{1f} + (2a_2 + a_6)\sigma_{33}\right]$$
$$\times \left[a_0 + (a_2 + a_6)\sigma_{1f} + (2a_2 + a_4)\sigma_{33}\right] = 0$$

or

$$\left[(a_0 + a_1) + a_2(\sigma_{1f} + 2\sigma_{33}) + a_3(\sigma_{1f} + 2\sigma_{33}) + a_4\sigma_{1f} + 2a_5\sigma_{1f} + a_6\sigma_{1f} \right]$$

$$\times \left[(2a_0 + a_1) + a_2(\sigma_{1f} + 2\sigma_{33}) + a_3(\sigma_{1f} + 2\sigma_{33}) + 2a_4\sigma_{33} + 2a_5\sigma_{33} + 2a_6\sigma_{33} \right]$$

$$- 2\left[a_0 + a_2(\sigma_{1f} + 2\sigma_{33}) + a_4\sigma_{1f} + a_6\sigma_{33} \right]$$

$$\times \left[a_0 + a_2(\sigma_{1f} + 2\sigma_{33}) + a_4\sigma_{33} + a_6\sigma_{1f} \right] = 0 \qquad (7\text{-}21)$$

Here σ_{1f} is the maximum or peak stress at "failure" defined by $d\sigma_{11}/d\epsilon_{11} = 0$.

Now Eq. (7-17a) can be used to define the slope of the volumetric strain versus axial strain relation at failure:

$$d\epsilon_{vf} = \left. \frac{d\epsilon_v}{d\epsilon_{11}} \right|_f = \frac{a_1 + (a_3 - 2a_6)\sigma_{1f} + 2(a_3 + a_5 + a_6)\sigma_{33}}{2a_0 + a_1 + (2a_2 + a_3)\sigma_{1f} + 2(2a_2 + a_3 + a_4 + a_5 + a_6)\sigma_{33}}$$

$$(7\text{-}22a)$$

where $d\epsilon_{vf}$ denotes the measured slope of ϵ_v versus ϵ_{11} curve at failure (Fig. 7-2). Hence,

$$2d\epsilon_{vf}a_0 + \left(d\epsilon_{vf} - 1 \right)a_1 + \left(2\sigma_{1f}d\epsilon_{vf} + 4\sigma_{33}d\epsilon_{vf} \right)a_2$$

$$+ \left(\sigma_{1f}d\epsilon_{vf} - \sigma_{1f} + 2\sigma_{33}d\epsilon_{vf} - 2\sigma_{33} \right)a_3 + \left(2\sigma_{33}d\epsilon_{vf} \right)a_4$$

$$+ \left(2\sigma_{33}d\epsilon_{vf} - 2\sigma_{33} \right)a_5 + \left(2\sigma_{1f} - 2\sigma_{33} + 2\sigma_{33}d\epsilon_{vf} \right)a_6 = 0 \qquad (7\text{-}22b)$$

If we define σ'_{11} as the axial stress at which the ratio $d\epsilon_v/d\epsilon_{11}$ is zero (Fig. 7-2), Eq. (7-17b) leads to one additional relation as

$$a_1 + \left(\sigma'_{11} + 2\sigma_{33} \right)a_3 + \left(2\sigma_{33} \right)a_5 - \left(2\sigma'_{11} - 2\sigma_{33} \right)a_6 = 0 \qquad (7\text{-}23)$$

Equations (7-20) to (7-23) can be used to obtain the required parameters from a set of CTC test data.

It is possible to specialize the grade-one hypoelastic model for other test configurations, such as conventional triaxial, one-dimensional compression, plane strain, axisymmetric, simple shear, and other general stress paths in the three-dimensional stress space (Fig. 5-10). Some of these are given in problems at the end of this chapter; we consider below the simple shear strain condition.

Simple shear strain.   The incremental strain tensor can be expressed as

$$d\epsilon_{ij} = \begin{pmatrix} 0 & d\epsilon_{12} & 0 \\ d\epsilon_{12} & 0 & 0 \\ 0 & 0 & 0 \end{pmatrix} \qquad (7\text{-}24)$$

Substitution of Eq. (7-24) in the general form [Eq. (7-9)] of the grade-one model gives

$$d\sigma_{11} = a_0\left(d\epsilon_{11} + d\varepsilon_{22}^{0} + d\epsilon_{33}\right)\delta_{11} + a_1\, d\varepsilon_{11}^{0}$$
$$+ a_2\left(\sigma_{11} + \sigma_{22} + \sigma_{33}\right)\left(d\epsilon_{11} + d\varepsilon_{22}^{0} + d\epsilon_{33}\right)\delta_{11}$$
$$+ a_3\left(\sigma_{11} + \sigma_{22} + \sigma_{33}\right)d\varepsilon_{v}^{0} + a_4\sigma_{11}\left(d\epsilon_{11} + d\varepsilon_{22}^{0} + d\epsilon_{33}\right)$$
$$+ a_5\left(\sigma_{1k}\, d\epsilon_{1k} + \sigma_{1k}\, d\epsilon_{1k}\right) + a_6\sigma_{kl}\, d\epsilon_{kl}\,\delta_{11}$$

or

$$d\sigma_{11} = a_5\left(\sigma_{11}\, d\varepsilon_{11}^{0} + \sigma_{12}\, d\epsilon_{12} + \sigma_{13}\, d\varepsilon_{13}^{0} + \sigma_{11}\, d\varepsilon_{11}^{0}\right.$$
$$\left. + \sigma_{12}\, d\epsilon_{12} + \sigma_{13}\, d\varepsilon_{13}^{0}\right) + a_6\left(\sigma_{21}\, d\epsilon_{21} + \sigma_{12}\, d\epsilon_{12}\right)$$

or

$$d\sigma_{11} = a_5\sigma_{12}\, d\epsilon_{12} + a_5\sigma_{12}\, d\epsilon_{12} + a_6\sigma_{12}\, d\epsilon_{12} + a_6\sigma_{12}\, d\epsilon_{12}$$
$$= 2\left(a_5 + a_6\right)\sigma_{12}\, d\epsilon_{12} \tag{7-25a}$$
$$d\sigma_{22} = d\sigma_{11} \tag{7-25b}$$
$$d\sigma_{33} = 2a_5\sigma_{12}\, d\epsilon_{12} \tag{7-25c}$$
$$d\sigma_{12} = a_1\, d\epsilon_{12} + a_3\left(\sigma_{11} + \sigma_{22} + \sigma_{33}\right)d\epsilon_{12} + a_5\sigma_{22}\, d\epsilon_{12} + a_5\sigma_{11}\, d\epsilon_{21}$$
$$= \left[a_1 + \left(\sigma_{11} + \sigma_{22} + \sigma_{33}\right)a_3 + \left(\sigma_{11} + \sigma_{22}\right)a_5\right]d\epsilon_{12} \tag{7-25d}$$

and

$$\sigma_{23} = \sigma_{13} = 0 \tag{7-25e}$$

Note that for grade-zero model, we have

$$d\sigma_{12} = a_1\, d\epsilon_{12} \tag{7-25f}$$

Hence $a_1 = 2G$.

Equation (7-25a) indicates that for inducing the simple shear condition, that is, no volume change, it is necessary that normal stresses $d\sigma_{11}$ and $d\sigma_{22}$ in Eqs. (7-25a) and (7-25b) and $d\sigma_{33}$ in Eq. (7-25c) must be applied to the boundaries of the specimen of the medium. Also, note that the required normal stresses are of different magnitudes and that they cause normal deviatoric stresses [Eqs. (7-25a) to (7-25c)] on the shearing planes. This effect is similar to the one noted for the hyperelastic models (Chapter 6). It is important to note that the first-order linear elastic laws are not capable of incorporating this effect. Hence, if the material exhibits second-order effects, it is necessary to use models that can account for them.

Comment. It becomes necessary to evaluate a greater number of parameters as the grade of the hypoelastic model increases. For grade-zero and grade-one models the required numbers of parameters are two and seven, respectively. For the grade-two model, the number becomes 18. It is difficult to determine such a large number of parameters and implement the models in solution techniques. Hence in this chapter we stop at the grade-one model. For details on higher-order models, the reader can consult various publications (1–6).

IMPLEMENTATION

Since the hypoelastic model is expressed in an incremental form, it can be applied readily in incremental nonlinear solution schemes (Chapter 5). For instance, [Eq. (7-9)] can be expressed in matrix notation as

$$\{d\sigma\} = [C_t]\{d\epsilon\} \tag{7-26}$$

where

$$\{d\sigma\}^T = [d\sigma_{11} \quad d\sigma_{22} \quad d\sigma_{33} \quad d\sigma_{12} \quad d\sigma_{23} \quad d\sigma_{13}]$$

$$\{d\epsilon\}^T = [d\epsilon_{11} \quad d\epsilon_{22} \quad d\epsilon_{33} \quad d\epsilon_{12} \quad d\epsilon_{23} \quad d\epsilon_{13}]$$

and $[C_t]$ denotes the constitutive or stress–strain matrix relevant to the given increments of stress and strain.

We now expand Eq. (7-26) for grade-one hypoelastic law; for example,

$$d\sigma_{11} = a_0(d\epsilon_{11} + d\epsilon_{22} + d\epsilon_{33})\delta_{11} + a_1 d\epsilon_{11} + a_2\sigma_{nn}(d\epsilon_{11} + d\epsilon_{22} + d\epsilon_{33})\delta_{11}$$

$$+ a_3\sigma_{nn} d\epsilon_{11} + a_4\sigma_{11}(d\epsilon_{11} + d\epsilon_{22} + d\epsilon_{33})$$

$$+ a_5(\sigma_{11} d\epsilon_{11} + \sigma_{12} d\epsilon_{12} + \sigma_{13} d\epsilon_{13} + \sigma_{11} d\epsilon_{11} + \sigma_{12} d\epsilon_{12} + \sigma_{13} d\epsilon_{13})$$

$$+ a_6(\sigma_{11} d\epsilon_{11} + \sigma_{12} d\epsilon_{12} + \sigma_{13} d\epsilon_{13} + \sigma_{21} d\epsilon_{21} + \sigma_{22} d\epsilon_{22}$$

$$+ \sigma_{23} d\epsilon_{23} + \sigma_{31} d\epsilon_{31} + \sigma_{32} d\epsilon_{32} + \sigma_{33} d\epsilon_{33})$$

$$= (a_0 + a_1 + a_2\sigma_{nn} + a_3\sigma_{nn} + a_4\sigma_{11} + 2a_5\sigma_{11}$$

$$+ a_6\sigma_{11}) d\epsilon_{11} + (a_0 + a_2\sigma_{nn} + a_4\sigma_{11} + a_6\sigma_{22}) d\epsilon_{22}$$

$$+ (a_0 + a_2\sigma_{nn} + a_4\sigma_{11} + a_6\sigma_{33}) d\epsilon_{33}$$

$$+ (2a_5 + 2a_6\sigma_{12}) d\epsilon_{12} + 2a_6\sigma_{23} d\epsilon_{23} + (2a_5\sigma_{13} + 2a_6\sigma_{13}) d\epsilon_{13}$$

$$\tag{7-27}$$

and so on. In matrix notation, this equation can be expressed as follows (page 146):

$$
\begin{Bmatrix} d\sigma_{11} \\ d\sigma_{22} \\ d\sigma_{33} \\ d\sigma_{12} \\ d\sigma_{23} \\ d\sigma_{13} \end{Bmatrix} =
$$

$$
\begin{bmatrix}
a_0 + a_1 \\ + (a_2 + a_3)\sigma_{nn} \\ + (a_4 + 2a_5 + a_6)\sigma_{11} &
a_0 + a_2\sigma_{nn} \\ + a_4\sigma_{22} \\ + a_6\sigma_{11} &
a_0 + a_2\sigma_{nn} \\ + a_4\sigma_{33} \\ + a_6\sigma_{11} &
(2a_5 + 2a_6)\sigma_{12} &
2a_6\sigma_{23} &
(2a_5 + 2a_6)\sigma_{13} \\[4pt]
a_0 + a_2\sigma_{nn} \\ + a_4\sigma_{22} \\ + a_6\sigma_{11} &
a_0 + a_1 \\ + (a_2 + a_3)\sigma_{nn} \\ + (a_4 + 2a_5 + a_6)\sigma_{22} &
a_0 + a_2\sigma_{nn} \\ + a_4\sigma_{33} \\ + a_6\sigma_{22} &
(2a_5 + 2a_6)\sigma_{12} &
(2a_5 + 2a_6)\sigma_{23} &
2a_6\sigma_{13} \\[4pt]
a_0 + a_2\sigma_{nn} \\ + a_4\sigma_{33} \\ + a_6\sigma_{11} &
a_0 + a_2\sigma_{nn} \\ + a_4\sigma_{33} \\ + a_6\sigma_{22} &
a_0 + a_1 \\ + (a_2 + a_3)\sigma_{nn} \\ + (a_4 + 2a_5 + a_6)\sigma_{33} &
2a_6\sigma_{12} &
2(a_5 + a_6)\sigma_{23} &
2(a_5 + a_6)\sigma_{13} \\[4pt]
(a_4 + a_5)\sigma_{12} &
(a_4 + a_5)\sigma_{12} &
a_4\sigma_{12} &
a_1 + a_3\sigma_{nn} \\ + a_5(\sigma_{11} + \sigma_{22}) &
a_5\sigma_{13} &
a_5\sigma_{23} \\[4pt]
a_4\sigma_{23} &
(a_4 + a_5)\sigma_{23} &
(a_4 + a_5)\sigma_{23} &
a_5\sigma_{31} &
a_1 + a_3\sigma_{nn} \\ + a_5(\sigma_{22} + \sigma_{33}) &
a_5\sigma_{12} \\[4pt]
(a_4 + a_5)\sigma_{31} &
a_4\sigma_{31} &
(a_4 + a_5)\sigma_{31} &
a_5\sigma_{23} &
a_5\sigma_{12} &
a_1 + a_2\sigma_{nn} \\ + a_5(\sigma_{11} + \sigma_{33})
\end{bmatrix}
\begin{Bmatrix} d\epsilon_{11} \\ d\epsilon_{22} \\ d\epsilon_{33} \\ d\epsilon_{12} \\ d\epsilon_{23} \\ d\epsilon_{13} \end{Bmatrix}
\qquad (7\text{-}28)
$$

Example 7-1: Artificial Soil

Parameters for a grade-one hypoelastic model for the artificial soil described in Chapter 5 are determined in this example. The incremental stress–strain relationship for this model is given as

$$d\sigma_{ij} = a_0 \, d\epsilon_{nn} \, \delta_{ij} + a_1 \, d\epsilon_{ij} + a_2 \sigma_{nn} \, d\epsilon_{kk} \, \delta_{ij} + a_3 \sigma_{nn} \, d\epsilon_{ij}$$

$$+ a_4 \sigma_{ij} \, d\epsilon_{kk} + a_5 \left(\sigma_{jk} \, d\epsilon_{ik} + \sigma_{ik} \, d\epsilon_{jk} \right) + a_6 \sigma_{kl} d\epsilon_{kl} \, \delta_{ij} \qquad (7\text{-}9)$$

As could be seen that there are two basic quantities on the right-hand side of the equation above; therefore, if the incremental stresses are to be predicted, we need to know the stress tensor and the incremental strain tensor along the desired stress path. In order to determine material parameters, let us consider some specific cases of Eq. (7-9).

As in the case of hyperelastic materials, we have assumed that the artificial soil is isotropic. If we select the principal axes as the frame of reference, the stress tensor, and the incremental strain tensor can be expressed as follows:

$$\sigma_{ij} = \begin{pmatrix} \sigma_1 & 0 & 0 \\ 0 & \sigma_2 & 0 \\ 0 & 0 & \sigma_3 \end{pmatrix} \qquad (7\text{-}29a)$$

$$d\epsilon_{ij} = \begin{pmatrix} d\epsilon_1 & 0 & 0 \\ 0 & d\epsilon_2 & 0 \\ 0 & 0 & d\epsilon_3 \end{pmatrix} \qquad (7\text{-}29b)$$

Substituting Eq. (7-29a) in Eq. (7-9), we obtain

$$d\sigma_{ij} = 0 \qquad \text{for } i \neq j \qquad (7\text{-}30a)$$

and

$$d\sigma_1 = a_0 \, d\epsilon_{nn} + a_1 \, d\epsilon_1 + a_2 \sigma_{nn} \, d\epsilon_{kk} + a_3 \sigma_{nn} \, d\epsilon_1 + a_4 \, d\epsilon_{kk} \sigma_1$$

$$+ a_5 \left(\sigma_1 \, d\epsilon_1 + \sigma_1 \, d\epsilon_1 \right) + a_6 \left(\sigma_1 \, d\epsilon_1 + \sigma_2 \, d\epsilon_2 + \sigma_3 \, d\epsilon_3 \right) \qquad (7\text{-}30b)$$

$$d\sigma_2 = a_0 \, d\epsilon_{nn} + a_1 \, d\epsilon_2 + a_2 \sigma_{nn} \, d\epsilon_{kk} + a_3 \sigma_{nn} \, d\epsilon_2 + a_4 \, d\epsilon_{kk} \sigma_2$$

$$+ a_5 \left(2\sigma_2 \, d\epsilon_2 \right) + a_6 \left(\sigma_1 \, d\epsilon_1 + \sigma_2 \, d\epsilon_2 + \sigma_3 \, d\epsilon_3 \right) \qquad (7\text{-}30c)$$

$$d\sigma_3 = a_0 \, d\epsilon_{nn} + a_1 \, d\epsilon_3 + a_2 \sigma_{nn} \, d\epsilon_{kk} + a_3 \sigma_{nn} \, d\epsilon_3 + a_4 \, d\epsilon_{kk} \sigma_3$$

$$+ a_5 \left(2\sigma_3 \, d\epsilon_3 \right) + a_6 \left(\sigma_1 \, d\epsilon_1 + \sigma_2 \, d\epsilon_2 + \sigma_3 \, d\epsilon_3 \right) \qquad (7\text{-}30d)$$

Let us now consider the hydrostatic compression (HC) state. For this case $\sigma_1 = \sigma_2 = \sigma_3$ and $d\epsilon_1 = d\epsilon_2 = d\epsilon_3$. By denoting stress as σ and strain increment as $d\epsilon$, Eq. (7-30d) can be written as

$$d\sigma = 3a_0 \, d\epsilon + a_1 \, d\epsilon + 9a_2 \sigma \, d\epsilon + 3a_3 \sigma \, d\epsilon + 3a_4 \sigma \, d\epsilon$$

$$+ 2a_5 \sigma \, d\epsilon + 3a_6 \sigma \, d\epsilon \qquad (7\text{-}31a)$$

or

$$d\sigma = \left(3a_0 + a_1 \right) d\epsilon + \left(9a_2 + 3a_3 + 3a_4 + 2a_5 + 3a_6 \right) \sigma \, d\epsilon \qquad (7\text{-}31b)$$

or

$$d\sigma = \alpha_1 \, d\epsilon + \alpha_2 \sigma \, d\epsilon \qquad (7\text{-}31c)$$

where

$$\alpha_1 = 3a_0 + a_1$$

$$\alpha_2 = 9a_2 + 3a_3 + 3a_4 + 2a_5 + 3a_6$$

Let us now consider the CTC, CTE, RTE, and TC tests (Chapter 5). For all those test conditions, the stress tensor takes the form

$$\sigma_{ij} = \begin{pmatrix} \sigma_1 & 0 & 0 \\ 0 & \sigma_3 & 0 \\ 0 & 0 & \sigma_3 \end{pmatrix} \qquad (7\text{-}32)$$

and the incremental strain tensor takes the form

$$d\epsilon_{ij} = \begin{pmatrix} d\epsilon_1 & 0 & 0 \\ 0 & d\epsilon_3 & 0 \\ 0 & 0 & d\epsilon_3 \end{pmatrix} \qquad (7\text{-}33)$$

Although the incremental strain tensor has the same form for all triaxial conditions, the sign of the increments are different. Table 7-1 summarizes the details of incremental quantities for the test conditions above.

For the foregoing test conditions, Eqs. (7-30) take the following forms:

$$d\sigma_1 = a_0(d\epsilon_1 + 2\,d\epsilon_3) + a_1\,d\epsilon_1 + a_2(\sigma_1 + 2\sigma_3)(d\epsilon_1 + 2\,d\epsilon_3) + a_3(\sigma_1 + 2\sigma_3)\,d\epsilon_1$$

$$+ a_4(d\epsilon_1 + 2\,d\epsilon_3)\sigma_1 + a_5(2\sigma_1\,d\epsilon_1) + a_6(\sigma_1\,d\epsilon_1 + 2\sigma_3\,d\epsilon_3) \qquad (7\text{-}34a)$$

By rearranging this equation, we have

$$d\sigma_1 = (a_0 + a_1)\,d\epsilon_1 + (a_2 + a_3 + a_4 + 2a_5 + a_6)\sigma_1\,d\epsilon_1 + (2a_2 + 2a_3)\sigma_3\,d\epsilon$$

$$+ 2a_0\,d\epsilon_3 + (2a_2 + 2a_4)\sigma_1\,d\epsilon_3 + (4a_2 + 2a_6)\sigma_3\,d\epsilon_3 \qquad (7\text{-}34b)$$

Equation (7-34b) can be written as

$$d\sigma_1 = \alpha_1\,d\epsilon_1 + \alpha_2\sigma_1\,d\epsilon_1 + \alpha_3\sigma_3\,d\epsilon_1 + \alpha_4\,d\epsilon_3 + \alpha_5\sigma_1\,d\epsilon_3 + \alpha_6\sigma_3\,d\epsilon_3 \qquad (7\text{-}35)$$

**TABLE 7-1   Sign of Incremental Quantities for Different
Testing Configurations (Fig. 5-10)**

Test	$d\sigma_1$	$d\sigma_2$	$d\sigma_3$	$d\epsilon_1$	$d\epsilon_2$	$d\epsilon_3$
CTC	+	0	0	+	−	−
CTE	0	+	+	−	+	+
RTE	−	0	0	−	+	+
TC	+	−	−	+	−	−
TE	−	+	+	−	+	+

where

$$\alpha_1 = a_0 + a_1$$
$$\alpha_2 = a_2 + a_3 + a_4 + 2a_5 + a_6$$
$$\alpha_3 = 2a_2 + 2a_3$$
$$\alpha_4 = 2a_0$$
$$\alpha_5 = 2a_2 + 2a_4$$
$$\alpha_6 = 4a_2 + 2a_6$$

Similarly, Eq. (7-30d) can be written as

$$d\sigma_3 = d\sigma_2 = \alpha_1 d\epsilon_1 + \alpha_2 \sigma_1 d\epsilon_1 + \alpha_3 \sigma_3 d\epsilon_1 + \alpha_4 d\epsilon_3 + \alpha_5 \sigma_1 d\epsilon_3 + \sigma_6 \sigma_3 d\epsilon_3 \qquad (7\text{-}36)$$

where

$$\alpha_1 = a_0$$
$$\alpha_2 = a_2 + a_6$$
$$\alpha_3 = 2a_2 + a_4$$
$$\alpha_4 = 2a_0 + a_1$$
$$\alpha_5 = 2a_2 + a_3$$
$$\alpha_6 = 4a_2 + 2a_3 + 2a_4 + 2a_5 + 2a_6$$

In the case of the simple shear strain test, the general stress tensor takes the form

$$\sigma_{ij} = \begin{pmatrix} \sigma_{11} & \sigma_{12} & 0 \\ \sigma_{12} & \sigma_{22} & 0 \\ 0 & 0 & \sigma_{33} \end{pmatrix} \qquad (7\text{-}37)$$

and the incremental strain tensor is given by

$$d\epsilon_{ij} = \begin{pmatrix} 0 & d\epsilon_{12} & 0 \\ d\epsilon_{12} & 0 & 0 \\ 0 & 0 & 0 \end{pmatrix} \qquad (7\text{-}38)$$

For this case Eq. (7-9) can be simplified to obtain

$$d\sigma_{11} = 2(a_5 + a_6)\sigma_{12} d\epsilon_{12} \qquad (7\text{-}39a)$$
$$d\sigma_{22} = d\sigma_{11} \qquad (7\text{-}39b)$$
$$d\sigma_{33} = 2a_6\sigma_{12} d\epsilon_{12} \qquad (7\text{-}39c)$$
$$d\sigma_{12} = \left[a_1 + (\sigma_{11} + \sigma_{22} + \sigma_{33})a_3 + (\sigma_{11} + \sigma_{22})a_5 \right] d\epsilon_{12} \qquad (7\text{-}39d)$$
$$d\sigma_{13} = \sigma_{23} = 0 \qquad (7\text{-}39e)$$

In the following section, numerical values of the parameters will be determined for the artificial soil described in Chapter 5. Here experimental data will be used in conjunction with the incremental relationships described in the foregoing section. The material is assumed to be isotropic, and average values of measured strains are used. For example, for the CTC test (Fig. 5-15) an average of the different extension strains is used. Also, only the loading envelope is considered.

Hydrostatic Compression Test

It can be observed that in the loading curve [Fig. 5-14(b)], a zero state of strain corresponds to an initial hydrostatic stress of 2.75 psi (19 kPa). Therefore, to model the loading behavior, let us assume that $\epsilon = 0$ when $\sigma_{oct} = 2.75$ psi. Table 7-2 gives details of the observed data for the hydrostatic case.

By using Eq. (7-31), the values of α_1 and α_2 were determined by using the least-squares method [Eq. (6-88)] as

$$\alpha_1 = 61.20 \text{ psi}$$
$$\alpha_2 = 36.20$$

Conventional Triaxial Compression Test

Figure 5-15 and Table 7-3 show observed data for a CTC test with an initial hydrostatic stress of 5.0 psi (35.0 kPa). By using Eq. (7-32) in conjunction with the least-squares fitting method [Eq. (6-88)], the values of α_i in Eq. (7-35) were obtained as

$$\alpha_1 = 175.0 \text{ psi}$$
$$\alpha_2 = -7.75$$
$$\alpha_3 = 7.77$$
$$\alpha_4 = -729.0 \text{ psi}$$
$$\alpha_5 = 76.1$$
$$\alpha_6 = 31.9$$

TABLE 7-2   HC Test Data for Artificial Soil [Fig. 5-14(b)]

σ	$\Delta\sigma$	ϵ	$\Delta\epsilon$	$\sigma\Delta\epsilon$
2.75		0.0		
	0.25		0.00125	0.00344
3.00		0.00125		
	1.00		0.005	0.015
4.00		0.00625		
	1.00		0.00525	0.021
5.00		0.0115		
	1.00		0.0045	0.0225
6.00		0.016		
	1.00		0.004	0.0240
7.00		0.020		
	1.00		0.003	0.0210
8.00		0.0230		
	1.00		0.003	0.024
9.00		0.0260		
	1.00		0.002	0.0180
10.00		0.0280		

TABLE 7-3 CTC Test Data for Artificial Soil (Fig. 5-15)

$\Delta\tau_{oct}$	$\Delta\sigma_1 = \dfrac{3}{\sqrt{2}}\tau_{oct}$	σ_1	$\sigma_2 = \sigma_3$	ϵ_1 (Coeff. of α_1)	$\Delta\epsilon_1$	$\epsilon_2 = \epsilon_3$	$\Delta\epsilon_2$ (Coeff. of α_4)	$\sigma_1\Delta\epsilon_1$ (Coeff. of α_2)	$\sigma_3\Delta\epsilon_1$ (Coeff. of α_3)	$\sigma_1\Delta\epsilon_3$ (Coeff. of α_5)	$\sigma_3\Delta\epsilon_3$ (Coeff. of α_6)
0.5	1.06	5.0	5.0	0.0	0.004	0.0	−0.0005	0.0200	0.0200	−0.0025	−0.0025
0.5	1.06	6.06	5.0	0.004	0.009	−0.0005	−0.0020	0.0545	0.0450	−0.0121	−0.0100
0.5	1.06	7.12	5.0	0.013	0.012	−0.0025	−0.0035	0.0854	0.0600	−0.0249	−0.0175
0.5	1.06	8.18	5.0	0.025	0.022	−0.0060	−0.0040	0.1800	0.1100	−0.0327	−0.0200
0.5	1.06	9.24	5.0	0.047	0.023	−0.0100	−0.0050	0.2125	0.1150	−0.0462	−0.0250
0.5	1.06	10.30	5.0	0.070	0.020	−0.0150	−0.0070	0.2060	0.1000	−0.0721	−0.0350
0.5	1.06	11.36	5.0	0.090		−0.0220					

TABLE 7-4 TC Test Data for Artificial Soil [Fig. 5-19(a)][a]

σ_1 psi	$\Delta\sigma_1$	$\sigma_2 = \sigma_3$	$\Delta\sigma_3$	ϵ_1	$\Delta\epsilon_1$ (Coeff. of α_1)	$\epsilon_2 = \epsilon_3$	$\Delta\epsilon_3$ (Coeff. of α_4)	$\sigma_1\Delta\epsilon_1$ (Coeff. of α_2)	$\sigma_3\Delta\epsilon_1$ (Coeff. of α_3)	$\sigma_1\Delta\epsilon_3$ (Coeff. of α_5)	$\sigma_3\Delta\epsilon_3$ (Coeff. of α_6)
10.0		10.0		0.0		0.0					
	1.414		−0.707		0.0005		0.0	0.005	0.005	0.0	0.0
11.414		9.293		0.0005		0.0					
	1.414		−0.707		0.0030		−0.0015	0.0342	0.0279	−0.0171	−0.0139
12.828		8.586		0.0035		−0.0015					
	1.414		−0.707		0.0040		−0.0010	0.0513	0.0343	−0.0128	−0.0086
14.242		7.879		0.0075		−0.0025					
	1.414		−0.707		0.0055		−0.0015	0.0783	0.0433	−0.0214	−0.0118
15.656		7.172		0.0130		−0.004					
	1.414		−0.707		0.0090		−0.0025	0.1409	0.0646	−0.0391	−0.0179
17.070		6.465		0.022		−0.0065					
	1.414		−0.707		0.0150		−0.0045	0.2561	0.0970	−0.0768	−0.0291
18.484		5.758		0.037		−0.011					
	1.414		−0.707		0.0290		−0.0110	0.5360	0.1670	−0.2033	−0.0633
19.898		5.051		0.066		−0.022					
	1.414		−0.707		0.0340		−0.0255	0.6765	0.1717	−0.5074	−0.1288
21.312		4.344		0.100		−0.0475					
	1.414		−0.707		0.0500		−0.0375	1.0656	0.2172	−0.7992	−0.1629
22.726		3.637		0.150		−0.085					

[a] Based on the same data as in Examples 6-1 and 6-2.

Triaxial Compression (TC) Test

Figure 5-19(a) and Table 7-4 show observed data for TC test with an initial hydrostatic stress of 10.0 psi (69.0 kPa). By using Eqs. (7-35) and (7-36), the values of α_i in the equations above were found as

α_i in Eq. (7-35):

$$\alpha_1 = 769.0 \text{ psi}$$

$$\alpha_2 = -54.2$$

$$\alpha_3 = 41.7$$

$$\alpha_4 = -4600.0 \text{ psi}$$

$$\alpha_5 = 123.0$$

$$\alpha_6 = 387.0$$

TABLE 7-5   Summary of Results for Hypoelastic Model for Artificial Soil

No.	Test	Equation	α	a_0	a_1	a_2	a_3	a_4	a_5	a_6	Right-Hand Side of Eq.
1	HC	(7-31)	α_1	3	1	0	0	0	0	0	61.2
2	HC	(7-31)	α_2	0	0	9	3	3	2	3	36.2
3[a]	CTC	(7-35)	α_1	1	1	0	0	0	0	0	175.0
4	CTC	(7-35)	α_2	0	0	1	1	1	2	1	−7.75
5	CTC	(7-35)	α_3	0	0	2	2	0	0	0	−7.77
6[a]	CTC	(7-35)	α_4	2	0	0	0	0	0	0	−729.0
7	CTC	(7-35)	α_5	0	0	2	0	2	0	0	76.1
8	CTC	(7-35)	α_6	0	0	4	0	0	0	2	31.9
9[a]	TC	(7-35)	α_1	1	1	0	0	0	0	0	769.0
10	TC	(7-35)	α_2	0	0	1	1	1	2	1	−54.2
11	TC	(7-35)	α_3	0	0	2	2	0	0	0	41.7
12	TC	(7-35)	α_4	2	0	0	0	0	0	0	−4600.0
13	TC	(7-35)	α_5	0	0	2	0	2	0	0	123.0
14	TC	(7-35)	α_6	0	0	4	0	0	0	2	387.0
15	TC	(7-36)	α_1	1	0	0	0	0	0	0	−399.0
16	TC	(7-36)	α_2	0	0	1	0	0	0	1	27.6
17	TC	(7-36)	α_3	0	0	2	0	1	0	0	−19.9
18	TC	(7-36)	α_4	2	1	0	0	0	0	0	2320.0
19	TC	(7-36)	α_5	0	0	2	1	0	0	0	−62.3
20	TC	(7-36)	α_6	0	0	4	2	2	2	2	−195.0

[a] These values were not used in the Least-squares Computations.

α_i in Eq. (7-36):

$$\alpha_1 = -399.0 \text{ psi}$$
$$\alpha_2 = 27.6$$
$$\alpha_3 = -19.9$$
$$\alpha_4 = 2320.0 \text{ psi}$$
$$\alpha_5 = -62.3$$
$$\alpha_6 = -195.0$$

A summary of the results from the analysis above is given in Table 7-5.

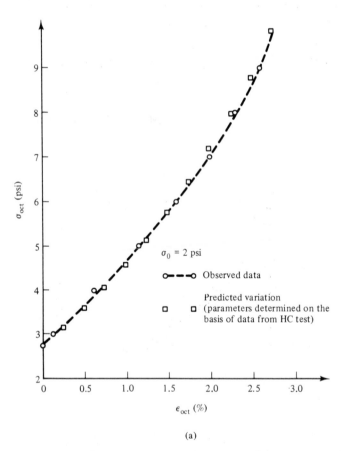

(a)

Figure 7-3   Comparison between predictions from grade-one hypoelastic model for artificial soil and observation for (a) HC test [Fig. 5-14(b)]: $\sigma_0 = 2$ psi (b) CTC test [Fig. 5-15)]: $\sigma_0 = 5$ psi (c) TC test [Fig. 5-19(a)]: $\sigma_0 = 10$ psi (d) HC test [Fig. 5-14(b)]: $\sigma_0 = 2$ psi, predictions based on average parameters (e) TC test [Fig. 5-19(a)]: $\sigma_0 = 10$ psi, predictions based on average parameters.

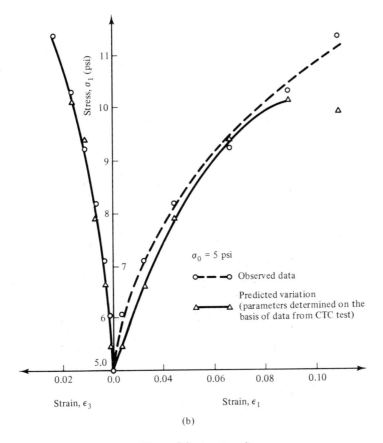

Figure 7-3 (*continued*)

Then by using the least-squares method, the following values of a_i were found:

$$a_0 = -1950.0 \text{ psi}$$
$$a_1 = 6070.0 \text{ psi}$$
$$a_2 = 35.7$$
$$a_3 = -57.5$$
$$a_4 = -16.1$$
$$a_5 = -39.1$$
$$a_6 = 10.3$$

It may be noted that the calculated negative value of a_0 may not be realistic. This is the consequence mainly of the averaging of the properties over a

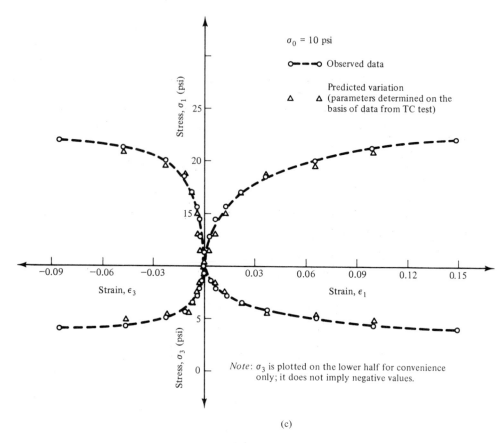

(c)

Figure 7-3 (*continued*)

number of stress paths. The details are included here for the sake of illustrating the procedure for finding the parameters of this model.

The values of α_i [Eqs. (7-31c) and (7-35)], determined on the basis of data from these individual stress paths, and the values of a_i, found by averaging results from various stress paths, were used to back predict the behavior of HC, CTC and TC tests. Comparisons of prediction and experimentally observed data are shown in Fig. 7-3. Predictions obtained by using the parameters, determined on the basis of data from specific stress paths, compare very well with observed curves for the HC, CTC, and TC tests, [Figs. 7-3(a), 7-3(b), and 7-3(c)].

Comparisons between observed data and predictions, based on the values of a_i obtained by averaging data from a number of stress paths, are shown in Figs. 7-3(d) and 7-3(e). The correlation between observed data and predictions is not satisfactory. It may be noted that the parameters thus determined represent weighted average values based on tests under a number of different

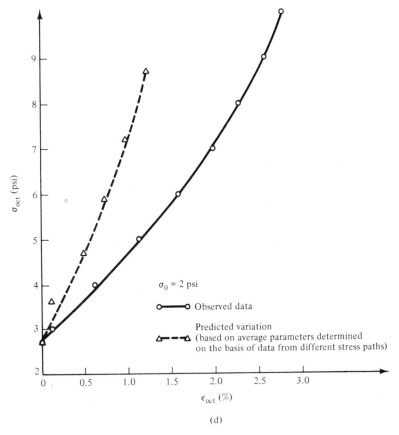

$\sigma_0 = 2$ psi

○——○ Observed data

Predicted variation
△– – –△ (based on average parameters determined
on the basis of data from different stress paths)

(d)

Figure 7-3 (*continued*)

stress paths. The deviation between predictions and observed data indicates that the grade-one hypoelastic model may not be appropriate for the artificial soil. Hence, the back predictions show different degrees of correlation with observed data for different stress paths.

Example 7-2   Colorado River Sand

Figure 7-4(a) and (b) shows four test data under CTC loading conditions for Colorado River sand under $\sigma_{33} = 2.32$, 4.60, 6.95, and 8.69 psi (16.00, 32.04, 48.00, and 60.00 kPa). The density of sand was 94 lb/ft^2 (1520 kg/m^3) and the specific gravity = 2.7. The sample was 1.4 in. (3.60 cm) in diameter and 3.2 in. (8.12 cm) high. Parameters a_0 to a_6 for the grade-one model were obtained by Mysore (10) using a curve-fitting scheme on three curves at $\sigma_{33} = 2.32$, 4.60, and 6.95 psi.

Table 7-6 gives values of $d\sigma_{11}/d\epsilon_1$ at the initial state $\sigma_{11} = \sigma_{33}$ [Eq. (7-16a)], $d\epsilon_v/d\epsilon_{11}$ at $\sigma_1 = \sigma_3$ [Eq. (7-15)], $d\epsilon_v/d\epsilon_{11}$ at failure, σ_1' at $d\epsilon_v/d\epsilon_{11} = 0$, and σ_{1f}. On the basis of these values and by using Eqs. (7-18), (7-19), (7-21), and (7-22), the following

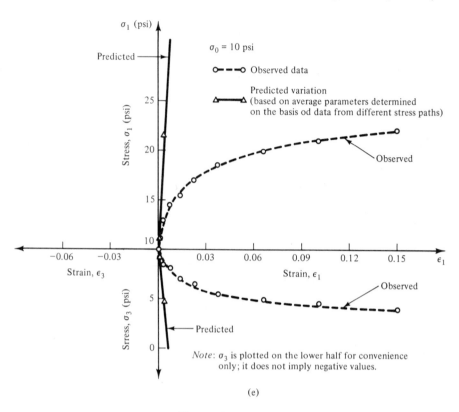

(e)

Figure 7-3 (*continued*)

values were determined (10):

$$a_0 = 1992.0500 \text{ psi}$$
$$a_1 = 392,3840 \text{ psi}$$
$$a_2 = -5.779$$
$$a_3 = 339.6080$$
$$a_4 = 315.2029$$
$$a_5 = 370.4618$$
$$a_6 = -262.0297$$

The dashed curves in Fig. 7-4(a) show predictions by using the foregoing parameters. The correlation with the curve for $\sigma_3 = 4.60$ psi, which was not used in the determination of the parameters, is also satisfactory. The predictions for ϵ_v versus ϵ_{11} curves are satisfactory [Fig. 7-4(b) and (c)]; however, the predictions are not close to the experimental results.

The foregoing parameters were used to extrapolate the behavior of sand at other (higher) confining stresses, $\sigma_3 = 12.0$, 20.0, and 30.0 psi (83, 138, and 207 kPa), by using Eqs. (7-14). Various parameters were obtained for the predicted curves (Table 7-7), and the failure stresses were obtained. Then Mohr circles were drawn as in Fig. 7-5. It can

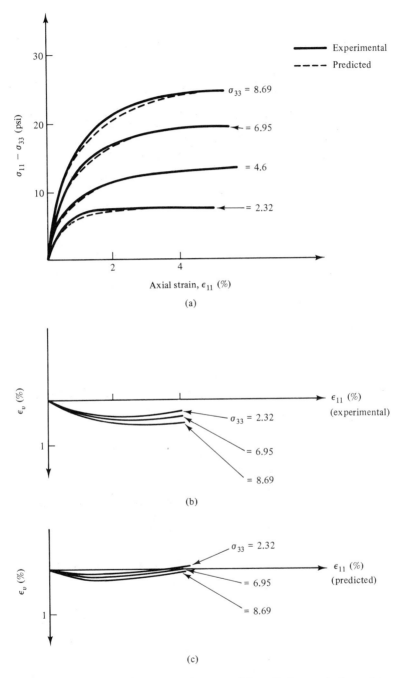

Figure 7-4 Comparison between grade-one model predictions and observation from CTC test for Colorado River sand (Ref. 10).

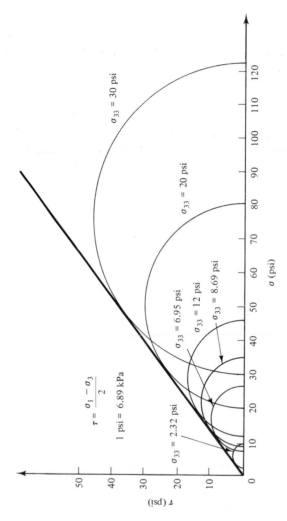

Figure 7-5 Comparison of failure stresses (Ref. 10).

$$\tau = \frac{\sigma_1 - \sigma_3}{2}$$

1 psi = 6.89 kPa

$\sigma_{33} = 30$ psi

$\sigma_{33} = 20$ psi

$\sigma_{33} = 6.95$ psi

$\sigma_{33} = 12$ psi

$\sigma_{33} = 8.69$ psi

$\sigma_{33} = 2.32$ psi

σ (psi)

τ (psi)

TABLE 7-6   Experimental Values for Colorado River Sand (10)[a]

Confining Pressure (psi)	Initial Tangent Modulus $\left.\dfrac{d\sigma_{11}}{d\epsilon_{11}}\right|_{\sigma_{11}=\sigma_{33}}$ (psi)	Initial Slope of $\epsilon_v-\epsilon_1$ Curve $\left.\dfrac{d\epsilon_v}{d\epsilon_{11}}\right|_{\sigma_{11}=\sigma_{33}}$	$\left.\dfrac{d\epsilon_v}{d\epsilon_{11}}\right|_{\text{failure}}$	Axial Stress at Which $\dfrac{d\epsilon_v}{d\epsilon_{11}}=0$ (psi)	Axial Stress at Failure, σ_{1f} (psi)
2.32	2055.55	0.33	0	9.92	9.92
4.60	2500.12	0.36	0	16.20	17.76
6.95	3264.71	0.41	0	25.1079	27.15
8.69	4269.23	0.48	0	31.8479	35.01

[a] 1 psi = 6.89 kPa.

TABLE 7-7   Predicted Values for Colorado River Sand (10)

Confining Pressure (psi)	Axial Failure Stress (psi)	Initial Tangent Modulus $\left.\dfrac{d\sigma_{11}}{d\epsilon_{11}}\right|_{\sigma_{11}=\sigma_{22}}$ (psi)	Initial slope of $\epsilon_v-\epsilon_1$ Curve
2.32	9.88	1444.95	0.214
4.60	17.72	2257.66	0.315
6.95	26.04	3021.03	0.400
8.69	32.56	3572.26	0.455
12.0	45.39	4576.96	0.544
20.0	77.66	6842.82	0.700
30.0	119.0	9486.44	0.827

be seen that the grade-one model can predict the failure behavior of the sand quite well for the CTC stress paths.

Example 7-3   Ottawa Sand

In this example, a grade-one hypoelastic model for Ottawa sand is presented; the experimental results were determined by Holubec (11) and were used by Coon and Evans (6) to derive grade-one model for this soil.

The experiments were conducted using a conventional triaxial device. Cylindrical test samples were 2.00 in. (5.00 cm) in diameter and 4.0 in. (10.0 cm) in height. Samples were prepared at three initial void ratios, 0.665, 0.555, and 0.465, and corresponding relative densities were 30%, 70%, and 94%. Further details of testing arrangement can be found in Refs. 6 and 11. Typical test results obtained by Holubec (11) are shown in Fig. 7-6.

Coon and Evans (6) expressed Eqs. (7-14a) and (7-14b) as

$$d\sigma_{11} = A_{11}\, d\epsilon_{11} + A_{12}\, d\epsilon_{22} \qquad (7\text{-}40a)$$

$$d\sigma_{22} = A_{21}\, d\epsilon_{11} + A_{22}\, d\epsilon_{22} \qquad (7\text{-}40b)$$

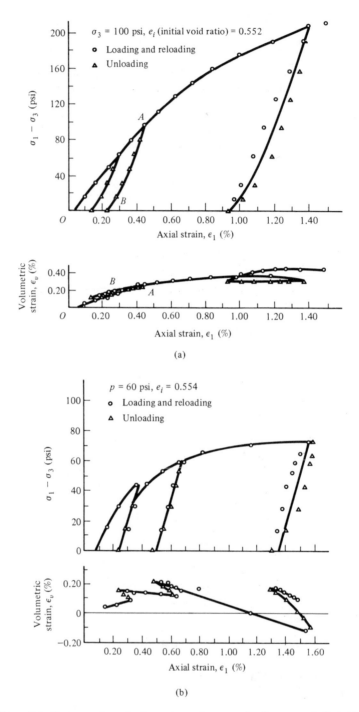

Figure 7-6 Stress–strain and volumetric–strain curves for Ottawa sand (Ref. 11): (a) CTC test results; (b) constant mean normal stress test results; (c) anisotropic consolidation tests (Ref. 11).

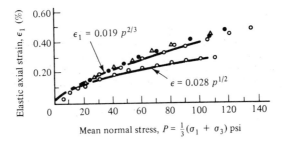

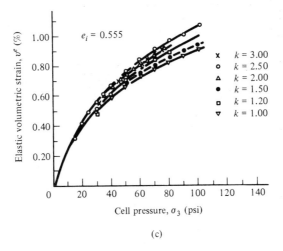

(c)

Figure 7-6 (*continued*)

where

$$A_{11} = (a_0 + a_1) + (a_2 + a_3 + a_4 + 2a_5 + a_6)\sigma_{11} + (2a_2 + 2a_3)\sigma_{22}$$
$$A_{12} = 2a_0 + (2a_2 + 2a_4)\sigma_{11} + (4a_2 + 2a_6)\sigma_{22}$$
$$A_{21} = a_0 + (a_2 + a_6)\sigma_{11} + (2a_2 + a_4)\sigma_{22}$$
$$A_{22} = (2a_0 + a_1) + (2a_2 + a_3)\sigma_{11} + (4a_2 + 2a_3 + 2a_4 + 2a_5 + 2a_6)\sigma_{22}$$

Parameters a_0 to a_6 have been determined (6) to fit Holubec's test data as shown in Fig. 7-7. The values of the parameters computed by Coon and Evans are given below.

$$a_0 = 2000 \text{ psi}$$
$$a_1 = 0 \text{ psi}$$
$$a_3 = -60$$
$$a_4 = 180$$
$$a_5 = 90$$
$$a_6 = -60$$

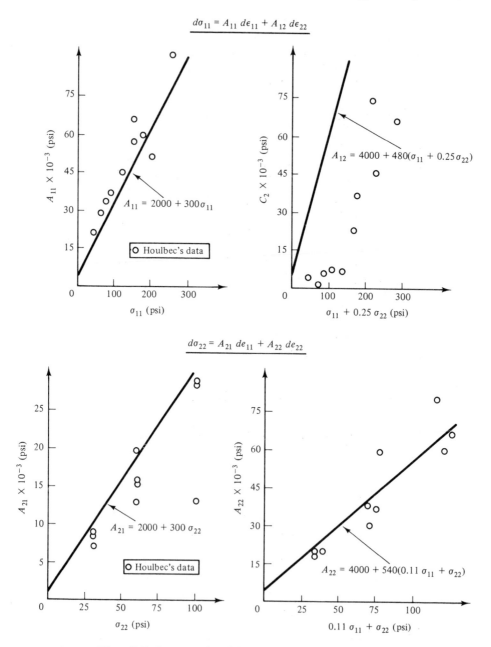

Figure 7-7   Incremental moduli for isotropic representation (Ref. 6).

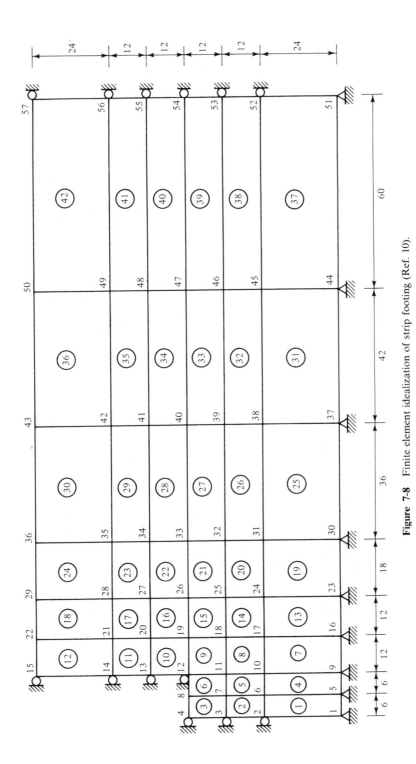

Figure 7-8 Finite element idealization of strip footing (Ref. 10).

It has been pointed out by Coon and Evans (6) that numerous combinations of constants could be found to give a reasonable fit to test data; the particular values given above have been chosen so that the triaxial behavior becomes integrable.

APPLICATIONS

The hypoelastic models described earlier in this chapter could describe certain aspects of stress-path-dependent behavior of geologic media. A grade-one hypoelastic model has been implemented in a numerical procedure for solving some boundary value problems in geotechnical engineering (10). The implementation of this model in a numerical technique such as the finite element method requires special attention because the resulting constitutive matrix, in general, is nonsymmetric.

The material constants used in the analysis were the same as those given in Example 7-2 for the Colorado River sand. One of the example problems reported in Ref. (10) will be described briefly in the following section to show the application of the hypoelastic model for evaluation of displacements and stresses in sand due to a strip footing.

The finite element mesh used in the analysis is shown in Fig. 7-8. The initial vertical stresses were computed based on the self-weight of the soil, and the initial horizontal stresses were calculated using $K_0 = 1 - \sin \phi$, the coefficient of lateral earth pressure at rest; here ϕ is the angle of friction. A vertical load of 60 psi (44 kN/m^2) was then applied on the footing, and the analysis was carried out for 60 load increments. Typical load–displacement relationships at a few nodal points are shown in Fig. 7-9. The predictions seem to be qualitatively reasonable. Although hypoelastic models can be implemented in numerical solution procedures, the computational effort can be relatively high

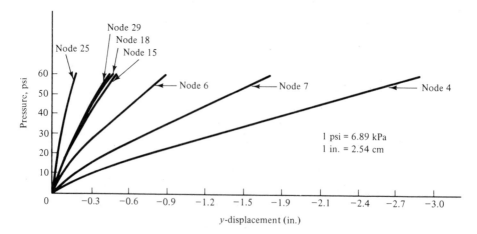

Figure 7-9 Load–displacement curves for typical nodes (Ref. 10).

in view of the nonsymmetric nature of the resulting constitutive matrix. One advantage is that the required constitutive parameters for grade-one model may often be determined from conventional triaxial tests. This model, in general, can be employed for situations with monotonically increasing loads.

PROBLEMS

7-1. Show that for grade-one model,

$$\alpha_0 = K - \frac{2G}{3} \qquad \text{and} \qquad \alpha_1 = 2G$$

[*Hint*: Use Eq. (7-7).]

7-2. Derive expressions for $d\sigma_{33}/d\epsilon_{33}$ and $d\epsilon_v/d\epsilon_{33}$ for grade-one hypoelastic model for conventional triaxial extension (CTE) test. (*Hint*: In this test, the stress σ_{33} is held constant and the axial stress σ_{11} is reduced.)

7-3. Derive the constitutive law for one-dimensional compression stress path for grade-one hypoelastic law. [*Hint*: Here $\sigma_{22} = \sigma_{33}$, $\sigma_{12} = \sigma_{23} = \sigma_{13} = 0$, and $\epsilon_{22} = \epsilon_{33} = 0$ and assume that $\sigma_{33} = K_0\sigma_{11}$, where $K_0 = 1 - \sin\phi$, the coefficient (lateral) earth pressure at rest, and ϕ is the angle of internal friction.]

Partial Solution: Two differential equations are

$$\frac{d\sigma_{11}}{d\epsilon_{11}} = a_0 + a_1 + (a_2 + a_3 + a_4 + 2a_5 + a_6)\sigma_{11} + 2(a_2 + a_3)\sigma_{33}$$

$$\frac{d\sigma_{33}}{d\epsilon_{11}} = a_0 + (a_2 + a_6)\sigma_{11} + (2a_2 + a_4)\sigma_{33}$$

With $\sigma_{33} = K_0\sigma_{11}$, we have

$$d\sigma_{11} = (A + B\sigma_{11})\,d\epsilon_{11}$$

where $A = a_0 + a_1$ and $B = (a_2 + a_3)(1 + 2K_0)$, and finally,

$$\sigma_{11} = \left(\frac{A}{B} + \sigma_0\right)e^{B\epsilon_{11}} - \frac{A}{B}$$

where σ_0 is the initial hydrostatic stress.

REFERENCES

1. Truesdell, C., "Hypoelasticity," *J. Ration. Mech. Anal.*, Vol. 4, 1955, pp. 83–133.

2. Truesdell, C., "Hypoelastic Shear," *J. Appl. Phys.*, Vol. 27, 1956, pp. 441–467.

3. Green, A. E., "Hypoelasticity and Plasticity, II," *J. Ration. Mech. Anal.*, Vol. 5, No. 5, 1956, pp. 725–734.

4. Eringen, A. C., *Nonlinear Theory of Continuous Media*, McGraw-Hill Book Company, New York, 1962.

5. Rivlin, R. S., and Ericksen, J. L., "Stress–Deformation Relation for Isotropic Materials," *J. Ration. Mech. Anal.*, Vol. 4, No. 2, Mar. 1955, pp. 323–425.

6. Coon, M. D., and Evans, R. J., "Recoverable Deformations of Cohesionless Soils," *J. Soil Mech. Found. Div., ASCE*, Vol. 97, No. SM2, Feb. 1971, pp. 375–390.

7. Desai, C. S., "Overview, Trends, and Projections: Theory and Applications of the Finite Element Method in Geotechnical Engineering," *State-of-the-Art Paper, Proc., Symp. on Appl. of the FEM in Geotech. Eng.*, Vicksburg, Miss., 1972.

8. Vagneron, J., Lade, P. V., and Lee, K. L., "Evaluation of Three Stress–Strain Models for Soils," *Proc. 2nd Int. Conf. Numer. Methods Geomech.*, C. S. Desai (Ed.), Blacksburg, Va., ASCE, June 1976.

9. Rohani, B., "Mechanical Constitutive Models for Engineering Materials," *Report S-77-19*, U.S. Army Corps of Engineers Waterway Exp. St., Vicksburg, Miss., 1977.

10. Mysore, R., "Finite Element Analysis of a Sand as a Hypoelastic Material," Ph.D. dissertation, State Univ. of New York, Buffalo, 1978.

11. Holubec, I., "Elastic Behavior of Cohesionless Soil," *J. Soil Mech. Found. Div., ASCE*, Vol. 94, No. SM6, Nov. 1968, pp. 1215–1231.

8

QUASILINEAR MODELS

A quasilinear model can be considered an approximation of actual nonlinear behavior by using essentially a series of linear behavior. Many of the models used in the past (1, 2) and based on *piecewise* linear behavior can fall into this category. In other words, a given nonlinear behavior is divided into pieces of linear elastic behavior, and very often the incremental Hooke's law or grade-zero hypoelastic law is assumed to describe the behavior. For each piece (Fig. 8-1), however, the material parameters, say K and G or E and ν, are revised by making them functions of the state of stress (strain). Thus a quasilinear model often approximates nonlinearity by including the nonlinear effects in essentially two parameters, K and G or E and ν.

BASIC EQUATION

As in Chapter 7, we express the constitutive equation for grade-zero incremental model as

$$d\sigma_{ij} = K \, d\epsilon_{nn} \delta_{ij} + 2G\left(d\epsilon_{ij} - \tfrac{1}{3} d\epsilon_{nn} \delta_{ij}\right) \tag{8-1a}$$

$$d\sigma_{ij}^m = C_{ijkl}^{tm} \, d\epsilon_{kl}^m \tag{8-1b}$$

or

$$\{d\sigma_m\} = [C_{tm}]\{d\epsilon_m\} \tag{8-1c}$$

where m denotes the mth increment of stress $\{d\sigma\}$ and strain $\{d\epsilon\}$, and $[C_{tm}]$

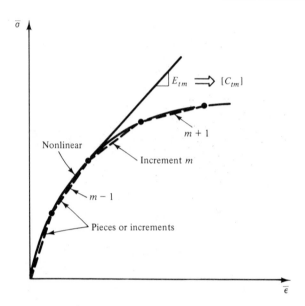

Figure 8-1   Piecewise linear or quasilinear approximation.

denotes the (tangent) constitutive matrix relevant to the mth increment (Fig. 8-1). The magnitude of $[\mathbf{C}_{tm}]$ is different for different increments, and it is often computed as a tangent value at a point within the mth increment. Since $[\mathbf{C}_{tm}]$ is composed of bulk and shear moduli or Young's modulus and Poisson's ratio, it is the tangent values of these parameters that we are interested in. For instance (Eq. 6-18b),

$$
[\mathbf{C}_{tm}] =
\begin{bmatrix}
K_t + \dfrac{4G}{3}t & K_t - \dfrac{4G}{3}t & K_t - \dfrac{4G}{3}t & 0 & 0 & 0 \\[2mm]
 & K_t + \dfrac{4G}{3}t & K_t - \dfrac{4G}{3}t & 0 & 0 & 0 \\[2mm]
 & & K_t + \dfrac{4G}{3}t & 0 & 0 & 0 \\[2mm]
\text{Symm.} & & & 2G_t & 0 & 0 \\[2mm]
 & & & & 2G_t & 0 \\[2mm]
 & & & & & 2G_t
\end{bmatrix}
$$

where G_t and K_t are the tangent shear and tangent bulk modulus, respectively.

Hence the idea of piecewise linear behavior is based on finding the changing or variable values of the parameters or moduli as the state of stress changes during the incremental loading. This is the reason the approach is often called variable parameter (VP) or variable moduli model (VM).

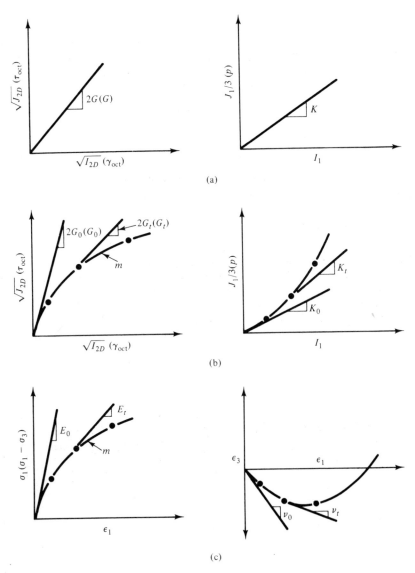

Figure 8-2 Variable parameter models: (a) linear elastic; (b) variable G and K; (c) variable E and ν.

We can let both parameters, say K_t and G_t or E_t and ν_t, vary during the incremental loading. It may also be possible to vary only one of them and keep the other constant. This will depend on the type of material. For example, for a saturated cohesive soil, the variation in the values of K_t and ν_t may be insignificant. Such a material can be characterized by keeping K_t or ν_t constant and varying G_t or E_t. That is, only one parameter is variable.

As stated earlier, the quasilinear models attempt to force some of the nonlinear effects in the two parameters. For example, K_t and G_t are expressed as functions of the state of stress

$$K_t = K_t\left(\sqrt{J_{2D}}, J_1\right) \tag{8-2a}$$

$$G_t = G_t\left(\sqrt{J_{2D}}, J_1\right) \tag{8-2b}$$

Coupled and Uncoupled Behavior

In general, both volumetric and shear response of a material are coupled. That is, a pure shear state of stress can cause volumetric deformations and a hydrostatic state of stress can cause shear deformations. Then we can write

$$\sqrt{J_{2D}} = F_1\left(\sqrt{I_{2D}}, I_1\right) \tag{8-3a}$$

$$\frac{J_1}{3} = p = F_2\left(\sqrt{I_{2D}}, I_1\right) \tag{8-3b}$$

If for certain materials, we can assume that the two responses are uncoupled, that is, the hydrostatic state of stress can cause only volumetric deformations and a state of shear stress only shear strains. Then we can specialize Eq. (8-3) as

$$\sqrt{J_{2D}} = F_1\left(\sqrt{I_{2D}}\right) \tag{8-4a}$$

$$\frac{J_1}{3} = p = F_2(I_1) \tag{8-4b}$$

Depending on the forms of F_1 and F_2, we can obtain various quasilinear models. If the functions F_1 and F_2 were constant, indicating linear relations, Eqs. (8-4a) and (8-4b) reduce to linear elastic laws as [Fig. 8-2(a)]

$$\sqrt{J_{2D}} = 2G\sqrt{I_{2D}} \tag{8-4c}$$

$$\frac{J_1}{3} = KI_1 \tag{8-4d}$$

VARIABLE PARAMETER OR VARIABLE MODULI MODELS

A variable parameter (VP) model is obtained if F_1 and F_2 provide simulation of a nonlinear stress–strain response as a series of linear increments or pieces [Fig. 8-2(b)]. Then

$$d\left(\sqrt{J_{2D}}\right)_m = 2G_{tm}d\left(\sqrt{I_{2D}}\right) \tag{8-5a}$$

$$d\left(\frac{J_1}{3}\right)_m = K_{tm}d(I_1) \tag{8-5b}$$

where m denotes the increment. The values of G_{tm} and K_{tm} change from increment to increment, but are usually assumed to remain constant within each increment.

Often, the stress–strain relations are expressed in terms of axial stress σ_1 or stress difference $(\sigma_1 - \sigma_3)$ versus axial strain, ϵ_1 [Fig. 8-2(c)], then Eq. (8-4a) can lead to

$$d(\sigma_1 - \sigma_3) = E_{tm}\,d\epsilon_1 \tag{8-6a}$$

and the second relation is expressed in terms of lateral strain ϵ_3 versus axial strain ϵ_1 as

$$d\epsilon_3 = -\nu_{tm}\,d\epsilon_1 \tag{8-6b}$$

where E_{tm} and ν_{tm} are tangent Young's modulus and Poisson's ratio, respectively, relevant to increment m.

Now we consider some examples of the VP models often used with solution techniques such as the finite element method. Consider a symbolic stress–strain curve (Fig. 8-3) expressed in terms of some measure of stress denoted by $\bar{\sigma}(\sigma_1, \sigma_1 - \sigma_3, \sqrt{J_{2D}}, \tau_{\text{oct}}, J_1 \text{ or } p)$ and of strain denoted by $\bar{\epsilon}(\epsilon_1, \epsilon_1 - \epsilon_3, \sqrt{I_{2D}}, \gamma_{\text{oct}}, I_1)$.

Functional Representation

As discussed in Chapter 5, for nonlinear analysis, the moduli are computed as tangents or (first) derivatives of the functions representing stress–strain response. Hence if the curve in Fig. 8-3 were expressed by using a mathematical function, we can evaluate the moduli as the first derivative of the function at a point relevant to a given increment. In the following, we have considered a number of possibilities as special cases of Eqs. (8-4).

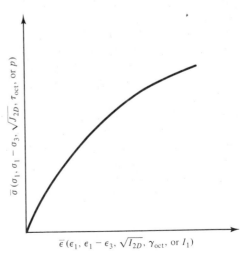

Figure 8-3   Symbolic stress–strain curve.

All stress–strain responses may not be amenable to any one mathematical function. For instance, often $(\sigma_1 - \sigma_3)$ versus ϵ_1 [Fig. 8-2(c)] curves may be simulated by using a hyperbola or parabola. The pressure–volume relations for materials are often obtained in forms shown in Fig. 8-2(b). Here it may be more appropriate to use exponential functions. In the following, we shall present some common functions used for simulating a given stress–strain curve.

Hyperbola.   Use of the hyperbola for representing stress–strain curves (for soils) (Fig. 8-4), was proposed by Kondner (3) and is given by Fig. 8-4(a)

$$\bar{\sigma} = \frac{\bar{\epsilon}}{a_1 + b_1 \bar{\epsilon}} \tag{8-7a}$$

where a_1 and b_1 are related to the initial slope or tangent modulus, E_{ti}, and the

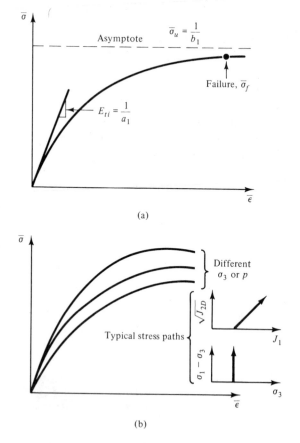

(a)

(b)

Figure 8-4   Hyperbolic simulation of stress–strain curves: (a) hyperbola; (b) effect of confinement (Refs. 3, 5).

asymptolic stress, $\bar{\sigma}_u$, to the curves as

$$E_{ti} = \frac{1}{a_1} \qquad (8\text{-}7b)$$

$$\bar{\sigma}_u = \frac{1}{b_1} \qquad (8\text{-}7c)$$

A modified form of the hyperbola, to account for the parabolic variation in the stress–strain curve at small strains was proposed by Hansen (4):

$$\bar{\sigma} = \left(\frac{\bar{\epsilon}}{a_2 + b_2\bar{\epsilon}} \right)^{1/2} \qquad (8\text{-}8)$$

The value of slope or tangent modulus at a point can be found by differentiating Eqs. (8-7) and (8-8) with respect to $\bar{\epsilon}$ as

$$E_t = \frac{d\bar{\sigma}}{d\bar{\epsilon}} = \frac{a_1}{(a_1 + b_1\bar{\epsilon})^2} \qquad (8\text{-}9a)$$

$$E_t = \frac{d\bar{\sigma}}{d\bar{\epsilon}} = \frac{1}{2\sqrt{\epsilon}} \frac{a_2}{(a_2 + b_2\bar{\epsilon})^{3/2}} \qquad (8\text{-}9b)$$

Since the response of many (geologic) media is dependent on mean pressure ($p = J_1/3$) or confining stress (σ_3), for a given stress path the behavior is often represented by a series of curves for different p or σ_3 [Fig. 8-4(b)]. To incorporate this aspect, Duncan and Chang (5) used the hyperbola in conjunction with the relation between initial modulus and confining pressure proposed by Janbu (6):

$$E_i = K_h'\sigma_3^n \qquad (8\text{-}10a)$$

$$E_i = K_h P_a \left(\frac{\sigma_3}{P_a} \right)^n \qquad (8\text{-}10b)$$

where K_h or K_h' and n represent material parameters and p_a is the atmospheric pressure used to express the results in a nondimensional form. With Eq. (8-10) and the Mohr–Coulomb failure criterion, the following expression for the tangent modulus can be obtained (5):

$$E_t = \left[1 - \frac{R_f(1 - \sin\phi)(\sigma_1 - \sigma_3)}{2c\cos\phi + 2\sigma_3\sin\phi} \right]^2 K_h P_a \left(\frac{\sigma_3}{P_a} \right)^n \qquad (8\text{-}11a)$$

Here c is the cohesive strength, ϕ the angle of friction, and R_f the ratio of the ultimate stress $\bar{\sigma}_u$ to failure stress $\bar{\sigma}_f$ [Fig. 8-4(a)].

A similar relation based on the hyperbolic concept was proposed by Kulhawy et al. (7) for the tangent Poisson's ratio as

$$\nu_t = \frac{G - F\log(\sigma_3/p_a)}{(1 - A)^2} \qquad (8\text{-}11b)$$

where G, F, and d are material parameters, and

$$A = \frac{(\sigma_1 - \sigma_3)d}{K_h p_a (\sigma_3/p_a)^n \left[1 - \dfrac{R_f(\sigma_1 - \sigma_3)(1 - \sin\phi)}{2c\cos\phi + 2\sigma_3 \sin\phi}\right]^2}$$

These models can be simple to use because only five parameters, having direct physical meanings, need to be determined. They can be obtained from laboratory test data with a cylindrical triaxial device conducted under a given stress path, say conventional triaxial compression (CTC) [Fig. 8-4(b)]. They can give satisfactory results for only a limited class of problems, however. For instance, if we are interested in stress–deformation analysis of a nonlinear semi-infinite medium under a monotonically increasing point or strip load, the results could be acceptable. However, for problems involving loading and unloading, and various stress paths in the soil, results from the hyperbolic simulation may not be reliable, one of the major limitations being that the model includes only one stress path, whereas the loading and/or unloading could cause a wide range of stress paths (Fig. 5-10). As indicated in Chapters 6 and 7, these models are not able to account for the (second-order) dilatancy effects. The expression for ν_t loses significance as soon as $\nu_t > 0.5$. Hence this and other models based on functional simulation of a given curve for a specific stress path should be used with care and essentially for cases involving monotonic loading.

Example 8-1   Hyperbolic Simulation

Figure 8-5 shows a set of stress–strain curves for a medium-dense sand tested by using a cylindrical triaxial device at a relative density of about 80% under different confining pressures (8, 9). The samples were first consolidated under a confining pressure, and then shear stress was applied under the constant confinement. Figure 8-6 shows a plot in terms of transformed parameters $\epsilon_1/(\sigma_1 - \sigma_3)$ and ϵ_1. It may be noted that the plots in the initial ranges are not linear, indicating deviation from the hyperbolic function in the early stage of the stress–strain response; such a deviation can occur for many materials.

     The transformed plot in Fig. 8-6 provides values of $a(E_i = 1/a)$, and $b[1/b = (\sigma_1 - \sigma_3)_u]$ for a given curve. The failure ratio, R_f, is found as the ratio of a (chosen) failure point, $(\sigma_1 - \sigma_3)_f$, in the ultimate range of the curve and $(\sigma_1 - \sigma_3)_u$, Fig. 8-4. The plot of E_i versus σ_3 (on logarithmic scale) provides values of K_h and n.

     The foregoing analysis was done for similar test data under two other densities, 60 and 100% (9). The values of the parameters for defining the model in Eq. (8-11a) were found to be

Initial Relative Density	K_h	n	R_f
60	909	0.490	0.895
80	1080	0.516	0.844
100	1530	0.600	0.889

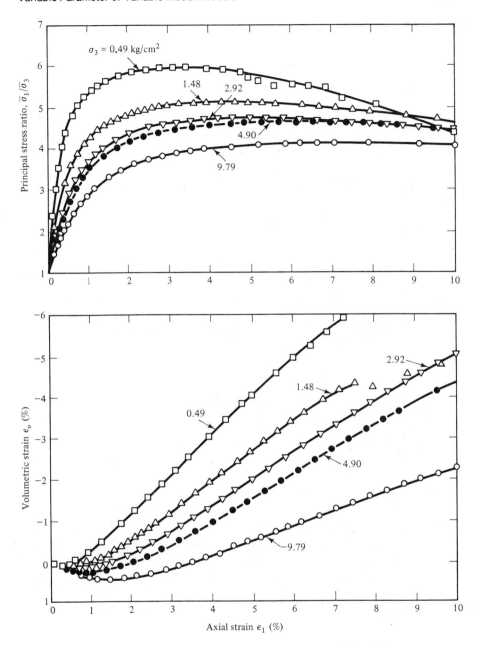

Figure 8-5 Stress ratio and volumetric strain versus axial strain for CTC tests on sand: σ_3 = effective confining pressure (Refs. 8, 9).

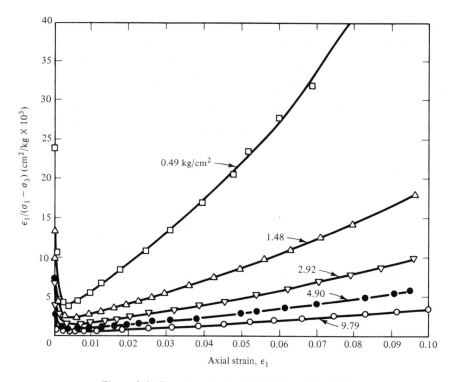

Figure 8-6 Transformed plots for CTC tests (Fig. 8-5).

The values of cohesion and the angle of friction can be found by drawing Mohr circles from the test data (Fig. 8-5); for the sand in question, their values were found to be $c = 0.00$, and $\phi = 31.5°$, respectively.

Hardin (10, 11) has proposed and used a variable shear modulus simulated by using a hyperbolic function; here shear modulus and the Poisson's ratio are made dependent on the maximum shear strain and hydrostatic pressure.

Resilient modulus (12).   The concept of resilient modulus, E_r, is often used to characterize materials such as gravel in pavements and ballast in railroad track support structures. The expression for the resilient modulus K_r is similar to that for the initial modulus E_i in Eq. (8-10a):

$$E_r = K_r (J_1)^{n_r} \tag{8-12a}$$

or

$$E_r = K_r (\sigma_3)^{n_r} \tag{8-12b}$$

in which K_r and n_r are material parameters.

The resilient modulus is often referred to the unloading (reloading) modulus during tests in which loading, unloading, and reloading are simulated under static or cyclic (repetitive) loading. A *resilient modulus* is often defined

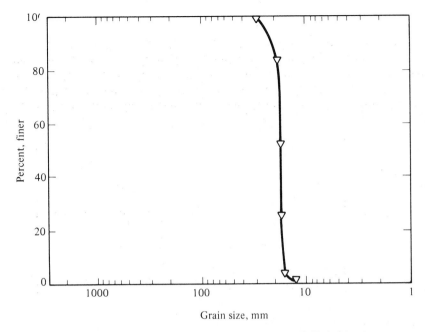

Figure 8-7 Grain size distribution curve for ballast II (Refs. 13–15).

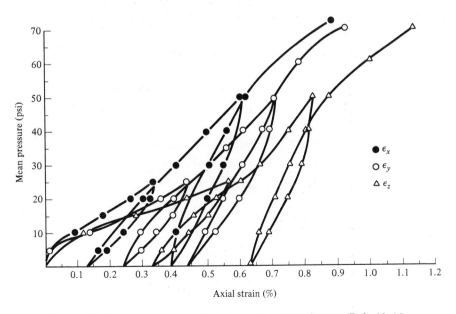

Figure 8-8 Stress–strain curves for hydrostatic compression test (Refs. 13, 14).

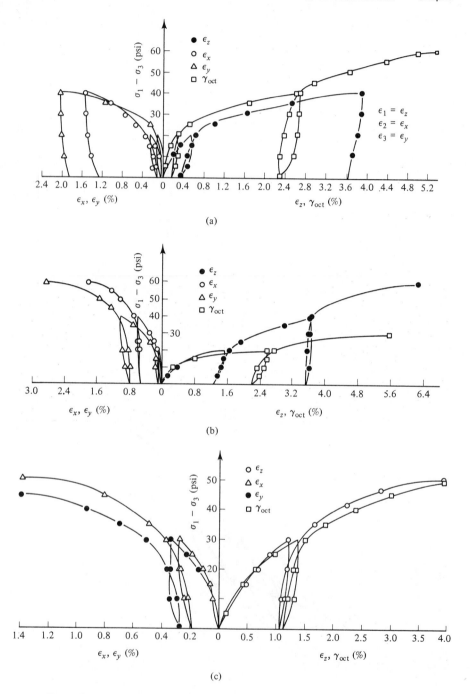

Figure 8-9 Stress–strain curves for conventional triaxial compression test: (a) $\sigma_0 = 10$ psi; (b) $\sigma_0 = 15$ psi; (c) $\sigma_0 = 20$ psi (Refs. 13, 14).

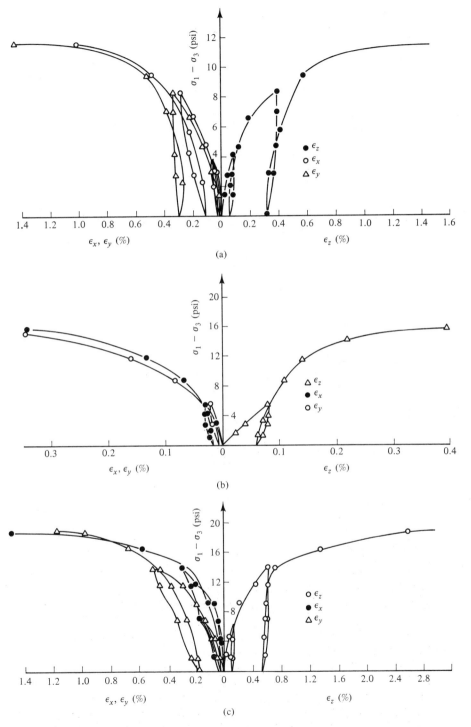

Figure 8-10 Stress–strain curves for reduced triaxial compression test: (a) $\sigma_0 = 30$ psi; (b) $\sigma_0 = 40$ psi; (c) $\sigma_0 = 50$ psi (Refs. 13, 14).

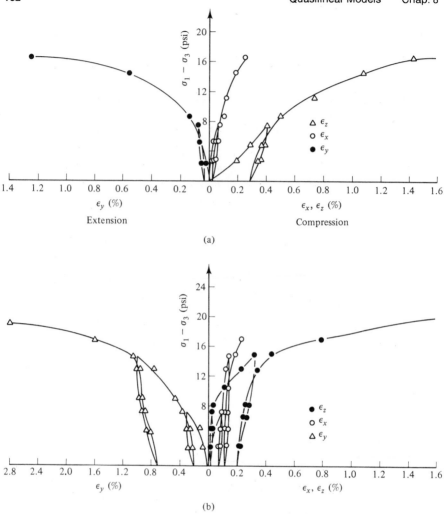

Figure 8-11 Stress–strain curves for simple shear test: (a) $\sigma_0 = 25$ psi; (b) $\sigma_0 = 35$ psi (Refs. 13, 14).

as the ratio of repeated deviatoric stress $(\sigma_1 - \sigma_3)$ to the recoverable part of the axial strain (in a conventional triaxial device).

Here we consider tests results for a ballast used in a railroad track support structure at the Transportation Test Center, Pueblo, Colorado; details of the problem are given in Refs. 13 to 15. Three different sizes of ballast were tested by using the truly triaxial device [Fig. 5-8(b)] and (initial) confining pressure. The initial test density was 90 lb/ft^3 (1.44 g/cm^3).

Figure 8-7 shows the grain size distribution for one of the sizes tested. Figures 8-8 to 8-12 show typical test results under HC, CTC, RTC, SS, and RTE stress paths.

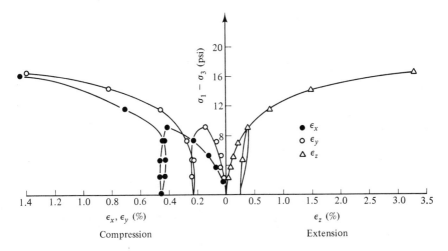

Figure 8-12 Stress–strain response curves for reduced triaxial extension test ($\sigma_0 =$ 40.00 psi) (Refs. 13, 14).

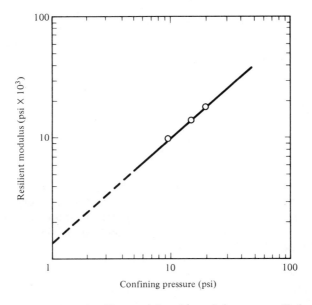

Figure 8-13 Variation of resilient modulus with confining pressure (Refs. 13–15).

Figure 8-13 shows a plot of the resilient modulus (average slope of the unloading–reloading curves) versus confining pressure σ_3. At $\sigma_3 = 1$, $K_r = E_r = 1300$ psi (8.97×10^3 kPa). The slope of the line gives the value of $n_r = 0.87$. Hence the expression for E_r is

$$E_r = 1300\sigma_3^{0.87} \tag{8-12c}$$

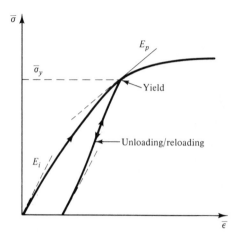

Figure 8-14 Ramberg–Osgood type model (Refs. 17, 18).

Ramberg–Osgood Type models. The Ramberg–Osgood (16) model has often been used for simulating curves for loading, unloading, and reloading behavior. A similar model was proposed by Richard and Abbott (17) and used in a general way by Desai and Wu (18). The expression for this model in terms of stress ($\bar{\sigma}$) and strain ($\bar{\epsilon}$), and for tangent modulus E_t, its first derivative at a point, can be written as (Fig. 8-14)

$$\bar{\sigma} = \frac{E_r \bar{\epsilon}}{\left[1 + \left(E_r \bar{\epsilon} / \bar{\sigma}_y \right)^m \right]^{1/m}} + E_p \bar{\epsilon} \tag{8-13a}$$

$$E_t = \frac{d\bar{\sigma}}{d\bar{\epsilon}} = \frac{E_r}{\left[1 + \left(E_r \bar{\epsilon} / \bar{\sigma}_y \right)^m \right]^{(m+1)/m}} + E_p \tag{8-13b}$$

where $E_r = E_i - E_p$, E_p is the modulus in plastic zones, E_i the initial modulus, $\bar{\sigma}_y$ the yield or plastic stress, and m a material parameter that defines the order of the curve. If $m = 1$, the simulation above can be shown to reduce to the hyperbola. The parameters E_i, E_p, $\bar{\sigma}_y$, and m can be determined using a curve-fitting procedure (18).

Example 8-2   Ramberg–Osgood Model

Consider the set of stress–strain curves in Fig. 8-15. The parameters for this model (Fig. 8-14) are as follows:

σ_3 (kg/cm^2)	E_i (kg/cm^2)	E_p (kg/cm^2)	m	$\bar{\sigma}_y$ (kg/cm^2)
0.49	262	56	8.65	1.90
2.92	1278	50	1.80	9.0
4.90	2500	50	1.73	14.6
9.79	4950	50	1.0	32.0

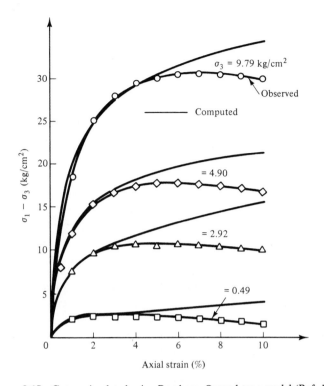

Figure 8-15   Curves simulated using Ramberg–Osgood type model (Ref. 18).

The value of E_i is found as the slope of the stress–strain curve at $(\sigma_1 - \sigma_3) = 0$. The slope at a convenient point in the ultimate zone can be chosen as E_p; here it is chosen as constant for all curves. The values of m can be obtained by solving the following equations using an iterative scheme (17, 18):

$$f(m) = A^m - \frac{1}{R^m} B^m + \left(\frac{1}{R^m} - 1 \right) = 0 \qquad (8\text{-}14\text{a})$$

where R is the ratio between strains at two selected points on the curve, $\bar{\epsilon}_2 = R\bar{\epsilon}_1$; $A = E_r/(E_1 - E_p)$, $B = E_r/(E_2 - E_p)$, $E_1 = \bar{\sigma}_1/\bar{\epsilon}_1$, and $E_2 = \bar{\sigma}_2/\bar{\epsilon}_2$, where 1 and 2 denote the two chosen points. Once m is found, $\bar{\sigma}_y$ can be evaluated from

$$\bar{\sigma}_y = \frac{E_r}{\left(A^m - 1 \right)^{1/m}} \qquad (8\text{-}14\text{b})$$

Spline functions.   The use of spline functions (19) for simulating stress–strain curves was proposed by Desai (20). A spline function approximates a given curve by a number of polynomials of a given degree spanning a number of nodes or data points (Fig. 8-16). The polynomials are fitted in such a way that simulation over the nodes of successive polynomials is as smooth as

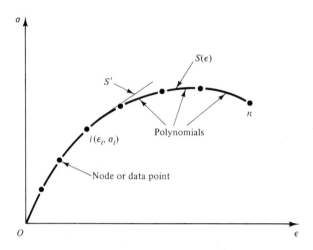

Figure 8-16   Simulation by using spline functions (Refs. 19, 20).

it can be made without resorting to a single polynomial over the complete range. The fitting by a spline provides an analytical curve similar to one obtained by using a mechanical spline or French curve.

A spline function provides a smoother approximation to the function it simulates and its derivatives. Hence it can provide better approximation to a curve and the tangent moduli than that given by use of the hyperbola (20). The major drawbacks of using a spline are that we need first to obtain a smooth curve from (scattered) test data, and a special subroutine is needed in the computer code to generate the derivatives or moduli.

As an example, consider a cubic spline (19, 20) expressed in terms of second derivatives, ϕ:

$$S(\bar{\epsilon}) = \sigma(\bar{\epsilon}) = \frac{\bar{\epsilon}_i - \bar{\epsilon}}{\bar{\epsilon}_i - \bar{\epsilon}_{i-1}} \bar{\sigma}_{i-1} + \frac{\bar{\epsilon} - \bar{\epsilon}_{i-1}}{\bar{\epsilon}_i - \bar{\epsilon}_{i-1}} \bar{\sigma}_i$$

$$+ \frac{1}{6(\bar{\epsilon}_i - \bar{\epsilon}_{i-1})} \left[(\bar{\epsilon}_i - \bar{\epsilon})^3 - (\bar{\epsilon}_i - \bar{\epsilon}_{i-1})^2 (\bar{\epsilon}_i - \bar{\epsilon}) \right] \phi_{i-1}$$

$$+ \frac{1}{6(\bar{\epsilon}_i - \bar{\epsilon}_{i-1})} \left[(\bar{\epsilon}_i - \bar{\epsilon}_{i-1})^3 - (\bar{\epsilon}_i - \bar{\epsilon}_{i-1})^2 (\bar{\epsilon} - \bar{\epsilon}_{i-1}) \right] \phi_i \quad (8\text{-}15)$$

where $\bar{\epsilon}_{i-1} \leqslant \bar{\epsilon} \leqslant \bar{\epsilon}_i$. A spline function can also be formed in terms of the first derivative. Further details of splines can be found in various references (19, 20).

Exponential functions.   Stress–strain curves such as those from pressure (p or $J_1/3$)–volume (I_1) (Fig. 8-17) relations may be more amenable to simulation by an exponential function. A curve shown in Fig. 8-17 can be

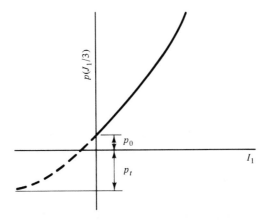

Figure 8-17   Exponential p–I_1 curve.

expressed as

$$p = (p_0 + p_t)e^{\alpha I_1} \tag{8-16}$$

where p_0 is the initial pressure, p_t the strength in volumetric tension, and α a material parameter.

Variable Moduli Models

A form of VP models used by Nelson et al. (21–24) is the variable moduli (VM) model; this form is based essentially on using K and G as variables. In the general case, we can assume that both parameters are variable. This will involve adopting functional representation for both K_t and G_t in Eq. (8-2).

First consider linear relations as

$$K_t = \alpha_0 + \alpha_1 p = K_0 + \alpha_1 p \tag{8-17a}$$

$$G_t = \beta_0 + \beta_1 \sqrt{J_{2D}} = G_0 + \beta_1 \sqrt{J_{2D}} \tag{8-17b}$$

Here $\alpha_0 = K_0$ and $\beta_0 = G_0$ are the initial values of the bulk and shear moduli, respectively [Fig. 8-2(b)], and α_1 and β_1 are other material parameters. Equations (8-17) are based on the assumption that the volumetric and shear behavior are uncoupled. Substitution of K_t and G_t above in Eq. (8-1a) leads to

$$d\sigma_{ij} = (\alpha_0 + \alpha_1 p)\, d\epsilon_{nn}\delta_{ij} + 2\Big(\beta_0 + \beta_1\sqrt{J_{2D}}\Big)\Big(d\epsilon_{ij} - \tfrac{1}{3}d\epsilon_{nn}\delta_{ij}\Big) \tag{8-18}$$

Now examine typical states of strain for Eq. (8-18). For simple shear strain state, we have $\epsilon_{12} = \epsilon_{21} = \epsilon$, and all other strain components are zero. Then for zero initial stress and strain, we have

$$d\sigma_{11} = d\sigma_{22} = d\sigma_{33} = d\sigma_{23} = d\sigma_{13} = 0 \tag{8-19a}$$

$$d\sigma_{12} = (\alpha_0 + \alpha_1 p)\, d\epsilon_{nn}\delta_{12}^{\,0} + 2(\beta_0 + \beta_1\sigma_{12})\Big(d\epsilon_{12} - \tfrac{1}{3}d\epsilon_{nn}\delta_{12}^{\,0}\Big)$$

$$= 2(\beta_0 + \beta_1\sigma_{12})\, d\epsilon_{12} \tag{8-19b}$$

Integration of Eq. (8-19b) gives

$$\epsilon_{12} = \frac{1}{2\beta_1} \ln \left(\frac{\beta_0 + \beta_1 \sigma_{12}}{\beta_0} \right) \tag{8-19c}$$

This means that for the simple shear strain state the model yields a nonlinear relation between the shear strain and shear stress.

Alternative forms of VM models.   For one of the alternatives, Nelson et al. (21–24) used Eq. (8-17a) for K_t but included the effect of mean pressure in the description of G_t as

$$G_t = \beta_0 + \beta_1 p + \beta_2 \sqrt{J_{2D}} \tag{8-20}$$

This assumes that volumetric strains are independent of shear stress but the shear behavior depends on both volumetric and shear stresses. Substitution of Eqs. (8-17a) and (8-20) in Eq. (8-1a) gives

$$d\sigma_{ij} = (\alpha_0 + \alpha_1 p)\, d\epsilon_{nn} \delta_{ij} + 2\left(\beta_0 + \beta_1 p + \beta_2 \sqrt{J_{2D}} \right)\left(d\epsilon_{ij} - \tfrac{1}{3} d\epsilon_{nn} \delta_{ij} \right) \tag{8-21}$$

Under simple shear state of strain, with zero initial state of strain, Eq. (8-21) gives

$$d\sigma_{11} = d\sigma_{22} = d\sigma_{33} = d\sigma_{23} = d\sigma_{13} = 0 \tag{8-22a}$$

and

$$d\sigma_{12} = (\alpha_0 + \alpha_1 p)\, d\epsilon_{nn} \delta_{12}^{\,0} + 2\left(\beta_0 + \beta_1 p + \beta_2 \sqrt{J_{2D}} \right)\left(d\epsilon_{12} - \tfrac{1}{3} d\epsilon_{nn} \delta_{12}^{\,0} \right)$$

$$= 2\left(\beta_0 + \beta_2 \sqrt{J_{2D}} \right) d\epsilon_{12} \tag{8-22b}$$

Here $p = 0$, since $d\sigma_{11} = d\sigma_{22} = d\sigma_{33} = 0$, and the result is the same as with Eq. (8-17b). Hence upon integration, we will obtain the same result as in Eq. (8-19c).

For a hydrostatic state of strain, we have $d\epsilon_{11} = d\epsilon_{22} = d\epsilon_{33} = d\epsilon$, and $\epsilon_{12} = \epsilon_{23} = \epsilon_{13} = 0$. Hence Eq. (8-21) leads to

$$d\sigma_{11} = (\alpha_0 + \alpha_1 p)(3d\epsilon)\, \delta_{11} + 2\left(\beta_0 + \beta_1 p + \beta_2 \sqrt{J_{2D}} \right)\left(d\epsilon - d\epsilon \right)^0$$

$$= 3(\alpha_0 + \alpha_1 p)\, d\epsilon$$

$$= d\sigma_{33} = d\sigma_{22} = d\sigma \tag{8-23a}$$

and

$$d\sigma_{12} = d\sigma_{22} = d\sigma_{13} = 0 \tag{8-23b}$$

Integration of Eq. (8-23a) leads to

$$p = \sigma = \frac{\alpha_0}{\alpha_1}\left(e^{\alpha_1 I_1} - 1 \right) \tag{8-24a}$$

or

$$I_1 = 3\epsilon = \frac{1}{\alpha_1} \ln\left(\frac{\alpha_0 + \alpha_1 p}{\alpha_0}\right) \tag{8-24b}$$

which represents the p–I_1 curve [Fig. 8-2(b)].

Thus the model predicts nonlinear behavior between σ_{12} and ϵ_{12}, and between p and I_1.

UNLOADING AND HYSTERETIC BEHAVIOR

Until now we have considered essentially monotonically increasing or decreasing loads, involving only elastic behavior. In reality, however, the behavior of geologic media is inelastic or plastic and involves hysteresis; this is represented symbolically in Fig. 8-18. This aspect is exhibited when the material, unloaded up to a certain point and then reloaded, does not follow the same paths (Fig. 8-18).

An improved treatment of the plastic or inelastic behavior would require use of the theory of incremental plasticity which is the subject of Chapters 9 to 12. In the meantime, we present here some details of schemes based on the quasilinear models to account approximately for the unloading and the hysteretic behavior.

The approximate models are based on the idea of assigning loading, unloading, and reloading moduli, α_L, α_U, and α_R, respectively [Fig. 8-18(a)]; here α's are used as a general notation. For instance, consider specific curves in Fig. 8-18(b) and (c). Here we can define loading, unloading, and reloading moduli as

$$\left.\frac{d\sigma_{11}}{d\epsilon_{11}}\right|_L = E_L, \quad \left.\frac{d\sigma_{11}}{d\epsilon_{11}}\right|_U = E_u, \quad \left.\frac{d\sigma_{11}}{d\epsilon_{11}}\right|_R = E_R \tag{8-25a}$$

$$-\left.\frac{d\epsilon_{33}}{d\epsilon_{11}}\right|_L = \nu_L, \quad -\left.\frac{d\epsilon_{33}}{d\epsilon_{11}}\right|_U = \nu_U, \quad -\left.\frac{d\epsilon_{33}}{d\epsilon_{11}}\right|_R = \nu_R \tag{8-25b}$$

$$\left.\frac{d(J_1/3)}{dI_1}\right|_L = \left.\frac{dp}{dI_1}\right|_L = K_L, \quad \left.\frac{d(J_1/3)}{dI_1}\right|_U = K_U, \quad \left.\frac{d(J_1/3)}{dI_1}\right|_R = K_R$$

$$\tag{8-25c}$$

$$\left.\frac{dS_{ij}}{dE_{ij}}\right|_L = \left.\frac{d\sqrt{J_{2D}}}{d\sqrt{I_{2D}}}\right|_L = 2G_L, \quad \left.\frac{dS_{ij}}{dE_{ij}}\right|_U = 2G_U, \quad \left.\frac{dS_{ij}}{dE_{ij}}\right|_R = 2G_R$$

$$\tag{8-25d}$$

If the material is in the loading stage, we use the loading moduli, E_L and ν_L or K_L and G_L; if it is unloading, the unloading moduli E_U and ν_U or K_U and G_U are used; and for reloading we use E_R and ν_R or G_R and K_R. The next

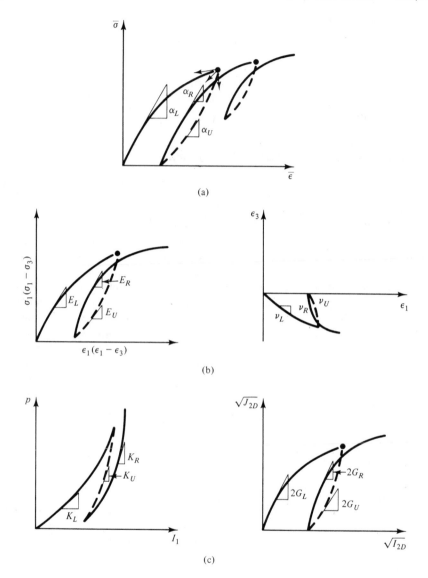

Figure 8-18 Hysteretic behavior.

question that arises relates to how to define the state at which unloading occurs. If the state of stress after the current increment, m, of load (stress) is lower than that at the end of the previous increment, $m - 1$, we can assume that unloading occurs. This is not as easy as it sounds. For instance, as indicated by a number of arrows in Fig. 8-18(a), unloading can occur along various directions. Furthermore, it is possible that the material may load in one mode of deformation, say shear, and unload in another, say volumetric.

One criteria often used is based on the rate of work performed by stresses during an increment of strain. We write rate of work dW as

$$dW = \sigma_{ij} d\epsilon_{ij} \qquad (8\text{-}26a)$$

The criterion is then expressed as

$$
\begin{array}{lll}
\text{Loading:} & dW > 0 & \\
\text{Unloading:} & dW < 0 & \qquad (8\text{-}26b) \\
\text{Neutral condition:} & dW = 0 &
\end{array}
$$

Substituting in Eq. (8-26a) and using the relation $J_{2D} = \frac{1}{2} S_{ij} S_{ij}$, we obtain

$$dW = \frac{dJ_{2D}}{2G} + \frac{(J_1/3)\, dJ_1}{3K} \qquad (8\text{-}27a)$$

Then the criterion in Eq. (8-26a) can be expressed separately as

$$
\begin{array}{lll}
\text{Loading:} & dJ_{2D} > 0 \quad \text{and} \quad dJ_1 > 0 & \qquad (8\text{-}27b) \\
\text{Unloading:} & dJ_{2D} < 0 \quad \text{and} \quad dJ_1 < 0 & \qquad (8\text{-}27c)
\end{array}
$$

Here it can be possible that we have loading in shear, that is, $dJ_{2D} > 0$, and unloading in mean pressure, that is, $dJ_1 < 0$, and vice versa. Other criteria for loading and unloading based on stress power are discussed in Chapter 12.

Comments. Although the VM models can account for behavior of a limited class of materials, loadings, and stress paths, they may not be treated or used as general models. For instance, because the behavior is decomposed into volumetric and deviatoric components, they cannot account for the volume change behavior under shear. The theory of VM models is rigorously valid for proportional loading in shear and a continuity problem may arise for complex stress paths where neutral loading in shear occurs (21-24).

Next we consider some examples of VM models and derivation of parameters for McCormick Ranch sand (21-24) and a ballast (13-15).

Example 8-3   McCormick Ranch Sand

A comprehensive series of uniaxial, hydrostatic, and conventional triaxial compression tests were performed under static and cyclic loading for McCormick Ranch sand; and various plasticity and variable moduli models were developed for the sand (21-24).

Figures 8-19 and 8-20 show test results from uniaxial and hydrostatic tests, respectively. Typical results from CTC tests are shown in Fig. 8-21.

Here we present one of the VM models proposed by Baron et al. (22-24) and Nelson (21). The pressure-volume relations are expressed through:

$$p_{\max} = \sum_{n=1}^{3} a_n \epsilon_{kk}^n \qquad \text{initial loading}$$

$$
K_U(p) =
\begin{cases}
K_{0U} + K_{1U} p & p \leqslant p_m \\
K_{\max} & p \geqslant p_m
\end{cases}
\qquad (8\text{-}28)
$$

$$p \geqslant 0 \qquad \text{tension cutoff}$$

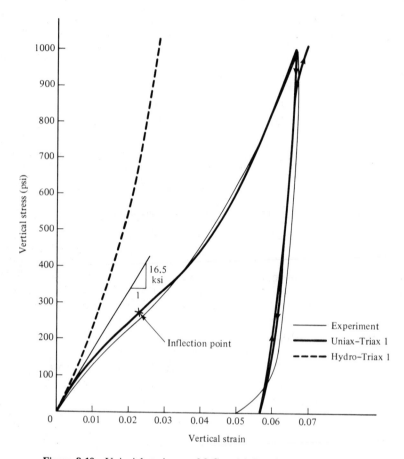

Figure 8-19 Uniaxial strain test: McCormick Ranch sand (Refs. 21–24).

Here p_m is a transition pressure and $p_{\max}$ is the maximum pressure. For hydrostatic behavior, the unloading and reloading (Fig. 8-20) are simulated by using the same modulus, K_{0U}.

For the deviatoric response, separate values for the shear modulus, G, are defined for loading, unloading and reloading.

Loading:

$$G_L\left(p,\sqrt{J_{2D}}\right) = G_0 + \bar{\gamma}\sqrt{J_{2D}} + \gamma_1 p + \gamma_2 p^2, \qquad p \leqslant p_c$$
$$= G_1 + \bar{\gamma}\sqrt{J_{2D}}, \qquad\qquad\qquad p \geqslant p_c \qquad (8\text{-}29a)$$

Unloading:

$$G_U\left(p,\sqrt{J_{2D}}\right) = G_{0U} + \bar{\gamma}_U\sqrt{J_{2D}} + \gamma_{1U} p + \gamma_{2U} p^2, \qquad p \leqslant p_c$$
$$= G_{1U} + \bar{\gamma}_U\sqrt{J_{2D}}, \qquad\qquad\qquad p \geqslant p_c \qquad (8\text{-}29b)$$

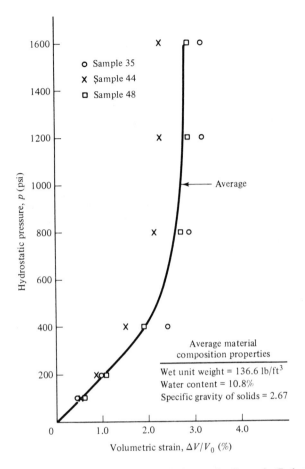

Figure 8-20 McCormick Ranch sand: static hydrostat, loading only (Refs. 21–24).

Reloading: Here the shear modulus G_R is expressed as the linear combination of G_L and G_U as

$$G_R\left(p,\sqrt{J_{2D}}\right) = \left(\frac{F}{F_{\max}}\right)^n G_L\left(p,\sqrt{J_{2D}}\right) + \left[1 - \left(\frac{F}{F_{\max}}\right)^n\right]G_U\left(p,\sqrt{J_{2D}}\right)$$

$$(8\text{-}30a)$$

where p_c is the critical or transition pressure, n a constant, F is defined as

$$F\left(p,\sqrt{J_{2D}}\right) = 1 - \frac{G_L\left(p,\sqrt{J_{2D}}\right)}{G_L(p,0)}$$

$$(8\text{-}30b)$$

and $F_{\max}$ is the maximum previous value of F. The values of F are within the range $0 \leqslant F \leqslant 1$; hence

$$G_L \leqslant G_R \leqslant G_U$$

$$(8\text{-}31)$$

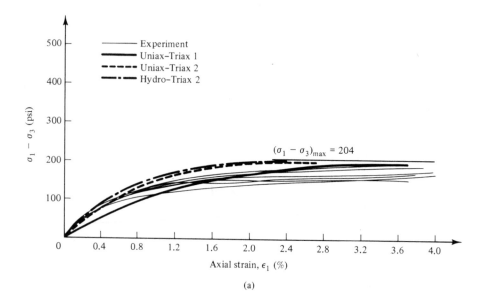

(a)

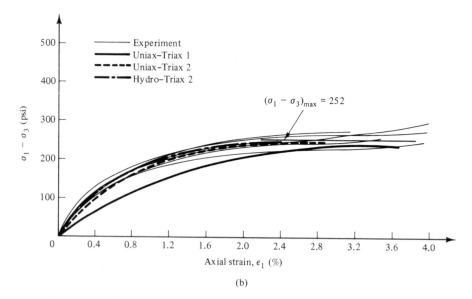

(b)

Figure 8-21 Triaxial compression test—stress difference versus axial strain (21):
(a) $\sigma_0 = 100$ psi; (b) $\sigma_0 = 200$ psi; (c) $\sigma_0 = 400$ psi; (d) $\sigma_0 = 800$ psi (Refs. 21, 24).

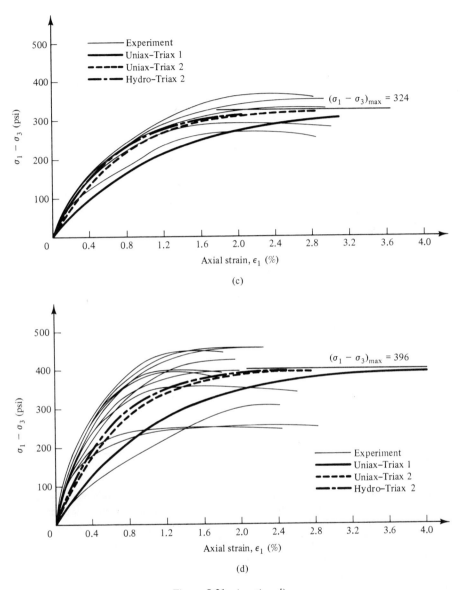

Figure 8-21 (*continued*)

Comprehensive details of the derivation of the parameters by using specializations of Eqs. (8-28) to (8-30) for various test configurations, and for the initial and final states of stresses with respect to observed data, using least-squares fitting techniques, are given by Nelson (21). Each test configuration provides all or a part of the parameters. Theoretically, such parameters from different tests must be the same. However, due to the deficiencies in the model and in the experimental data, different

TABLE 8-1   Material Constants for Variable Moduli Model of McCormick Ranch Sand[a] (21, 22)

Density, ρ	131.1 lb/ft^3
Initial loading, p_{max}	
$\quad a_1 \equiv K_0$	5.83 ksi
$\quad a_2$	13.333 ksi
$\quad a_3$	1111.1 ksi
Loading shear modulus, G_L	
$\quad G_0$	8.0 ksi
$\quad \bar{\gamma}$	-110.0
$\quad \gamma_1$	32.4
$\quad \gamma_2$	-15.0 (ksi)$^{-1}$
$\quad G_1$	25.496 ksi
$\quad p_c$	1080.00 psi
Initial moduli	
$\quad \nu_0$	0.0292
$\quad M_0$	16.497 ksi
$\quad E_0$	16.468 ksi
Unloading bulk modulus	
$\quad K_{0U}$	32.0 ksi
$\quad K_{1U}$	143.0
$\quad K_{max}$	300. ksi
$\quad p_m$	1.8741 ksi
Unloading shear modulus, G_U	
$\quad G_{0U}$	8.0 ksi
$\quad \bar{\gamma}_U$	500.00
$\quad \gamma_{1U}$	33.0
$\quad \gamma_{2U}$	-15.278 (ksi)$^{-1}$
$\quad G_{1U}$	25.820 ksi
$\quad G_{max\,U}$	141.71 ksi
$\quad G_R$	
$\quad n$	4

[a] 1 kip = 1000 lb; 1 lb = 4.45 N; 1 in. = 2.54 cm.

values can arise from different data. Then a weighted averaging scheme can be used to find the final set of parameters that can be used for other major configurations.

Parameters for the foregoing VM model for the McCormick sand as given by Nelson (21) and Baron et al. (22) are reproduced in Table 8-1.

Example 8-4   Ballast in Rail Track Support Structures

Ballast (two different sizes) used in the railroad track support structure at the Transportation Test Section, Pueblo, Colorado (13–15), was tested using the truly triaxial device [Fig. 5-8(b)]. The grain size distribution of one of the sizes used is shown in Fig. 8-7. Tests were conducted under a variety of stress paths (Fig. 5-10).

Figures 8-8 to 8-12 show typical data from different tests: HC, CTC, RTC, SS, and RTE stress paths. The following variable moduli model was proposed for the loading behavior:

$$K = K_0 + K_1 I_1 + K_2 I_2 \tag{8-32a}$$

$$G = G_0 + \gamma_1 J_1 + \gamma_2 \sqrt{J_{2D}} \tag{8-32b}$$

By assuming initial values K_0 and G_0 to represent unloading and reloading moduli, and by a process of least-squares fitting, the following values were found for the parameters in Eq. (8-32):

$$K_0 = 4 \times 10^3 \text{ psi } \left(27.6 \times 10^3 \text{ kN/m}^2\right)$$

$$K_1 = -5.4917 \times 10^3 \text{ psi } \left(-37.89 \times 10^3 \text{ kN/m}^2\right)$$

$$K_2 = 2.536 \times 10^7 \text{ psi } \left(17.5 \times 10^7 \text{ kN/m}^2\right)$$

$$G_0 = 1.714 \times 10^3 \text{ psi } \left(11.83 \times 10^3 \text{ kN/m}^2\right)$$

$$\gamma_1 = 2.5593$$

$$\gamma_2 = -60.172$$

APPLICATIONS

Quasilinear models have been commonly used in nonlinear analysis of geotechnical engineering problems. A few representative examples have been selected for the purpose of illustrating practical application to boundary value problems.

Example 8-5   Analysis of Piles

Application of the hyperbolic stress–strain model is described in Ref. 8 for design analysis of piles in sands. The model has been used in conjunction with a finite element formulation of this problem. A typical finite element mesh, made up with quadrilateral isoparametric elements, used in the analysis is shown in Fig. 8-22. Interface behavior between pile and sand has been modeled using interface elements.

Two field pile load tests were considered in the study: Arkansas Lock and Dam No. 4 (LD4) on the Arkansas River in Arkansas; and Jonesville Lock (JL) on the Black River in Louisiana. Typical material parameters for the JL site were given in Example

TABLE 8-2   **Parameters Adopted for Sands at Arkansas LD4 Piles (8)**

(a) For E_t:

Layer	Parameter			
	K_h	n	R_f	ϕ
1	1500	0.6	0.9	32
2	1200	0.5	0.8	31
3	1500	0.6	0.9	32

(b) For ν_t:

Layer	Parameter		
	G	F	d
All three layers	0.54	0.24	4.0

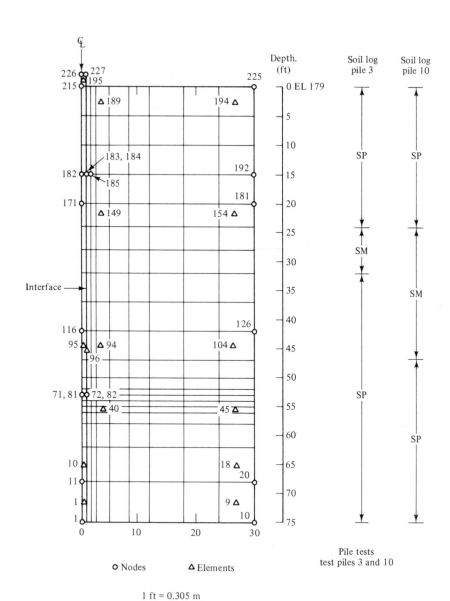

1 ft = 0.305 m

Figure 8-22 Finite element mesh (Ref. 8).

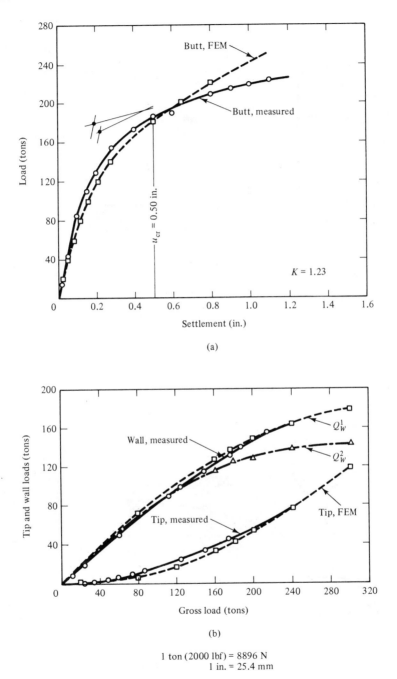

Figure 8-23 Comparisons for pile 10, LD4: (a) load–settlement curve; (b) gross load versus tip and wall friction loads (Ref. 8).

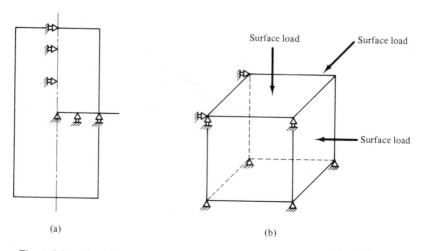

Figure 8-24 Simulation of triaxial test specimens: (a) cylindrical triaxial specimen; (b) cubical specimen (Ref. 25).

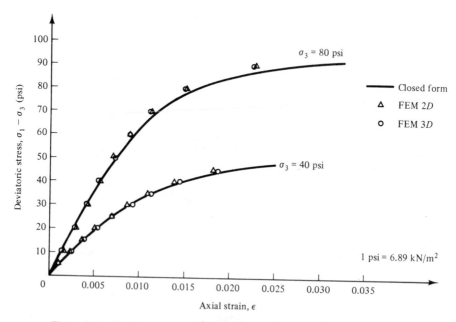

Figure 8-25 Verification of predictions from variable moduli model (Ref. 25).

8-1. Approximate values of the parameters adopted for the LD4 site are shown in Table 8-2.

The analysis was performed by using an incremental-iterative procedure. A comparison of the computed and measured load–displacement curves for a typical pile is shown in Fig. 8-23(a). Measured tip load and wall friction are compared with corresponding computed quantities as shown in Fig. 8-23(b). The comparisons are considered satisfactory; other cases reported in Refs. 8 and 9 indicate similar trends. Although the hyperbolic model is seen to be satisfactory for this example involving monotonic loading, general applicability of the model does not appear to have been sufficiently well founded.

Example 8-6   Analysis of Laboratory Tests

The variable moduli model described in Example 8-4 and Eq. (8-32) has been implemented in conjunction with a soil–structure interaction analysis (25); this has been applied to investigate load-transfer mechanisms in multicomponent systems such as track support structures (26). Details of this application are given in a later chapter because several constitutive models in addition to the variable moduli model are involved. Here we give an example of simulating a triaxial test.

A cylindrical (axisymmetric) specimen [Fig. 8-24(a)] is analyzed using the two-dimensional code, while a cubical triaxial specimen [Fig. 8-24(b)] is analyzed using a three-dimensional code. Figure 8-24 shows the finite element meshes containing only one element for each, and the boundary conditions. The following parameters were used for the variable moduli model (24) [Eq. (8-32)]:

$$K_0 = 1000 \text{ psi } (6890 \text{ kPa})$$
$$G_0 = 462 \text{ psi } (3183.2 \text{ kPa})$$
$$K_1 = -10^5 \text{ psi } (-6.89 \times 10^5 \text{ kPa})$$
$$K_2 = 4 \times 10^6 \text{ psi } (27.56 \times 10^6 \text{ kPa})$$
$$\gamma_1 = 60.0$$
$$\gamma_2 = -133.0$$

Conventional triaxial compression (CTC) stress paths ($\sigma_2 = \sigma_3 =$ constant while σ_1 is increased) have been simulated for two confining pressures. One iteration per increment has been used in the analysis. A comparison of finite element predictions and closed-form solutions is shown in Fig. 8-25. Predictions from both two- and three-dimensional codes compare very well with the closed-form solutions given by Nelson and Baron (24).

REFERENCES

1. Desai, C. S., and Abel, J. F., *Introduction to the Finite Element Method*, Van Nostrand Reinhold Company, New York, 1972.
2. Desai, C. S., and Christian, J. T. (Eds.), *Numerical Methods in Geotechnical Engineering*, McGraw-Hill Book Company, New York, 1977.

3. Kondner, R. L., "Hyperbolic Stress–Strain Response: Cohesive Soils," *J. Soil Mech. Found. Div., ASCE*, Vol. 89, No. SM1, Jan. 1963, pp. 115–143.

4. Hansen, J. B., "Discussion of 'Hyperbolic Stress–Strain Response: Cohesive Soils'," by R. L. Kondner, *J. Soil Mech. Found. Div., ASCE*, Vol. 89, No. SM4, July 1963, pp. 241–242.

5. Duncan, J. M., and Chang, C. Y., "Nonlinear Analysis of Stress and Strain in Soils," *J. Soil Mech. Found. Div., ASCE*, Vol. 96, No. SM5, Sept. 1970, pp. 1629–1653.

6. Janbu, N., "Soil Compressibility as Determined by Odometer and Triaxial Test," *Proc. Eur. Conf. Soil Mech. Found. Eng.*, Wiesbaden, Vol. 1, 1963, pp. 19–25.

7. Kulhawy, F. H., Duncan, J. M., and Seed, H. B., "Finite Element Analysis of Stresses and Movements in Embankments during Construction," *Rep. 569-8*, U.S. Army Corps of Engineers Waterways Exp. Stn., Vicksburg, Miss., 1969.

8. Desai, C. S., "Numerical Design—Analysis for Piles in Sands," *J. Geotech. Eng. Div., ASCE*, Vol. 100, No. GT6, June 1974, pp. 613–635.

9. Desai, C. S., and Holloway, D. M., "Load-Deformation Analysis of Deep Pile Foundations," *Proc., Symp. on Appl. of FEM in Geotech. Eng.*, Vicksburg, Miss., Sept. 1972.

10. Hardin, B. O., and Black, W. L., "Sand Stiffness under Various Triaxial Stresses," *J. Soil Mech. Found. Div., ASCE*, Vol. 92, No. SM2, Mar. 1966, pp. 27–42.

11. Hardin, B. O., "Effect of Strain Amplitude on the Shear Modulus of Soils," *Tech. Rep. AFWL-7R-72-201*, Air Force Weapons Lab., Kirtland AFB, N.Mex., 1972.

12. Hicks, R. G., and Monismith, C. L., "Factors Influencing the Resilient Response of Granular Materials," *Highway Res. Rec. 345*, 1971.

13. Desai, C. S., Siriwardane, H. J., and Janardhanam, R., "Interaction and Load Transfer in Track Guideway Systems," Final Report No. DOT/RSPA/DMA-50/83/11, *DOT,Office of Univ. Res.*, Washington, D.C., 1983.

14. Janardhanam, R., "Constitutive Laws of Materials in Track Guideway Support Structures," Ph.D. dissertation, Va. Polytech. Inst. and State Univ., Blacksburg, 1981.

15. Janardhanam, R., and Desai, C. S., "Three-Dimensional Testing and Modelling of a Ballast," *J. Geotech. Eng. Div., ASCE*, Vol. 108, No. GT6, 1983.

16. Ramberg, W., and Osgood, W. R., "Description of Stress–Strain Curves by Three Parameters," *Tech. Note 902*, National Advisory Comm. Aeronaut., Washington, D.C., 1943.

17. Richard, R. M., and Abbott, B. J., "Versatile Elastic–Plastic Stress–Strain Formula," *Tech. Note, J. Eng. Mech. Div., ASCE*, Vol. 101, No. EM4, Aug. 1975, pp. 511–515.

18. Desai, C. S., and Wu, T. H., "A General Function for Stress–Strain Curves," *Proc. 2nd Int. Conf. Numer. Methods Geomech.*, C. S. Desai (Ed.), Blacksburg, Va., ASCE, June 1976.

19. Ahlberg, J. H., Nilsen, E. N., and Walsh, J. L., *The Theory of Splines and Their Applications*, Academic Press, Inc., New York, 1967.

20. Desai, C. S., "Nonlinear Analysis Using Spline Functions," *J. Soil Mech. Found. Div.*, *ASCE*, Vol. 97, No. SM10, Oct. 1971, pp. 1461–1480.

21. Nelson, I., "Investigation of Ground Shock Effects in Nonlinear Hysteretic Media: Modelling the Behavior of a Real Soil," *Rep. 2 to* U.S. Army Corps of Engineers Waterways Exp. Stn., Vicksburg, Miss., by Weidlinger Associates, New York, 1970.

22. Baron, M. L., Nelson, I., and Sandler, I., "Influence of Constitutive Models on Ground Motion Predictions," *Contract Rep. S-71-10, Rep. 2*, U.S. Army Corps of Engineers Waterways Exp. Stn., Vicksburg, Miss., Nov. 1971.

23. Nelson, I., and Baron, M. L., "Investigation of Ground Shock Effects in Nonlinear Hysteretic Media," *Contract Rep. S-68-1*, U.S. Army Corps of Engineers Waterways Exp. Stn., Vicksburg, Miss., by Weidlinger Associates, New York, 1968.

24. Nelson, I., and Baron, M. L., "Application of Variable Moduli Models to Soil Behavior," *Int. J. Solids Struc.*, Vol. 7, 1971, pp. 399–417.

25. Siriwardane, H. J., and Desai, C. S., "Implementation of Some Constitutive Laws for Three- and Two-Dimensional Analysis," *Proc. Symp. Implementation of Computer Procedures and Stress–Strain Laws in Geotech. Eng.*, Chicago, Aug. 1981.

26. Desai, C. S., and Siriwardane, H. J., "Analysis of Track Support Structures," *J. Geotech. Eng. Div.*, *ASCE*, Vol. 108, No. GT3, 1982, pp. 461–480.

9

PLASTICITY:
AN INTRODUCTION

The literature on the theory of plasticity and its applications is wide and available in various text books and technical publications. In this chapter, we have attempted to present a review of these classical developments adopted from the available literature. For additional details the reader may consult the publications cited in the reference list.

An external load causes strains and stresses in a body. When the external load is removed, the body may or may not return to its original configuration. So far in the text, we have dealt essentially with materials that undergo only recoverable deformations, that is, they return to their initial state when the load is removed; they are called *elastic*. In this chapter we consider materials that retain a part of the deformation on unloading. These types of materials are called *inelastic* or *plastic*.

Experimental observations show that certain materials such as steel behave elastically up to a certain level of stress. This observation is the foundation of the theory of elasticity. A typical stress–strain curve of a metal under simple tension is shown in Fig. 9-1. When the load is increased gradually, the material behaves elastically up to point *A*, and regains the original state if the load is removed. If the metal specimen is stressed beyond point *A*, say up to point *B*, and then unloaded, there will be some permanent or irrecoverable deformations in the body, and the material is said to have undergone plastic deformations. It is interesting to note that if the specimen is reloaded from point *C* (Fig. 9-1), it will often behave elastically until the stress level reaches point *B*. The stress–strain response depicted in Fig. 9-1 is called *hardening* behavior.

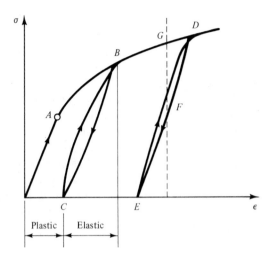

Figure 9-1 Typical stress–strain curve for metal under simple (uniaxial) tension.

When the specimen is loaded from A to B, both elastic and plastic deformations will occur, and this is known as _elastoplastic_ behavior. Because the reloading paths do not follow the original loading path, the strains will be dependent on the history of stress applications when plastic deformations occur. For instance, points F and G represent different states of stress for the same state of strain. As will be seen later, the plastic behavior is characterized by the history-dependent deformation.

Study and analysis of the behavior of materials that experience plastic deformation constitutes the theory of plasticity. The subject of plasticity can be classified into two categories: _physical_ theories and _mathematical_ theories (1). The physical theories attempt to describe why the material behaves plastically. This involves study at the microscopic level. On the other hand, mathematical theories make an attempt to formalize the experimental observations at a macroscopic level. Engineers are more interested in the second category; hence emphasis will be placed on the mathematical theory.

REVIEW

The historical review given in this section is based on information given in several texts on the theory of plasticity (1–4). Perhaps the first reference in the history of plasticity can be attributed to Coulomb (1773), who proposed a yield criterion for solids such as soils. Subsequently, Rankine (1853) applied Coulomb's concept to problems of the calculation of earth pressures on retaining walls. However, Tresca is regarded as the first one to perform a scientific study of the plasticity of metals. He published the results on punching and extrusion of metals in 1864 and formulated his famous yield criterion. Subsequently, in 1870, Saint-Venant applied Tresca's yield criterion to determine stresses in a partly plastic cylinder subjected to torsion or bending. Levy, in 1871, adopting Saint-Venant's concept of ideal plasticity, proposed a three-dimensional relationship between stress and plastic strain rate. von Mises, who independently proposed equations similar to Levy's, suggested a yield criterion in 1913 on the basis of mathematical considerations. Ruess (1930) made an allowance for elastic strains following an earlier suggestion by Prandlt (1–4). However, a unified theory began to evolve only around 1945. Since that time a considerable amount of work has been done in this area. For a detailed history of this review, the reader is referred to the text by Hill (1).

There are two major aspects that constitute the theory of plasticity: (a) the yield criterion and (b) post-yield behavior. In subsequent sections these two aspects are considered in detail.

YIELD CRITERIA

The yield criterion can be defined as the limit of elastic deformations expressed by a combination of states of stress. For a one-dimensional state of stress, the yield criterion can be easily visualized: for instance, quantities such as measures of uniaxial compressive stress or uniaxial tensile stress. However, under multiaxial states of stress, this becomes complicated, and a mathematical expression involving all the stresses is often required. Establishment of this mathematical expression, known as the *yield criterion*, has to be done based on experimental observations. Original experimental work was done on metals, and hence the development of this subject started with metal plasticity. In this chapter some of the yield criteria developed for metals will be discussed first.

In view of the complexities involved in the yielding of materials under three-dimensional states of stress, it is convenient to define a scalar function, F, as the yield criterion. That is,

$$F = F(\sigma_{11}, \sigma_{22}, \sigma_{33}, \sigma_{12}, \sigma_{23}, \sigma_{13}) \qquad (9\text{-}1)$$

In the most general case, F will be a function of all six components of the stress tensor. If the material is assumed to be homogeneous, this function is valid everywhere in the material. Equation (9-1) can also be expressed in terms of the principal stresses and their directions (5). That is,

$$F = F(\sigma_1, \sigma_2, \sigma_3, n_1, n_2, n_3) \qquad (9\text{-}2)$$

Here σ_1, σ_2, and σ_3 are the principal stresses, and n_1, n_2, and n_3 are their direction cosines (Chapter 4).

Equation (9-1) can be complicated, as the yield criterion is dependent on six quantities. This can be simplified if an assumption is made regarding material properties. If we assume that the material is isotropic, it does not have any preferred directions. Therefore, the yield criterion in Eq. (9-2) can be expressed only in terms of principal stresses. That is, for an isotropic material,

$$F = F(\sigma_1, \sigma_2, \sigma_3) \tag{9-3}$$

This can also be expressed more conveniently in terms of the invariants of the stress tensor as follows:

$$F = F(J_1, J_2, J_3) \tag{9-4}$$

where J_1, J_2, and J_3 are the invariants of the stress tensor as defined in Chapter 4.

The influence of hydrostatic stress on the plastic deformations has been found to be negligible for many metals. Bridgeman's experiments (6) have shown that the increase in shear modulus when the mean pressure was increased from zero to 10^5 atm was only 2.5% for coiled steel, and 1.8% for nickel. The experimental results of Bridgeman (6) have also shown that volumetric deformations of a nonporous solid can be assumed to be elastic. In fact, the volumetric strains experienced by most metals have been found to be very small in the working range of pressures.

Experimental evidence shows that the yielding of a metal is not affected significantly by moderate hydrostatic pressure, either applied alone or superimposed on a state of combined stress. Some description of the original work supporting this fact can be found in the text by Hill (1). This assumption leads to the conclusion that the yield criterion depends essentially only on the state of the deviatoric stress. Therefore, the yield function in Eq. (9-4) can be expressed in terms of the invariants of the deviatoric stress tensor as

$$F(J_{2D}, J_{3D}) = 0 \tag{9-5}$$

Here J_{2D} and J_{3D} are the second and third invariants of the deviatoric stress tensor, respectively. In a subsequent section, some yield criteria in metal plasticity will be discussed, and it will be shown that those criteria are related to the quantities of the deviatoric stress tensor. In the next section, representation of yield criteria in a three-dimensional frame of reference is described.

Haigh–Westergard Stress Space

As discussed previously, the yield function for metals can depend only on the state of deviatoric stress. This can be written as

$$F(S_1, S_2, S_3) = k^2 \tag{9-6}$$

where S_1, S_2, and S_3 are the principal quantities of the deviatoric stress tensor. More generally, Eq. (9-3) or (9-6) represents a surface in the three-dimensional stress space. For cases of biaxial states of stress, the functions above can be represented as curves in a two-dimensional space. Figure 9-2 shows a three-dimensional stress space where the principal directions have been selected as

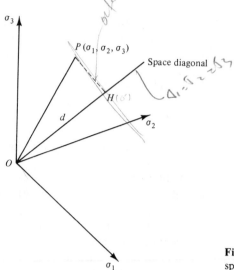

Figure 9-2 Haigh–Westergard stress space.

the coordinate axes. This is known as the Haigh–Westergard stress space (1, 2).

With reference to Fig. 9-2, the length OP to a stress point $(\sigma_1, \sigma_2, \sigma_3)$ can be written as

$$OP^2 = \sigma_1^2 + \sigma_2^2 + \sigma_3^2 \tag{9-7}$$

Along the space diagonal,

$$\sigma_1 = \sigma_2 = \sigma_3 = \frac{J_1}{3} \tag{9-8}$$

and the direction cosines of the space diagonal are

$$n_1 = n_2 = n_3 = \frac{1}{\sqrt{3}} \tag{9-9}$$

Hence the equation of a plane normal to the space diagonal can be written as

$$\sigma_1 + \sigma_2 + \sigma_3 = c = \text{constant} \tag{9-10}$$

The normal distance d from the origin to this plane can be found using coordinate geometry as

$$d = \frac{c}{\sqrt{1 + 1 + 1}} = \frac{c}{\sqrt{3}} \tag{9-11}$$

Substituting the value of c into Eq. (9-10), we obtain

$$\sigma_1 + \sigma_2 + \sigma_3 = \sqrt{3}\, d \tag{9-12}$$

Equation (9-12) shows that the hydrostatic stress at a point is increasing in a linear manner with the distance from the origin to the plane that is normal to the space diagonal and which passes through the stress point. Such a plane will be called an *octahedral* or *normal plane* (see Chapter 4). If we assume that the yield criterion is independent of hydrostatic stress, the shape and size of the yield function on any normal plane will be the same. In other words, the projection of the yield surface on any plane normal to the space diagonal will be the same. This projection is known as the *yield locus.* When the yield locus is the same on any normal plane, it is easier to represent this on a plane passing through the origin. Such a plane passing through the origin normal to the space diagonal is known as the *Π-plane*. A projection of a typical yield surface on the Π-plane is shown in Fig. 9-3; on this plane, the hydrostatic stress is zero. Since there are no preferred directions for an isotropic body, the behavior in all directions is the same. If we assume that the yield limits in compression and tension are equal, then the yield locus on the Π-plane must pass through the six points $T_1, T_2, T_3, \ldots, T_6$ (Fig. 9-3) on the principal axes equidistant from the origin. Therefore, the yield locus must be symmetrical, and has the same shape in each of the six 60° sectors on the Π-plane.

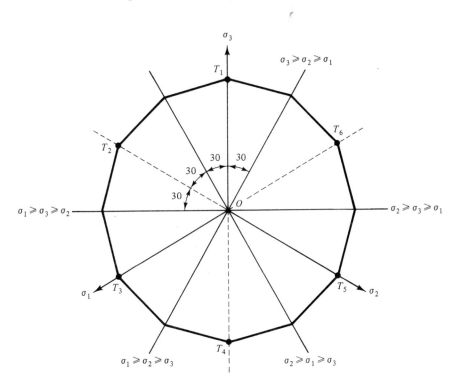

Figure 9-3 Typical yield locus on Π-plane. Note that the six points T_1, T_2, T_3, T_4, T_5, and T_6 on the axes 1, 2, 3 are equidistant from the origin. The yield curve consists of 12 equal areas.

Projection of Points on the Π-Plane

It will be seen later in this chapter and later in the book that yield criteria are frequently depicted on the Π-plane, and hence it becomes necessary to locate the projection of any given state of stress on the Π-plane in order to verify whether a stress point is inside or outside the yield surface. Moreover, in order to experimentally establish yield loci on Π-plane, it becomes necessary to project the experimentally observed yield points on the Π-plane. In the following section, we describe the procedure for finding the projections on the Π-plane.

Let us consider a point $P(\sigma_1, \sigma_2, \sigma_3)$ on the yield surface (Fig. 9-2). Let us suppose that it is required to find P_1 (Fig. 9-4), the projection of point P on the Π-plane. The three components $\sigma_{1\Pi}$, $\sigma_{2\Pi}$, and $\sigma_{3\Pi}$ (Figs. 9-4 and 9-5) on the Π-plane can be found by using trigonometry. Referring to Fig. 9-5, we have

$$\sigma_{1\Pi} = \sigma_1 \cos(90 - \alpha) \tag{9-13}$$

Since the σ_1, σ_2, and σ_3 axes make equal angles with the space diagonal

$$\cos \alpha = \frac{1}{\sqrt{3}} \tag{9-14}$$

$$\sin \alpha = \sqrt{\tfrac{2}{3}} \tag{9-15}$$

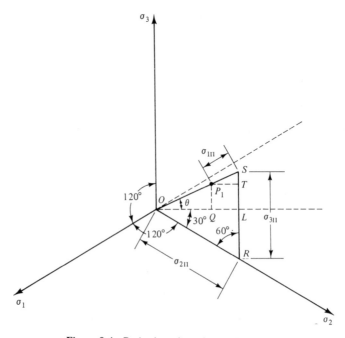

Figure 9-4   Projection of a point on the Π-plane.

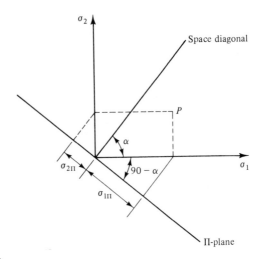

Figure 9-5 Projected components on the
Π-plane.

$\cos \alpha = 1/\sqrt{3}, \sin \alpha = \sqrt{2/3}$

Hence Eq. (9-13) becomes

$$\sigma_{1\Pi} = \sqrt{\tfrac{2}{3}}\,(\sigma_1) \qquad (9\text{-}16)$$

Similarly,

$$\sigma_{2\Pi} = \sqrt{\tfrac{2}{3}}\,(\sigma_2) \qquad (9\text{-}17)$$

$$\sigma_{3\Pi} = \sqrt{\tfrac{2}{3}}\,(\sigma_3) \qquad (9\text{-}18)$$

It should be noted that in the sector where P_1 is located in Fig. 9-4, the major principal stress is σ_2, the minor principal stress is σ_1, and the intermediate principal stress is σ_3.

In order to locate the projection of point P on the Π-plane (the projected point is denoted by P_1), two quantities are required (Fig. 9-4): the angle θ and the distance OP_1. Actually, these quantities are the polar coordinates of point P_1. To reach P_1 in Fig. 9-4, one can travel a distance $\sigma_{2\Pi}$ along σ_2, then a distance $\sigma_{3\Pi}$ in the σ_3-direction, and a distance $\sigma_{1\Pi}$ in σ_1-direction. Now

$$OQ = OL - P_1T$$

$$= OR \cos 30 - P_1 S \sin 60$$

$$= (\sigma_{2\Pi} - \sigma_{1\Pi})\frac{\sqrt{3}}{2}$$

$$= \frac{1}{\sqrt{2}}(\sigma_2 - \sigma_1) \qquad (9\text{-}19)$$

Similarly, the length P_1Q can be found as

$$P_1Q = SR - RL - ST$$

$$= \sigma_{3\Pi} - \sigma_{2\Pi} \sin 30 - \sigma_{1\Pi} \cos 60$$

$$= \frac{1}{\sqrt{6}} (2\sigma_3 - \sigma_1 - \sigma_2) \tag{9-20}$$

Therefore,

$$\theta = \tan^{-1} \left(\frac{P_1Q}{OQ} \right)$$

or

$$\theta = \tan^{-1} \left[\frac{2\sigma_3 - \sigma_1 - \sigma_2}{\sqrt{3} (\sigma_2 - \sigma_1)} \right] \tag{9-21}$$

Here σ_3, σ_1, and σ_2 are intermediate, minor, and major principal stresses, respectively.

The length of OP_1 can be written as

$$|OP_1|^2 = |OQ|^2 + |P_1Q|^2$$

$$= \tfrac{1}{2}(\sigma_1 - \sigma_2)^2 + \tfrac{1}{6}(2\sigma_3 - \sigma_1 - \sigma_2)^2 \tag{9-22}$$

By simplification of Eq. (9-22), it can be shown that

$$|OP_1|^2 = \tfrac{1}{3}\left[(\sigma_1 - \sigma_2)^2 + (\sigma_2 - \sigma_3)^2 + (\sigma_1 - \sigma_3)^2 \right] \tag{9-23}$$

By comparing Eqs. (9-23) and (4-6f), the length of OP_1 can be expressed in terms of J_{2D} as

$$R = |OP_1| = \sqrt{2J_{2D}} \tag{9-24}$$

With reference to Fig. 9-2,

$$|PH|^2 = |OP|^2 - |OH|^2 \tag{9-25}$$

Using Eq. (9-12), we find that

$$|PH|^2 = \sigma_1^2 + \sigma_2^2 + \sigma_3^2 - \tfrac{1}{3}(\sigma_1 + \sigma_2 + \sigma_3)^2$$

$$= \tfrac{1}{3}\left[(\sigma_1 - \sigma_2)^2 + (\sigma_2 - \sigma_3)^2 + (\sigma_1 - \sigma_3)^2 \right] \tag{9-26}$$

Comparing Eqs. (9-23) and (9-26), it can be seen that $|PH|$ in Fig. 9-2 and $|OP_1|$ in Fig. 9-4 have the same magnitudes. Actually, the projection of PH on the Π-plane should be the same as PH, since it is parallel to the Π-plane. The length of PH can be related to octahedral shear stress using Eqs. (5-14a) and (4-6h) as

$$|PH| = \sqrt{3}\, \tau_{\text{oct}} \tag{9-27}$$

As discussed previously, there are six sectors in the Π-plane (Fig. 9-3), and the principal stresses are different in each sector. For example, consider the sector $T_5 0 T_6$, where the major principal stress is σ_2, the intermediate principal stress is σ_3, and the minor principal stress is σ_1. As another example, consider the sector between the σ_1 and $-\sigma_3$ axes in Fig. 9-3. In this sector the major principal stress is σ_1, the intermediate principal stress is σ_2, and the minor principal stress is σ_3. Depending on the sector, the expression for θ in Eq. (9-21) will take different forms.

SOME YIELD CRITERIA USED IN METAL PLASTICITY

In this section some of the yield criteria developed for metals are reviewed. Most of the experimental work was carried out using thin metal tubes under biaxial states of stress. Hence the yield locus on a two-dimensional space and its counterpart on the Π-plane will be presented for some of the widely used yield criteria.

von Mises Yield Criterion

According to this theory, yielding will initiate when the second invariant of the deviatoric stress tensor reaches a certain value. Actually, this criterion assumes that yielding begins when the distortional energy reaches a value that is equal to the distortional energy at yield in simple tension. This criterion was suggested by von Mises in 1913 (1, 2), and is also known as the *distortional energy theory*. According to this criterion,

$$J_{2D} < k^2 \qquad \text{if the material is in the elastic state}$$
$$J_{2D} = k^2 \qquad \text{if the material is yielding} \tag{9-28a}$$

Here k is a material constant to be determined from experiments. Using Eq. (4-6e), one can express the von Mises yield criterion as

$$\tfrac{1}{6}\left[(\sigma_{11} - \sigma_{22})^2 + (\sigma_{22} - \sigma_{33})^2 + (\sigma_{33} - \sigma_{11})^2\right] + \sigma_{12}^2 + \sigma_{23}^2 + \sigma_{13}^2 = k^2 \tag{9-28b}$$

In the principal stress space, Eq. (9-28b) can be written as

$$\tfrac{1}{6}\left[(\sigma_1 - \sigma_2)^2 + (\sigma_2 - \sigma_3)^2 + (\sigma_3 - \sigma_1)^2\right] = k^2 \tag{9-28c}$$

As an example, consider a simple tension test. Under this state,

$$\sigma_2 = \sigma_3 = 0 \tag{9-29a}$$

and

$$\sigma_1 = -\sigma_{yt} \qquad \text{at yield} \tag{9-29b}$$

where σ_{yt} is the yield stress under simple tension. For the case of isotropic materials,

$$\sigma_{yt} = \sigma_{yc} = \sigma_y \tag{9-29c}$$

where σ_{yc} is the yield stress in uniaxial compression. In the following section, yield stress under uniaxial conditions will be denoted by σ_y. Therefore, from Eq. (9-28c), the value of k can be determined as

$$\tfrac{1}{3}\sigma_1^2 = k^2 = \tfrac{1}{3}\sigma_y^2 \tag{9-30a}$$

Hence

$$k = \frac{\sigma_y}{\sqrt{3}} \tag{9-30b}$$

Substituting the value of k into Eq. (9-28b), the yield criterion can be written as

$$\tfrac{1}{6}\left[(\sigma_{11}-\sigma_{22})^2 + (\sigma_{22}-\sigma_{33})^2 + (\sigma_{33}-\sigma_{11})^2\right] + \sigma_{12}^2 + \sigma_{23}^2 + \sigma_{13}^2 = \frac{\sigma_y^2}{3} \tag{9-31}$$

As another example, consider a pure shear state of stress at yield, that is,

$$\sigma_{11} = \sigma_{22} = \sigma_{33} = \sigma_{23} = \sigma_{13} = 0$$

and

$$\sigma_{12} = s$$

Substituting these values in Eq. (9-31), one can find that

$$s = \frac{\sigma_y}{\sqrt{3}} \tag{9-32}$$

This shows that for a material obeying the von Mises yield criterion, the yield stress in pure shear is $1/\sqrt{3}$ times the yield stress in uniaxial tension. For the special case of plane stress (i.e., $\sigma_3 = 0$), von Mises yield criterion will be an ellipse on σ_1–σ_2 stress plane. For this case Eq. (9-28c) reduces to the form

$$(\sigma_1 - \sigma_2)^2 + (\sigma_2 - 0)^2 + (0 - \sigma_1)^2 = 6k^2 \tag{6-33a}$$

Substituting the value of k, this can be simplified to obtain

$$2\sigma_1^2 - 2\sigma_1\sigma_2 + 2\sigma_2^2 = 6k^2 = 2\sigma_y^2 \tag{6-33b}$$

which is an ellipse in σ_1–σ_2 space. Equation (6-33b) can be rearranged to obtain the form

$$\left(\frac{\sigma_1}{\sigma_y}\right)^2 - \left(\frac{\sigma_1}{\sigma_y}\right)\left(\frac{\sigma_2}{\sigma_y}\right) + \left(\frac{\sigma_2}{\sigma_y}\right)^2 = 1 \tag{9-33c}$$

This represents an ellipse in the nondimensional (σ_1/σ_y)–(σ_2/σ_y) space, as shown in Fig. 9-6. This is known as the *von Mises ellipse*. In order to find the yield locus on the Π-plane, let us consider Eq. (9-24). Then the von Mises

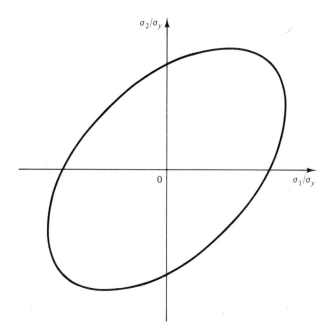

Figure 9-6 Von Mises ellipse for plane stress condition ($\sigma_3 = 0$).

criterion can be written as

$$R = \sqrt{2}\sqrt{J_{2D}} = \frac{\sqrt{2}}{\sqrt{3}}\sigma_y \tag{9-33d}$$

This represents a circle of radius $\sqrt{\frac{2}{3}}\sigma_y$ on the Π-plane (Fig. 9-7).

Tresca Yield Criterion

According to this criterion, yielding will initiate when the maximum value of the extreme shear stress is reached (Fig. 9-8). Any state of stress located inside the yield locus is considered to be under elastic state. In this theory, yield strength in tension and compression has been assumed to be equal, which is a limitation for many materials (2).

As explained in the section on the Π-plane, there are six sectors (Fig. 9-3) in which the principal stresses are different. In the sector T_5OT_6,

$$\sigma_2 \geqslant \sigma_3 \geqslant \sigma_1 \tag{9-34a}$$

Therefore, the yield criterion reduces to

$$\sigma_2 - \sigma_1 = \sigma_y \tag{9-34b}$$

for that sector, and it is independent of the value of σ_3. Hence Eq. (9-34b) represents a line parallel to σ_3-axis. As shown in Fig. 9-4 and Eq. (9-19), the

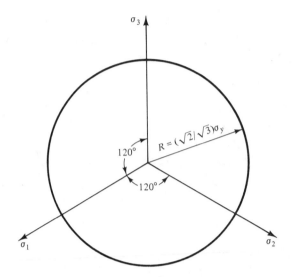

Figure 9-7   Von Mises criterion on the Π-plane.

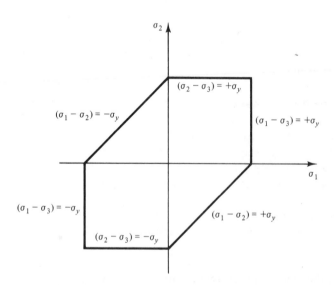

Figure 9-8   Tresca yield criterion under plane stress–state ($\sigma_3 = 0$).

normal distance to this line from the origin is a critical value which is proportional to the maximum shear stress attained under uniaxial conditions. The maximum shear stress under uniaxial conditions is equal to $\sigma_y/2$, where σ_y is the uniaxial yield stress. The Tresca criterion can be expressed in terms of principal stresses as follows:

$$\tfrac{1}{2}(\sigma_1 - \sigma_2) = \frac{\pm\sigma_y}{2} \tag{9-35a}$$

$$\tfrac{1}{2}(\sigma_2 - \sigma_3) = \frac{\pm\sigma_y}{2} \tag{9-35b}$$

$$\tfrac{1}{2}(\sigma_3 - \sigma_1) = \frac{\pm\sigma_y}{2} \tag{9-35c}$$

where σ_1, σ_2, and σ_3 are major, intermediate, and minor principal stresses, respectively.

Equations (9-35) can also be written as

$$\sigma_1 - \sigma_2 = \pm\sigma_y \tag{9-35d}$$

$$\sigma_2 - \sigma_3 = \pm\sigma_y \tag{9-35e}$$

$$\sigma_3 - \sigma_1 = \pm\sigma_y \tag{9-35f}$$

Let us consider a uniaxial case as an example to illustrate the Tresca criterion. For the uniaxial case,

$$\sigma_3 = \sigma_2 = 0$$

Therefore, $\sigma_1 = \pm\sigma_y$.

As another example, consider the case of a pure shear state. Here the stress tensor in the principal stress space can be expressed as

$$\begin{pmatrix} s & 0 & 0 \\ 0 & -s & 0 \\ 0 & 0 & 0 \end{pmatrix} \tag{9-36a}$$

That is,

$$\sigma_1 = s$$

$$\sigma_2 = -s \tag{9-36b}$$

$$\sigma_3 = 0$$

The maximum value of the shear stress is given by Eq. (9-35a), which produces

$$s - (-s) = \sigma_y$$

or $\tag{9-36c}$

$$s = \frac{\sigma_y}{2}$$

Hence, according to the Tresca yield criterion, the yield stress in the pure shear state is equal to half of the yield stress under uniaxial conditions. It is interesting to compare Eq. (9-36c) with Eq. (9-32) to see how the Tresca and von Mises criteria are different for these two states of stresses.

A comparison of the von Mises and Tresca criteria is shown in Figs. 9-9 and 9-10. The value of $\sqrt{J_{2D}}$ for the pure shear state in Eq. (9-36a) is equal to $\sigma_y/2$. Now using Eq. (9-24), length OB in Fig. 9-9(a) can be found as (2)

$$OB = \frac{\sigma_y}{\sqrt{2}} \ (\sec 30) = \sqrt{\tfrac{2}{3}} \ \sigma_y \qquad (9\text{-}37\text{a})$$

Hence it can be seen that both criteria predict the same yield stress under a uniaxial state of stress. However, predictions are different for any other state of stress. This deviation is maximum under a pure shear state of stress. The ratio of yield stress under pure shear conditions predicted by the Tresca criterion to that of von Mises can be found as (Fig. 9-9)

$$\text{Ratio} = \frac{\sigma_y/\sqrt{2}}{\sqrt{\tfrac{2}{3}} \ \sigma_y} = \frac{\sqrt{3}}{2} \qquad (9\text{-}37\text{b})$$

One of the limitations of the Tresca criterion is that the yield strength in tension and compression has been assumed to be equal.

POST-YIELD BEHAVIOR: PLASTIC STRESS–STRAIN RELATIONS

Before discussing mathematical aspects of plastic behavior, we describe certain idealizations of stress–strain behavior; these idealizations are:

1. Perfectly elastic
2. Rigid, perfectly plastic
3. Rigid, linear strain hardening
4. Elastic, perfectly plastic
5. Elastic, linear strain hardening
6. Strain softening

Physical and graphical representations of these idealized models are shown in Fig. 9-11. An explanation of strain hardening is given subsequently.

Strain softening is depicted in the symbolic curve (Fig. 9-11); however, the physical model shown does not include softening. Many materials exhibit strain softening and efforts have been made to incorporate it in the context of various models described in subsequent chapters. However, we feel that considerable work will be needed before viable models for this phenomenon will be available, and hence this text does not include many details of strain softening.

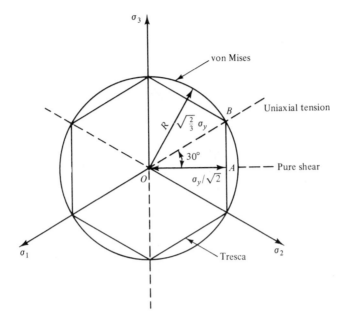

Figure 9-9 Von Mises and Tresca criteria on the Π-plane.

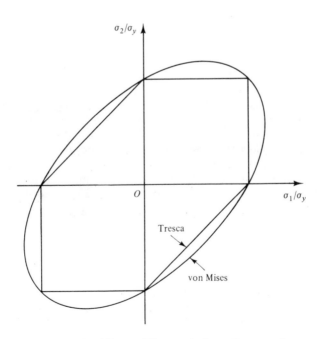

Figure 9-10 Von Mises and Tresca criteria on the σ_1–σ_2 plane.

Physical Model   Stress–Strain Idealization

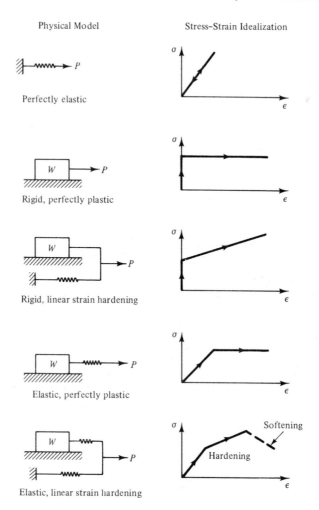

Perfectly elastic

Rigid, perfectly plastic

Rigid, linear strain hardening

Elastic, perfectly plastic

Elastic, linear strain hardening

Figure 9-11   Idealized stress–strain models (Ref. 2).

As the reader knows by now, the elastic deformations do not depend on how and along what stress path the state of stress was reached. They depend only on the final state of stress. However, in the case of plastic deformations, the strains depend on the history of the stress states, and the stress–strain relationships are generally nonlinear.

Influence of Stress Path

Simple explanations for the influence of a stress path on plastic deformations are given below (Fig. 9-12); this explanation is similar to that given by Mendelson (2). Let us assume that a test specimen is loaded with an increasing normal stress with zero shear stress, that is, along stress path OCA (σ-axis) (Fig. 9-12). The initial yield surface for this specimen is shown by AQG, and is denoted as "yield surface 1" in Fig. 9-12. The material will behave elastically until it reaches a state represented by point A, which is on the initial yield surface. Upon further loading it will behave elastoplastically, and due to hardening the yield surface will expand to a new position. The new yield locus passing through point B, is denoted as "yield surface 2" in Fig. 9-12.

The plastic strains that occur during deformations along the stress path above will have only nonzero normal components, since the applied state of stress is uniaxial. Hence the normal plastic strain along the x-direction (Fig. 9-12) is denoted as

$$\epsilon_x^p = \epsilon^p \qquad (9\text{-}38a)$$

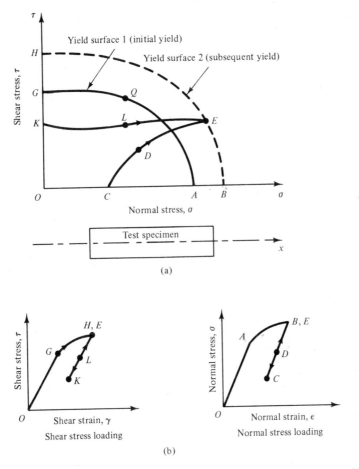

Figure 9-12 History-dependent nature of plastic strains: (a) yielding behavior; (b) stress–strain response (Ref. 2) (note that H, E; and B, E are different stress points).

Here the superscript p denotes plastic quantities. Since the volumetric plastic strain $(\epsilon_x^p + \epsilon_y^p + \epsilon_z^p)$ is equal to zero,

$$\epsilon_y^p = \epsilon_z^p = -\frac{\epsilon^p}{2} \qquad (9\text{-}38b)$$

Also,

$$\gamma_{xy}^p = \gamma_{yz}^p = \gamma_{zx}^p = 0 \qquad (9\text{-}38c)$$

Here γ represents shear strains.

Now let the specimen be unloaded along path BC, and then sheared along path CDE until state E is reached. During unloading along path BC, material will undergo elastic deformations only; this is because the states of stress along BC are inside the current (subsequent) yield surface, i.e. yield surface 2. Upon reloading along CDE the material deforms only elastically, and no plastic deformations occur because the state of stress does not go outside the current yield surface. Hence plastic strains at point E will be the same as those at point B. That is, the accumulated plastic strains along path $OBACDE$ (Fig. 9-12) are equal to those given in Eqs. (9-38).

Now let us look at shear loading of a similar specimen along the stress path OGH (Fig. 9-12). During loading, the material will deform elastically until the state of stress reaches point G, which is on the initial yield surface. When it is loaded beyond stress point G along the same stress path, the material will deform elastoplastically, and the yield surface will expand up to point H due to hardening; this new yield surface is denoted as "yield surface 2" in Fig. 9-12. The plastic strains that occur along this stress path will have only nonzero shear components. That is,

$$\epsilon_x^p = \epsilon_y^p = \epsilon_z^p = 0.0 \qquad (9\text{-}39a)$$

$$\gamma_{xy}^p = \gamma^p \qquad (9\text{-}39b)$$

$$\gamma_{yz}^p = \gamma_{xz}^p = 0.0 \qquad (9\text{-}39c)$$

Now if material is unloaded from point H to point K along the stress path HK, the deformations will have only elastic components. Upon reloading along path KLE up to point E, the material deforms elastically since the states of stress do not cross the current yield surface. Therefore, plastic strains along path $OHGKLE$ will be the same as those at point H [Eq. (9-39)].

The discussion above shows that plastic strains ϵ_x^p and γ^p at point E, reached via two different stress paths, are different. Thus the behavior is dependent on the stress path or stress history. This stress-path-dependent nature requires us to consider incremental plastic strains throughout the history and to accumulate them to determine the total plastic strains. However, if a proportional stress path (all the stresses increase in the same ratio) is followed, the plastic strains will be independent of the stress history and will depend only on the final stress state. In the following sections some of the incremental theories of ideal plasticity are considered; detailed descriptions of the theories are available in several other references $(1, 2, 5, 7)$.

Levy–Mises Stress–Strain Relations

Perhaps the first attempt for determination of plastic stress–strain relationship was made by Saint-Venant in 1870. He proposed that the principal axes of incremental plastic strains coincide with the current principal axes of the stresses. In the Levy–Mises approach it is assumed that total strain increments are equal to the plastic strain increments; that is, elastic strains are negligible. This leads to

$$\frac{d\epsilon_{11}}{S_{11}} = \frac{d\epsilon_{22}}{S_{22}} = \frac{d\epsilon_{33}}{S_{33}} = \frac{d\epsilon_{12}}{S_{12}} = \frac{d\epsilon_{23}}{S_{23}} = \frac{d\epsilon_{13}}{S_{13}} = \lambda \qquad (9\text{-}40\text{a})$$

or

$$d\epsilon_{ij} = \lambda S_{ij} \qquad (9\text{-}40\text{b})$$

where S_{ij} is the deviatoric stress tensor, $d\epsilon_{ij}$ the incremental (plastic) strain tensor, and λ a nonnegative scalar factor which may vary throughout the loading history. The equations above are known as the *flow rule* of the von Mises criterion. The concept of flow rule will be discussed later. It is also assumed that the material is incompressible under plastic flow or yielding. That is, the plastic volumetric strain ϵ_{kk}^{p} is equal to zero. Hence

$$E_{ij}^{p} = \epsilon_{ij}^{p} - \tfrac{1}{3}\delta_{ij}\epsilon_{kk}^{p\,0} \qquad (9\text{-}41)$$

Therefore, deviatoric strain tensor is equal to the total strain tensor. Equation (9-40) shows that the incremental plastic strain is related to the deviatoric stress and not to the total stress, which indicates that the relationships above do not depend on the hydrostatic stress. Although this assumption holds good for many metals, Levy–Mises equations, in general, are not valid for geologic materials. Because of the first assumption, $d\epsilon = d\epsilon^{p}$, Levy–Mises equations can only be applied to problems in which only the plastic strains are significant, that is, when elastic strains are negligible compared to the plastic strains. Generalization of the Levy–Mises equation to include both elastic and plastic components of the strain is due to Prandtl and Ruess (1), and this is the subject of the next section.

Prandtl–Reuss Equations

These equations describe an ideal plastic law $(1, 2, 4)$. It is assumed that the total strains can be decomposed into elastic and plastic components as

$$d\epsilon = d\epsilon^{e} + d\epsilon^{p} \qquad (9\text{-}42)$$

where $d\epsilon^{e}$ is the incremental elastic strain and $d\epsilon^{p}$ is the incremental plastic strain. It is also assumed that the plastic strain increment is proportional to the deviatoric stress tensor at any instant of loading. That is,

$$\frac{d\epsilon_{11}^{p}}{S_{11}} = \frac{d\epsilon_{22}^{p}}{S_{22}} = \frac{d\epsilon_{33}^{p}}{S_{33}} = \frac{d\epsilon_{12}^{p}}{S_{12}} = \frac{d\epsilon_{23}^{p}}{S_{23}} = \frac{d\epsilon_{13}^{p}}{S_{13}} = \lambda \qquad (9\text{-}43a)$$

or

$$d\epsilon_{ii}^{p} = \lambda S_{ij} \qquad (9\text{-}43b)$$

where λ is a nonnegative scalar factor which may vary during the loading history. In this flow rule it is implied that principal axes of incremental plastic strains coincide with the principal axes of stresses. In principal axes of reference, this equation takes the form

$$\frac{d\epsilon_{1}^{p}}{S_{1}} = \frac{d\epsilon_{2}^{p}}{S_{2}} = \frac{d\epsilon_{3}^{p}}{S_{3}} = \lambda \qquad (9\text{-}43c)$$

Hence

$$d\epsilon_{1}^{p} = \lambda S_{1}$$
$$d\epsilon_{2}^{p} = \lambda S_{2} \qquad (9\text{-}43d)$$
$$d\epsilon_{3}^{p} = \lambda S_{3}$$

It is also possible to write

$$d\epsilon_{1}^{p} - d\epsilon_{2}^{p} = \lambda(S_{1} - S_{2})$$
$$d\epsilon_{2}^{p} - d\epsilon_{3}^{p} = \lambda(S_{2} - S_{3}) \qquad (9\text{-}44a)$$
$$d\epsilon_{1}^{p} - d\epsilon_{3}^{p} = \lambda(S_{1} - S_{3})$$

or

$$\frac{d\epsilon_{1}^{p} - d\epsilon_{2}^{p}}{S_{1} - S_{2}} = \frac{d\epsilon_{2}^{p} - d\epsilon_{3}^{p}}{S_{2} - S_{3}} = \frac{d\epsilon_{1}^{p} - d\epsilon_{3}^{p}}{S_{1} - S_{3}} = \lambda \qquad (9\text{-}44b)$$

Expression for λ.   Equations (9-43) can be rearranged to obtain

$$d\epsilon_{11}^{p} = \lambda S_{11} = \lambda\left[\sigma_{11} - \left(\frac{\sigma_{11} + \sigma_{22} + \sigma_{33}}{3}\right)\right]$$
$$= \tfrac{2}{3}\lambda\left[\sigma_{11} - \tfrac{1}{2}(\sigma_{22} + \sigma_{33})\right] \qquad (9\text{-}44c)$$

Similarly,

$$d\epsilon_{22}^{p} = \tfrac{2}{3}\lambda\left[\sigma_{22} - \tfrac{1}{2}(\sigma_{11} + \sigma_{33})\right] \qquad (9\text{-}44d)$$
$$d\epsilon_{33}^{p} = \tfrac{2}{3}\lambda\left[\sigma_{33} - \tfrac{1}{2}(\sigma_{11} + \sigma_{22})\right] \qquad (9\text{-}44e)$$
$$d\epsilon_{12}^{p} = \lambda\sigma_{12} \qquad (9\text{-}44f)$$
$$d\epsilon_{23}^{p} = \lambda\sigma_{23} \qquad (9\text{-}44g)$$
$$d\epsilon_{13}^{p} = \lambda\sigma_{13} \qquad (9\text{-}44h)$$

which leads to

$$d\epsilon_{11}^p - d\epsilon_{22}^p = \lambda(S_{11} - S_{22}) \tag{9-45a}$$

$$d\epsilon_{22}^p - d\epsilon_{33}^p = \lambda(S_{22} - S_{33}) \tag{9-45b}$$

$$d\epsilon_{11}^p - d\epsilon_{33}^p = \lambda(S_{11} - S_{33}) \tag{9-45c}$$

Hence

$$\left(d\epsilon_{11}^p - d\epsilon_{22}^p\right)^2 + \left(d\epsilon_{22}^p - d\epsilon_{33}^p\right)^2$$
$$+ \left(d\epsilon_{11}^p - d\epsilon_{33}^p\right)^2 + 6\left(d\epsilon_{12}^p\right)^2 + 6\left(d\epsilon_{23}^p\right)^2 + 6\left(d\epsilon_{13}^p\right)^2 \tag{9-46}$$
$$= \left[(\sigma_{11} - \sigma_{22})^2 + (\sigma_{22} - \sigma_{33})^2 + (\sigma_{11} - \sigma_{33})^2 + 6\sigma_{12}^2 + 6\sigma_{23}^2 + 6\sigma_{13}^2\right]\lambda^2$$

or in terms of invariants, this can be written as [Eq. (3-15d)]

$$9\left(d\gamma_{oct}^p\right)^2 = 9(\tau_{oct})^2(\lambda)^2 \tag{9-47a}$$

or

$$\lambda = \frac{d\gamma_{oct}^p}{\tau_{oct}} = \sqrt{\frac{3}{2}}\frac{d\gamma_{oct}^p}{\sqrt{J_{2D}}} \tag{9-47b}$$

Plastic Work

The concept of plastic work plays an important role in the plastic stress–strain laws, and it is presented briefly in this section. The total work done per unit volume of a deformable body during a strain increment can be written as (2, 7)

$$dW = \sigma_{ij}d\epsilon_{ij} \tag{9-48}$$

where $d\epsilon_{ij}$ is the total strain increment. This consists of both elastic and plastic components. It can be decomposed as

$$d\epsilon_{ij} = d\epsilon_{ij}^e + d\epsilon_{ij}^p \tag{9-49}$$

where the superscripts e and p denote elastic and plastic components, respectively. Therefore, dW is given by

$$dW = \sigma_{ij}\left(d\epsilon_{ij}^e + d\epsilon_{ij}^p\right)$$
$$= \sigma_{ij}d\epsilon_{ij}^e + \sigma_{ij}d\epsilon_{ij}^p$$
$$= dW^e + dW^p \tag{9-50}$$

The quantity dW^e is the elastic energy which is recoverable. The quantity dW^p is the plastic work and cannot be recovered, as the plastic deformations are irreversible. Since the work done in metals due to hydrostatic stress is zero, the plastic work dW^p can also be written as

$$dW^p = S_{ij}\epsilon_{ij}^p \tag{9-51a}$$

or in terms of principal stresses as

$$dW^P = S_1 \, d\epsilon_1^p + S_2 \, d\epsilon_2^p + S_3 \, d\epsilon_3^p \qquad (9\text{-}51\text{b})$$

Here it was assumed that the principal axes of stress coincide with the principal axes of incremental plastic strains. Now, let us consider a representation on the Π-plane (Fig. 9-13). On the Π-plane, $\sigma_1 + \sigma_2 + \sigma_3 = 0$, and hence

$$S_1 = \sigma_1$$
$$S_2 = \sigma_2 \qquad (9\text{-}52)$$
$$S_3 = \sigma_3$$

For metals, since the plastic volumetric strain, $d\epsilon_1^p + d\epsilon_2^p + d\epsilon_3^p$, is zero, the incremental plastic strains can also be represented on the same plot. However, the strain quantities have to be multiplied by a stress quantity in order to obtain the same dimensions, and usually the constant is taken as $2G$, G being the shear modulus. The deviatoric stress vector is represented by OP, and the incremental plastic strain vector OR can be represented on the same diagram, as shown in Fig. 9-13. The plastic work can be written as $(2,4)$

$$dW^P = OP \cdot \frac{OR}{2G}$$

$$= |OP| \cdot |OR| \frac{\cos(\gamma - \theta)}{2G} \qquad (9\text{-}53)$$

Here

$$|OP| = \left(S_1^2 + S_2^2 + S_3^2 \right)^{1/2} = \sqrt{2J_{2D}}$$

$$|OR| = 2G\left[\left(d\epsilon_1^p \right)^2 + \left(d\epsilon_2^p \right)^2 + \left(d\epsilon_3^p \right)^2 \right]^{1/2}$$

$$= 2G\left(\sqrt{3d\gamma_{\text{oct}}^p} \right)$$

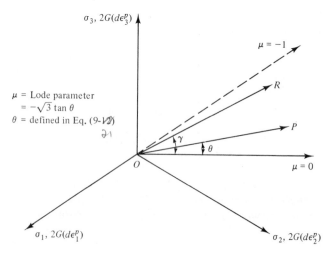

Figure 9-13 Stress and incremental plastic strain vectors on the Π-plane.

If it is assumed that the direction of the stress vector and the plastic strain increment vector coincide, then $\gamma = \theta$. Therefore,

$$dW^p = \sqrt{2J_{2D}}\left(\sqrt{3}\,d\gamma^p_{\text{oct}}\right) = 2J_{2D}\lambda \tag{9-54}$$

Hardening Behavior

Due to plastic flow, work hardening or strain hardening occurs in certain materials. Two hypotheses have been proposed to define the degree of hardening. One hypothesis, proposed by Hill (1), assumes that hardening depends only on the plastic work, and is independent of the strain path. This implies that the resistance to further yielding depends only on the total plastic work that has been done on the material. According to this hypothesis, the yield criterion can be written as

$$f = f\left(\sigma_{ij}, W^p\right) \tag{9-55a}$$

The second hypothesis assumes that plastic strain, ϵ^p, is a measure of hardening. According to this hypothesis, the yield function can be written as

$$f = f\left(\sigma_{ij}, \epsilon^p\right) \tag{9-55b}$$

In the literature, Eq. (9-55a) is referred to as the *work hardening hypothesis*, and Eq. (9-55b) is referred to as the *strain hardening hypothesis*. The implications of work hardening are discussed subsequently.

GENERALIZED PLASTIC STRESS–STRAIN RELATIONS

The definition of plastic flow behavior of materials is important in developing plastic stress–strain relations. The plastic flow is defined through a flow rule which is discussed in the following section.

Plastic Flow Behavior

When the state of stress reaches the yield criterion f, the material undergoes plastic deformations; this is also called plastic flow. In the theory of plasticity (1–3) the direction of plastic strain vectors is defined through a flow rule by assuming the existence of a plastic potential function, to which the incremental strain vectors are orthogonal. Then the increments of the plastic strain can be expressed as

$$d\epsilon^p_{ij} = \lambda \frac{\partial Q}{\partial \sigma_{ij}} \tag{9-56}$$

which is referred to as the *normality rule*; here, Q is the plastic potential function, and λ is a positive scalar factor of proportionality. For some materials, the plastic potential function, Q, and the yield function, f, can be assumed to be the same. Such materials are considered to follow the *associa-*

tive flow rule of plasticity. However, for many geologic materials, the yield function f and the plastic potential function Q are often different. These materials are considered to follow *nonassociative flow* rules of plasticity. The Levy–Misses and Prandtl–Ruess plastic stress–strain relations [Eq. (9-40a) and (9-43a)] are special cases of the flow rule expressed in Eq. (9-56).

A generalized approach for determining plastic stress–strain relations for any yield criterion has been suggested by Drucker (8–10), and details are given later in this chapter. The derivation of plastic stress–strain relations is based on certain assumptions and postulates, and it is useful to review them and their implications on shape of the yield surface and stress–strain relations, prior to detailed derivations. In the following section, we present the concept of work hardening and its implications.

Implications of Work Hardening

The meaning of work hardening in simple tension is depicted in Fig. 9-14; here tension is taken as positive. However, for the more general states of stresses and stress paths, the concept of work hardening can be expressed in terms of the work done by an additional or incremental set of stresses caused by an external agency. The work referred to is only the work done by the added set of forces (stresses) on the displacements (strains) that result, *not* the total work done by all the forces acting (8). Work hardening means that for all such added stresses the material will remain in stable equilibrium. Furthermore, it is governed by the following postulates:

1. During the application of stresses, the work done by the external agency will be positive.
2. Over a cycle of application and removal of stresses, the work done by the external agency will be zero or positive.

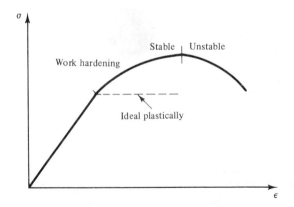

Figure 9-14 Stress–strain behavior (Ref. 8).

In other words, work hardening implies that useful net energy cannot be extracted from material subjected to a set of external forces in such a cycle.

To establish the mathematical relationships of these postulates, let us consider a material whose current states of stress and strain are given by σ_{ij} and ϵ_{ij}, respectively. Suppose that an external agency alters the stress by $d\sigma_{ij}$ at all points, and causes a small strain increment $d\epsilon_{ij}$. Then according to postulate 1,

$$d\sigma_{ij} d\epsilon_{ij} > 0 \qquad (9\text{-}57a)$$

or

$$d\sigma_{ij}\left(d\epsilon_{ij}^{e} + d\epsilon_{ij}^{p}\right) > 0 \qquad (9\text{-}57b)$$

The second postulate implies that

$$d\sigma_{ij} d\epsilon_{ij}^{p} \geqslant 0 \qquad (9\text{-}57c)$$

At this stage, two basic assumptions are made (8, 9):

1. There exists a surface, called the *yield surface*, which represents the yield limit associated with a state of stress along any stress path. Only elastic changes of deformation will occur for changes in stresses inside the yield surface, while plastic deformations occur for all stress paths directed toward the exterior of the yield surface.
2. The relation between the infinitesimal changes of the stress and plastic strain is linear. This implies that the sum of plastic strain increments obtained by two sets of stress increments, $d\sigma_{ij}'$ and $d\sigma_{ij}''$, will be the same as the plastic strain increments resulting from $d\sigma_{ij} = d\sigma_{ij}' + d\sigma_{ij}''$.

Furthermore, some conditions should be satisfied to ensure an appropriate description of the physical process involved in plastic deformations. There are four conditions formulated by Prager (11), and these are given below.

1. **Condition of continuity.** Let us consider a state of stress σ_{ij} which lies on the yield surface. An infinitesimal change of stress $d\sigma_{ij}$ causes unloading or loading, or neutral loading depending on whether the stress path is directed toward the interior or exterior of the yield surface or is tangential to it, respectively. To avoid awkward discontinuities in stress–strain relations, the condition of continuity requires that neutral loading does not cause plastic deformations.
2. **Condition of uniqueness.** This condition ensures that for a given mechanical state of a body and a system of infinitesimal increments of surface tractions, the resulting increments of stresses and strains (elastic and plastic) are unique.
3. **Condition of irreversibility.** This condition requires that because of the irreversible nature of plastic deformations, the work done on the plastic

deformations, dW^p [Eq. (9-50)], will be positive; that is,

$$\sigma_{ij}\,d\epsilon_{ij}^{P} > 0 \tag{9-58}$$

4. **Condition of consistency.** Loading from a plastic state will lead to another plastic state. The condition of consistency requires that a yield criterion will be satisfied as long as the material is in a plastic state.

The implications of these four conditions together with the assumptions made will lead to important restraints on the plastic stress–strain behavior. To investigate the implications of the condition of continuity, let us consider a stress state $\sigma_{ij}(\sigma)$ that lies on the yield surface. Let us choose an increment in stress $d\sigma_{ij}(d\sigma)$ whose stress path is directed toward the exterior of the yield surface. In order to give a geometric representation, let us consider the vector quantities of stress and incremental stress as shown in Fig. 9-15. The incremental vector $d\sigma$ can be decomposed into a tangential and a normal component. That is,

$$d\sigma = d\sigma^{(n)} + d\sigma^{(t)} \tag{9-59}$$

where the superscripts n and t denote normal and tangential components, respectively. According to the assumption of linearity (assumption 2), the incremental plastic strains caused by $d\sigma$ will be equal to the vector sum of incremental plastic strains caused by $d\sigma^{(n)}$ and $d\sigma^{(t)}$ separately. However, the condition of continuity requires that the plastic strains due to $d\sigma^{(t)}$ be zero. Thus the incremental plastic strain $d\epsilon^{p}$ due to $d\sigma$ is dependent only on the normal component $d\sigma^{(n)}$. This leads to the fact that the incremental plastic strain vector is normal to the yield surface. This is the well-known *normality rule* in the theory of plasticity, Eq. (9-56).

In view of the path dependence of plastic stress–strain relations, it is not reasonable to assume that state of stress at the end of any stress path will give

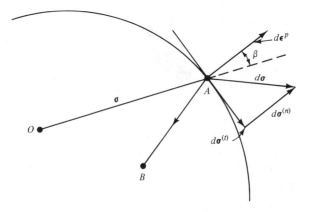

Figure 9-15   Stress increment vector.

a unique solution to stresses and strains. In other words, it is required to know the complete history of applied surface tractions, body forces, and boundary conditions. Drucker (12) has shown that the uniqueness is assured for linear incremental theories of plasticity if the two postulates given above are satisfied.

Shape of Yield Surface

From a given plastic state of stress, point A (Fig. 9-15), it is possible to reach a stress-free state, point O, (zero state of stress) by following an appropriate (unloading) stress path such as AB (Fig. 9-15). This implies that the origin of the stress coordinate system must lie inside the yield surface.

The condition of irreversibility [Eq. (9-58)] requires that (Fig. 9-15)

$$\boldsymbol{\sigma} \cdot d\boldsymbol{\epsilon}^p > 0 \qquad (9\text{-}60a)$$

or

$$|\boldsymbol{\sigma}| \cdot |d\boldsymbol{\epsilon}^p| \cos \beta > 0 \qquad (9\text{-}60b)$$

This means that the radius vector denoted by $\boldsymbol{\sigma}$ in Fig. 9-15 should make an acute angle (β) with the incremental plastic strain vector. However, the incremental plastic strain vector is normal to the yield surface, and is acting toward the exterior of the closed surface. In other words, irreversibility condition requires that any radius vector to a point on the yield surface must make an acute angle with its exterior normal at that point. That is, the yield surface has to be convex. Mathematical proof of convexity is given below.

Let us consider a state of stress σ_{ij}^* as represented by point A in Fig. 9-16(a), and say that the state of stress is brought to σ_{ij} (point P) on the yield surface along a stress path AP. Now let us increase the stress by $d\sigma_{ij}$ [Fig. 9-16(a)], which will cause plastic deformations, and then release $d\sigma_{ij}$ and return to the original state at point A. The plastic work over the above cycle A-P-A can be written as (9)

$$dW^p = \left(\sigma_{ij} - \sigma_{ij}^*\right) d\epsilon_{ij}^p + d\sigma_{ij} d\epsilon_{ij}^p > 0 \qquad (9\text{-}61)$$

As σ_{ij}^* can be chosen to be equal to σ_{ij} itself, $d\sigma_{ij}$ and $d\epsilon_{ij}^p$ must satisfy

$$d\sigma_{ij} d\epsilon_{ij}^p \geqslant 0 \qquad (9\text{-}62)$$

This indicates that $d\sigma_{ij}$ and $d\epsilon_{ij}^p$ must make an acute angle with each other. In fact, this is the second postulate given by Drucker [Eq. (9-57c)]. Moreover, by considering the loading cycle starting at σ_{ij}^*, it can be shown that the second postulate leads to

$$\left(\sigma_{ij} - \sigma_{ij}^*\right) d\epsilon_{ij}^p \geqslant 0 \qquad (9\text{-}63)$$

That is,

$$|\sigma_{ij} - \sigma_{ij}^*| \cdot |d\epsilon_{ij}^p| \cos \gamma \geqslant 0$$

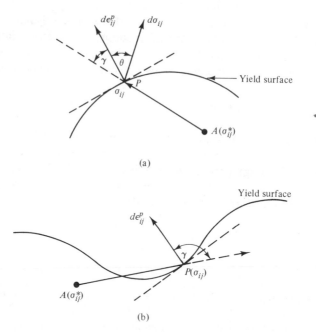

Figure 9-16 Convexity of the yield surface: (a) yield locus which is convex; (b) yield locus which is not convex (Refs. 8–10).

or $-\pi/2 \leqslant \gamma \leqslant \pi/2$. This implies that all points σ_{ij}^* must be located on one side of a plane perpendicular to $d\epsilon_{ij}^P$ at point P [Fig. 9-16(b)]. That is, all points σ_{ij}^* must lie on one side of the tangential plane at point P. Hence no vector $(\sigma_{ij} - \sigma_{ij}^*)$ should intersect the yield surface twice. Therefore, the condition of irreversibility implies that the yield surface should be convex.

Derivation of Plastic Stress–Strain Relation

According to the condition of consistency, loading from a plastic state should reach another plastic state, which satisfies the yield criterion.

For plastic deformations to occur in a work-hardening material, the following condition should be satisfied:

$$df = \frac{\partial f}{\partial \sigma_{ij}} d\sigma_{ij} > 0 \qquad \text{loading} \qquad (9\text{-}64a)$$

$$df < 0 \qquad\qquad\qquad \text{unloading} \qquad (9\text{-}64b)$$

However, for an ideal plastic material, the plastic flow will occur when $df = 0$.

To obtain a generalized form of plastic stress–strain relationships, let us consider an increment of stress $d\sigma_{kl}$ from an existing state of stress σ_{kl} which lies on the yield surface. This stress increment $d\sigma_{kl}$ can be decomposed into two components, $d\sigma_{kl}^{(n)}$ and $d\sigma_{kl}^{(t)}$, which are normal and tangential to the yield surface, respectively. According to the foregoing discussion, plastic deformations will occur when $df > 0$. Therefore,

$$\left(\frac{\partial f}{\partial \sigma_{kl}}\right) d\sigma_{kl} = \frac{\partial f}{\partial \sigma_{kl}}\left(d\sigma_{kl}^{(n)} + d\sigma_{kl}^{(t)}\right) > 0 \qquad (9\text{-}65)$$

However, the tangential component of the stress increment $d\sigma_{kl}^{(t)}$ does not cause any plastic deformations according to the condition of continuity. Hence

$$\left(\frac{\partial f}{\partial \sigma_{kl}}\right) d\sigma_{kl}^{(t)} = 0 \qquad (9\text{-}66)$$

The normal component $d\sigma_{kl}^{(n)}$ is proportional to the gradient of the yield function, that is,

$$d\sigma_{kl}^{(n)} = r\left(\frac{\partial f}{\partial \sigma_{kl}}\right) \qquad (9\text{-}67)$$

where r is a positive scalar quantity. Hence

$$\left(\frac{\partial f}{\partial \sigma_{kl}}\right) d\sigma_{kl}^{(n)} = \left(\frac{\partial f}{\partial \sigma_{kl}}\right) r\left(\frac{\partial f}{\partial \sigma_{kl}}\right) > 0 \qquad (9\text{-}68)$$

From Eqs. (9-65), (9-66), and (9-68), it follows that

$$\left(\frac{\partial f}{\partial \sigma_{kl}}\right) d\sigma_{kl} = \left(\frac{\partial f}{\partial \sigma_{kl}}\right) d\sigma_{kl}^{(n)} = r\left(\frac{\partial f}{\partial \sigma_{kl}}\right)\left(\frac{\partial f}{\partial \sigma_{kl}}\right) > 0 \qquad (9\text{-}69)$$

Therefore,

$$r = \frac{(\partial f/\partial \sigma_{kl})\, d\sigma_{kl}}{(\partial f/\partial \sigma_{mn})(\partial f/\partial \sigma_{mn})} = D_{kl}\, d\sigma_{kl} \qquad (9\text{-}70)$$

where D_{kl} is a quantity that does not depend on incremental stress $d\sigma_{kl}$.

According to the linearity assumption, however, if all components of $d\sigma_{kl}$ are increased in a given ratio, the corresponding components of $d\epsilon_{ij}^{p}$ are also increased in the same ratio (8). That is,

$$d\epsilon_{ij}^{p} = C_{ijkl}\, d\sigma_{kl} \qquad (9\text{-}71a)$$

where C_{ijkl} may be a function of stress, strain, and the history of loading, but is assumed to be independent of incremental stress, $d\sigma_{kl}$. Comparing Eq. (9-71a) with Eq. (9-70), it can be observed that $d\epsilon_{ij}^{p}$ is proportional to the scalar quantity r. That is,

$$d\epsilon_{ij}^{p} = h_{ij} r \qquad (9\text{-}71b)$$

Substituting for r from Eq. (9-70), we can write

$$d\epsilon_{ij}^{p} = g_{ij}\left(\frac{\partial f}{\partial \sigma_{kl}}\right) d\sigma_{kl} \qquad (9\text{-}71c)$$

where g_{ij} may depend on stress, strain, and history of loading, but does not involve incremental stress, $d\sigma_{kl}$. The second postulate by Drucker [Eq. (9-57c)] can be written as

$$d\sigma_{ij}\,d\epsilon_{ij}^p = \left(d\sigma_{ij}^{(n)} + d\sigma_{ij}^{(t)}\right)d\epsilon_{ij}^p \geqslant 0 \tag{9-72}$$

Since $d\sigma_{ij}^{(t)}$ does not produce any plastic deformations, the stress increment $d\sigma_{ij} = d\sigma_{ij}^{(n)} + C\,d\sigma_{ij}^{(t)}$ also constitutes loading for all values of C, and will produce the same plastic deformations, $d\epsilon_{ij}^p$. Therefore, the condition in Eq. (9-72) becomes

$$C\,d\sigma_{ij}^{(t)}\,d\epsilon_{ij}^p + d\sigma_{ij}^{(n)}\,d\epsilon_{ij}^p \geqslant 0 \tag{9-73}$$

But $d\sigma_{ij}^{(t)}\,d\epsilon_{ij}^p = 0$ and using Eq. (9-71c) we have

$$d\sigma_{ij}^{(t)}\,g_{ij}\frac{\partial f}{\partial \sigma_{kl}}d\sigma_{kl} = 0 \tag{9-74}$$

However, it was stated in Eq. (9-64a) that $(\partial f/\partial \sigma_{kl})\,d\sigma_{kl} > 0$. Therefore, the condition above reduces to

$$d\sigma_{ij}^{(t)}g_{ij} = 0 \tag{9-75}$$

Equation (9-75) must be satisfied by all $d\sigma_{ij}^{(t)}$ that satisfy Eq. (9-66). Therefore,

$$g_{ij} = G\frac{\partial f}{\partial \sigma_{ij}} \tag{9-76}$$

where G is a scalar which may depend on stress, strain, and history of loading. Hence the plastic strain increment $d\epsilon_{ij}^p$ [Eq. (9-71c)] can be written as

$$d\epsilon_{ij}^p = G\frac{\partial f}{\partial \sigma_{ij}}\left(\frac{\partial f}{\partial \sigma_{kl}}\right)d\sigma_{kl} \tag{9-77a}$$

or

$$d\epsilon_{ij}^p = G\frac{\partial f}{\partial \sigma_{ij}}df \tag{9-77b}$$

or

$$d\epsilon_{ij}^p = \lambda\frac{\partial f}{\partial \sigma_{ij}} \tag{9-77c}$$

where $\lambda = G\,df$. Equation (9-77c) is also referred to as the *flow rule associated with the yield criterion f*. The quantity λ is a positive scalar factor of proportionality. Note that Eq. (9-77c) is a special case of Eq. (9-56).

Stress–Strain Relations for Ideal Plastic Materials

Incremental plastic stress–strain models for various materials have been derived and implemented in numerical procedures such as the finite element method; some of the initial works and their reviews are available in Refs.

13–22. In this chapter we present such a derivation for perfectly plastic materials.

For an ideal plastic material the yield function f does not move in the stress space. That is,

$$f(\sigma_{ij}) = k \qquad (9\text{-}78)$$

where k is a constant defining the yield limit. During plastic flow

$$df = \frac{\partial f}{\partial \sigma_{ij}} d\sigma_{ij} = 0 \qquad (9\text{-}79)$$

As shown previously,

$$d\epsilon_{ij}^{p} = \lambda \frac{\partial f}{\partial \sigma_{ij}} \qquad (9\text{-}77\text{c})$$

A basic assumption made in the development of stress–strain relations for elastic–plastic materials is that for each load increment the corresponding strain increment can be decomposed into elastic and plastic components. That is,

$$d\epsilon_{ij} = d\epsilon_{ij}^{e} + d\epsilon_{ij}^{p} \qquad (9\text{-}80)$$

Substituting Eqs. (6-19) and (9-77c) in Eq. (9-80), one can write

$$d\epsilon_{ij} = \left(\frac{dS_{ij}}{2G} + \frac{dJ_1}{9K} \delta_{ij} \right) + \lambda \frac{\partial f}{\partial \sigma_{ij}} \qquad (9\text{-}81\text{a})$$

The inverse relationship corresponding to Eq. (9-81a) can be obtained as

$$dS_{ij} + \frac{dJ_1}{3} \delta_{ij} = 2G \, d\epsilon_{ij} - \frac{2G \, dJ_1}{9K} \delta_{ij} - 2G\lambda \frac{\partial f}{\partial \sigma_{ij}} + \frac{dJ_1}{3} \delta_{ij}$$

or

$$d\sigma_{ij} = 2G \, d\epsilon_{ij} - 2G\lambda \frac{\partial f}{\partial \sigma_{ij}} + \left(\frac{1}{3} - \frac{2G}{9K} \right) dJ_1 \delta_{ij} \qquad (9\text{-}81\text{b})$$

Substituting the value of $d\sigma_{ij}$ from Eq. (9-81b) into Eq. (9-79), it is possible to write

$$2G \frac{\partial f}{\partial \sigma_{ij}} d\epsilon_{ij} - 2G\lambda \frac{\partial f}{\partial \sigma_{ij}} \left(\frac{\partial f}{\partial \sigma_{ij}} \right) + \left(\frac{1}{3} - \frac{2G}{9K} \right) dJ_1 \frac{\partial f}{\partial \sigma_{ij}} \delta_{ij} = 0 \qquad (9\text{-}82)$$

Now, from Eq. (9-81a), we obtain

$$dI_1 = \frac{dJ_1}{3K} + \lambda \frac{\partial f}{\partial \sigma_{ij}} \delta_{ij} \qquad (9\text{-}83\text{a})$$

which gives

$$dJ_1 = 3K\left(dI_1 - \lambda\frac{\partial f}{\partial \sigma_{ij}}\delta_{ij}\right)$$

(9-83b)

Substituting Eq. (9-83b) into Eq. (9-82) and simplifying, we obtain

$$\lambda = \frac{\dfrac{\partial f}{\partial \sigma_{ij}}d\epsilon_{ij} + \dfrac{3K-2G}{6G}dI_1\dfrac{\partial f}{\partial \sigma_{ij}}\delta_{ij}}{\dfrac{\partial f}{\partial \sigma_{ij}}\left(\dfrac{\partial f}{\partial \sigma_{ij}}\right) + \dfrac{3K-2G}{6G}\left(\dfrac{\partial f}{\partial \sigma_{ij}}\delta_{ij}\right)^2}$$

(9-84)

The value of λ can now be substituted into Eq. (9-81a), and we rewrite it as

$$d\epsilon_{ij} = \left(\frac{dS_{ij}}{2G} + \frac{dJ_1}{9K}\delta_{ij}\right) + \left[\frac{\dfrac{\partial f}{\partial \sigma_{mn}}d\epsilon_{mn} + \dfrac{3K-2G}{6G}dI_1\dfrac{\partial f}{\partial \sigma_{mn}}\delta_{mn}}{\dfrac{\partial f}{\partial \sigma_{mn}}\left(\dfrac{\partial f}{\partial \sigma_{mn}}\right) + \dfrac{3K-2G}{6G}\left(\dfrac{\partial f}{\partial \sigma_{mn}}\delta_{mn}\right)^2}\right]\frac{\partial f}{\partial \sigma_{ij}}$$

(9-85)

Substituting Eqs. (9-83b) and (9-84) into Eq. (9-81b), the inverse relationships can be written as

$$d\sigma_{ij} = 2G\,dE_{ij} + K\,dI_1\delta_{ij} - \left[\frac{\dfrac{\partial f}{\partial \sigma_{mn}}d\epsilon_{mn} + \dfrac{3K-2G}{6G}dI_1\dfrac{\partial f}{\partial \sigma_{mn}}\delta_{mn}}{\dfrac{\partial f}{\partial \sigma_{nm}}\left(\dfrac{\partial f}{\partial \sigma_{nm}}\right) + \dfrac{3K-2G}{6G}\left(\dfrac{\partial f}{\partial \sigma_{mn}}\delta_{mn}\right)^2}\right]$$
$$\times \left[\left\{\frac{3K-2G}{3}\frac{\partial f}{\partial \sigma_{mn}}\delta_{mn}\right\}\delta_{ij} + 2G\frac{\partial f}{\partial \sigma_{ij}}\right]$$

(9-86a)

or in matrix notation,

$$\{d\sigma\} = [[\mathbf{C}^e] - [\mathbf{C}^p]]\cdot\{d\epsilon\}$$

(9-86b)

where $[\mathbf{C}^e]$ and $[\mathbf{C}^p]$ are the elastic and plastic stress–strain matrices, respectively. Once the yield function f is known for any material, Eqs. (9-85) and (9-86) could be used to evaluate elastoplastic constitutive relationships. However, for most materials, particularly geological, the yield function can be often expressed in terms of the invariants J_1 and $\sqrt{J_{2D}}$ if isotropic properties are assumed. Therefore, the stress–strain relations given in Eqs. (9-85) and (9-86) can be further simplified as given below.

The yield function can be expressed as

$$f = f\left(J_1, \sqrt{J_{2D}}\right) = k$$

(9-87)

Hence

$$\frac{\partial f}{\partial \sigma_{ij}} = \frac{\partial f}{\partial J_1}\frac{\partial J_1}{\partial \sigma_{ij}} + \frac{\partial f}{\partial \sqrt{J_{2D}}}\frac{\partial \left(\sqrt{J_{2D}}\right)}{\partial \sigma_{ij}}$$

$$= \frac{\partial f}{\partial J_1}\delta_{ij} + \frac{\partial f}{\partial \sqrt{J_{2D}}}\left(\frac{1}{2\sqrt{J_{2D}}}S_{ij}\right) \tag{9-88}$$

Substituting Eq. (9-88) into Eq. (9-85) will yield

$$d\epsilon_{ij} = \frac{dS_{ij}}{2G} + \frac{dJ_1}{9K}\delta_{ij} + \left[\frac{3K\,dI_1\dfrac{\partial f}{\partial J_1} + \dfrac{G}{\sqrt{J_{2D}}}\dfrac{\partial f}{\partial\sqrt{J_{2D}}}S_{mn}\,dE_{mn}}{9K\left(\dfrac{\partial f}{\partial J_1}\right)^2 + G\left(\dfrac{\partial f}{\partial\sqrt{J_{2D}}}\right)^2}\right]$$

$$\times \left[\frac{\partial f}{\partial J_1}\delta_{ij} + \frac{1}{2\sqrt{J_{2D}}}\frac{\partial f}{\partial\sqrt{J_{2D}}}S_{ij}\right] \tag{9-89}$$

and the inverse relationship can be found as

$$d\sigma_{ij} = 2G\,dE_{ij} + K\,dI_1\,\delta_{ij} - \left[\frac{3K\,dI_1\dfrac{\partial f}{\partial J_1} + \dfrac{G}{\sqrt{J_{2D}}}\dfrac{\partial f}{\partial\sqrt{J_{2D}}}S_{mn}\,dE_{mn}}{9K\left(\dfrac{\partial f}{\partial J_1}\right)^2 + G\left(\dfrac{\partial f}{\partial\sqrt{J_{2D}}}\right)^2}\right]$$

$$\times \left[3K\frac{\partial f}{\partial J_1}\delta_{ij} + \frac{G}{\sqrt{J_{2D}}}\frac{\partial f}{\partial\sqrt{J_{2D}}}S_{ij}\right] \tag{9-90}$$

Example 9-1 Tresca Criterion

Let us assume that the maximum and minimum principal stresses are known where $\sigma_1 \geqslant \sigma_2 \geqslant \sigma_3$. The Tresca criterion becomes

$$f = \tfrac{1}{2}(\sigma_1 - \sigma_3) - k \tag{9-91}$$

and from Eq. (9-77c),

$$d\epsilon_{ij}^p = \lambda\frac{\partial f}{\partial \sigma_{ij}}$$

Therefore,

$$\frac{\partial f}{\partial \sigma_1} = \frac{1}{2}, \qquad \frac{\partial f}{\partial \sigma_3} = -\frac{1}{2}, \qquad \frac{\partial f}{\partial \sigma_2} = 0 \tag{9-92}$$

Therefore,

$$d\epsilon_1^p = \tfrac{1}{2}\lambda$$
$$d\epsilon_2^p = 0 \tag{9-93}$$
$$d\epsilon_3^p = -\tfrac{1}{2}\lambda$$

Example 9-2   Prandtl–Reuss Material

Here the yield function is assumed to be the von Mises criterion. That is,

$$f = \sqrt{J_{2D}} - k \tag{9-94}$$

Hence

$$\frac{\partial f}{\partial J_1} = 0 \tag{9-95a}$$

$$\frac{\partial f}{\partial \sqrt{J_{2D}}} = 1 \tag{9-95b}$$

Substituting Eqs. (9-95) into Eq. (9-89), one can write

$$d\epsilon_{ij} = \frac{dS_{ij}}{2G} + \frac{dJ_1}{9K}\delta_{ij} + \frac{S_{mn}\, dE_{mn}}{2J_{2D}} S_{ij} \tag{9-96a}$$

Substituting $J_{2D} = k^2$ into Eq. (9-96a) yields

$$d\epsilon_{ij} = \frac{dS_{ij}}{2G} + \frac{dJ_1}{9K}\delta_{ij} + \frac{S_{mn}\, dE_{mn}}{2k^2} S_{ij} \tag{9-96b}$$

Similarly, Eq. (9-90) can be simplified and written as

$$d\sigma_{ij} = \left(2G\, dE_{ij} + K\, dI_1 \delta_{ij}\right) - G\frac{S_{mn}\, dE_{mn}}{k^2} S_{ij} \tag{9-97}$$

REFERENCES

1. Hill, R., *The Mathematical Theory of Plasticity*, Oxford University Press, London, 1950.

2. Mendelson, A., *Plasticity: Theory and Applications*, Macmillan Publishing Co., Inc., New York, 1968.

3. Kachanov, L. M., *Foundations of Theory of Plasticity*, North-Holland Publishing Company, Amsterdam, 1971.

4. Slater, R. A. C., *Engineering Plasticity and Applications to Metal Forming*, Halsted Press, New York, 1979.

5. Malvern, L. E., *Introduction to Mechanics of a Continuous Medium*, Prentice-Hall, Inc., Englewood Cliffs, N.J., 1969.

6. Bridgeman, P. W., *Studies in Large Plastic Flow and Fracture with Special Emphasis on the Effects of Hydrostatic Pressure*, McGraw-Hill Book Company, New York, 1952.

7. Fung, Y. C., *Foundations of Solid Mechanics*, Prentice-Hall, Inc., Englewood Cliffs, N.J., 1965.

8. Drucker, D. C., "Some Implications of Work Hardening and Ideal Plasticity," *Quart. Appl. Math.*, Vol. 7, 1950, pp. 411–418.

9. Drucker, D. C., "A More Fundamental Approach to Plastic Stress–Strain Relations," *Proc. First U.S. Natl. Congr. Appl. Mech.*, 1951, pp. 487–491.

10. Drucker, D. C., "A Definition of Stable Inelastic Material," *ASME Trans.*, Vol. 81, 1959, pp. 101–106.

11. Prager, W., "Recent Developments in Mathematical Theory of Plasticity," *J. Appl. Phys.*, Vol. 20, No. 3, Mar. 1949, pp. 235–241.

12. Drucker, D. C., "On Uniqueness in the Theory of Plasticity," *Quart. Appl. Math.*, Vol. 14, 1956, pp. 35–42.

13. Reyes, S. F. and Deere, D. U., "Elastic-Plastic Analysis of Underground Openings by the Finite Element Method," *Proc., 1st Int. Cong. Rock Mech.*, Vol. II, Lisbon, 1966, pp. 477–486.

14. Marcal, P. V. and King, I. P., "Elastic-Plastic Analysis of Two Dimensional Stress Analysis Systems by the Finite Element Method," *Int. J. Mech. Sci.*, Vol. 9, 1967, pp. 143–155.

15. Yamada, Y., Kawai, T. and Yoshimura, N., "Analysis of Elastic-Plastic Problems by the Matrix Displacement Method," *Proc., 2nd Conf. on Matrix Meth. in Struct. Mech.*, Wright-Patterson Air Force Base, Ohio, Dec. 1969, pp. 1271–1299.

16. Shieh, W. Y. J. and Sandhu, R. S., "Application of Elasto-Plastic Analysis in Earth Structures," *Proc., Nat. Meeting on Water Resources Eng., ASCE*, Memphis, Jan. 1970.

17. Rohani, R., "Mechanical Constitutive Models for Engineering Materials," *Misc. Paper S-77-19, U.S. Army Engineers Waterways Expt. St.*, Vicksburg, Miss., 1977.

18. Zienkiewicz, O. C., *The Finite Element Method*, 3rd Edition, McGraw-Hill Book Company, New York, 1977.

19. Desai, C. S. and Abel, J. F., *Introduction to the Finite Element Method*, Van Nostrand Reinhold Co., New York, 1972.

20. Desai, C. S. and Christian, J. T. (Eds.), *Numerical Methods in Geotechnical Engineering*, McGraw-Hill Book Company, New York, 1977.

21. Bathe, K. H. and Wilson, E. L., *Numerical Methods in Finite Element Analysis*, Prentice-Hall, Inc., Englewood Cliffs, N.J., 1977.

22. Owen, D. R. J. and Hinton, E., *Finite Elements in Plasticity: Theory and Practice*, Pineridge Press, Swansea, U.K., 1980.

10

PLASTICITY MODELS: I

In Chapter 9 we dealt with materials whose yielding or plastic flow behavior was independent of the first invariant of the stress tensor or mean pressure; these types of materials are called nonfrictional materials. A physical model involving a mass resting on a frictionless surface can be considered equivalent to this model. Here, the frictional force or resistance will be zero for any normal load acting on the mass. On the other hand, if there were friction between the mass and the surface, the frictional resistance will be proportional to the normal load. Similar ideas can be extended to describe the behavior of the internal structure of some materials. Many geologic materials exhibit this type of behavior.

The yield criteria described in Chapter 9 assume that the strength of the material is independent of the spherical or hydrostatic stress. This assumption has been shown to be valid for materials, such as metals, which are considered as frictionless. The behavior of most geologic media can be quite different from that of metals, and their strength is dependent on the hydrostatic stress. Under fully or partially drained conditions, the strength of soil often increases with mean pressure and exhibits frictional characteristics. There are certain exceptions. For instance, the undrained behavior of clays can be similar to that of metals. In this chapter we consider frictional materials and their stress–strain behavior based on the postulates described in Chapter 9.

MOHR–COULOMB FAILURE CRITERION

According to the Mohr–Coulomb criterion, the shear strength increases with increasing normal stress on the failure plane:

$$\tau = c + \sigma \tan \phi \qquad (10\text{-}1)$$

where τ is the shear stress on the failure plane, c the cohesion of the material, σ the normal effective stress on the failure surface, and ϕ the angle of internal friction. This failure criterion is shown graphically in Fig. 10-1. The concept of Mohr circle can be used to express the criterion in terms of principal

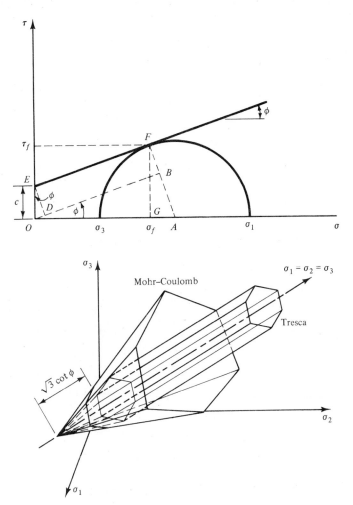

Figure 10-1 Mohr–Coulomb criterion.

stresses. With reference to Fig. 10-1, we can write

$$\frac{\sigma_1 - \sigma_3}{2} = AB + BF \tag{10-2a}$$

That is,

$$\frac{\sigma_1 - \sigma_3}{2} = OA \sin \phi + c \cos \phi \tag{10-2b}$$

Hence

$$\frac{\sigma_1 - \sigma_3}{2} = \frac{\sigma_1 + \sigma_3}{2} \sin \phi + c \cos \phi \tag{10-2c}$$

Here σ_1 and σ_3 are the major and minor principal stresses, respectively. As can be seen in Eq. (10-2c), the Mohr–Coulomb criterion ignores the effects of intermediate principal stress. Equation (10-2c) can also be written as

$$\sqrt{\left(\frac{\sigma_{11} - \sigma_{33}}{2}\right)^2 + \sigma_{12}^2} = \frac{\sigma_{11} + \sigma_{33}}{2} \sin \phi + c \cos \phi \tag{10-3}$$

Equation (10-2c) represents an irregular hexagonal pyramid in the stress space; the projection of this surface on the Π-plane is shown in Fig. 10-2; appropriate quantities (major and minor principal stresses) are selected and substituted in Eq. (10-2c) to obtain these plots. In a subsequent section we consider a generalization of the Mohr–Coulomb criterion, and the corresponding constitutive law.

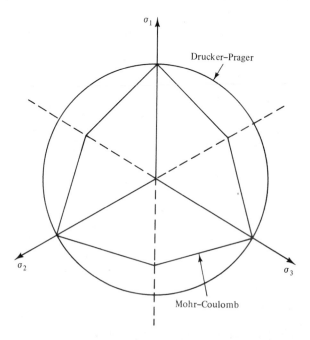

Figure 10-2   Mohr–Coulomb and Drucker–Prager criteria on the Π-plane.

According to the Mohr–Coulomb criterion, the yield strength in compression is higher than that in extension; this shows the dependence of the behavior on the third invariant of the stress tensor. At the same time, the Mohr–Coulomb criterion is expressed in terms of maximum and minimum principal stresses, and hence does not incorporate the effects of intermediate principal stresses. Therefore, it is inconvenient to express the Mohr–Coulomb criterion in terms of a general three-dimensional state of stress defined by six components of stress vector. Hence description of the Mohr–Coulomb criterion in terms of conventional forms of stress invariants (Chapter 4) becomes difficult.

An alternative form of invariants (1) can be used to define failure and yield criteria such as the Mohr–Coulomb. The alternative set of invariants include a quantity θ defined as (1,2)

$$\theta = -\tfrac{1}{3} \sin^{-1}\left(-\frac{3\sqrt{3}}{2} \frac{J_{3D}}{J_{2D}^{3/2}} \right) \tag{10-4a}$$

and

$$-\frac{\pi}{6} \leqslant \theta \leqslant \frac{\pi}{6} \tag{10-4b}$$

where J_{2D} and J_{3D} are invariants of the deviatoric stress tensor defined in Chapter 4. The alternative set of invariants J_1, J_{2D}, and θ could be used in expressing the Mohr–Coulomb criterion conveniently in a three-dimensional stress space as

$$f = J_1 \sin\phi + \sqrt{J_{2D}}\, \cos\theta - \frac{\sqrt{J_{2D}}}{3} \sin\phi \sin\theta - c\cos\phi = 0 \tag{10-5}$$

It should be noted that when $\phi = 0$, the Mohr–Coulomb criterion reduces to the Tresca criterion described in Chapter 9.

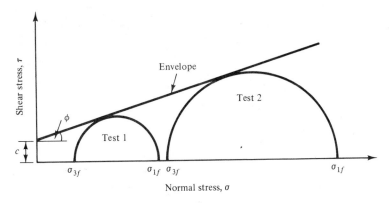

Figure 10-3 Parameters of Mohr–Coulomb model, where σ_{3f} is the minor principal stress at failure and σ_{1f} is the major principal stress at failure.

Determination of Parameters

The two parameters associated with the Mohr–Coulomb failure criterion, angle of internal friction ϕ and cohesion c, can be determined from laboratory tests on the material conducted up to the ultimate or failure conditions. The Mohr–Coulomb failure line is the envelope of Mohr circles at failure, and the values of ϕ and c could be determined as shown in Figs. 10-1 and 10-3. Conventional triaxial compression tests with cylindrical specimens are commonly used in determining the cohesion and angle of internal friction. A minimum of two tests is required to determine c and ϕ. A numerical example of determination of c and ϕ is given subsequently in this chapter.

DRUCKER–PRAGER MODEL

A generalization to account for the effects of all principal stresses was suggested by Drucker and Prager (3) by using the invariants of the stress tensor. This generalized criterion can be written as

$$f = \sqrt{J_{2D}} - \alpha J_1 - k \tag{10-6}$$

where α and k are positive material parameters, J_1 is the first invariant of the stress tensor, and J_{2D} is the second invariant of the deviatoric stress tensor. Equation (10-6) represents a straight line on a J_1 versus $\sqrt{J_{2D}}$ plot (Fig. 10-4). In the three-dimensional stress space, the criterion plots as a right circular cone, and the projection on the Π-plane is a circle, as shown in Fig. 10-2. When the state of stress reaches the failure surface [Eq. (10-6)], the material undergoes plastic deformations. According to the criterion, a state of stress

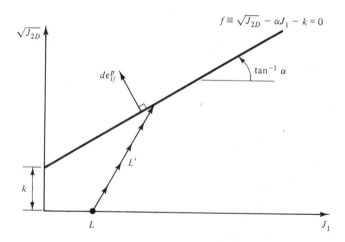

Figure 10-4   Drucker–Prager criterion.

outside the surface is not stable. The material could undergo plastic deformations while the stress point is moving on the failure surface.

Determination of Parameters

The two material parameters α and k for the Drucker–Prager model can be determined from the slope and intercept of failure envelope plotted on the $J_1-\sqrt{J_{2D}}$ space (Fig. 10-4). In order to establish the failure envelope for a material it is necessary to perform laboratory tests up to the ultimate or failure conditions. For this purpose, laboratory equipment such as the conventional triaxial device, plane strain device, or truly triaxial device described in Chapter 5 can be used.

In this text, ultimate condition is usually defined as the asymptotic value(s) of stress to the final range of a stress-strain curve. This is also considered as the state of ultimate yield. Failure is often assumed to be a state in the final range such as peak stress and value of stress corresponding to a chosen strain state. The critical state (Chapter 11) refers to the condition in the final range when no volume change occurs. The failure and critical states may coincide with the ultimate state or may correspond to a lower state of stress.

The values of α and k can be expressed in terms of angle of internal friction ϕ and cohesion c. However, the values of ϕ and c determined by using conventional triaxial compression tests are different from those determined under plane strain conditions. For these conditions the values of α and k could be expressed as follows (4–6):

Conventional triaxial compression:

$$\alpha = \frac{2\sin\phi}{\sqrt{3}\,(3-\sin\phi)} \tag{10-7a}$$

$$k = \frac{6c\cos\phi}{\sqrt{3}\,(3-\sin\phi)} \tag{10-7b}$$

Plane strain condition:

$$\alpha = \frac{\tan\phi}{\left(9+12\tan^2\phi\right)^{1/2}} \tag{10-8a}$$

$$k = \frac{3c}{\left(9+12\tan^2\phi\right)^{1/2}} \tag{10-8b}$$

A numerical example for derivation of material parameters for the Drucker–Prager model is given later in this chapter.

For use of the von Mises, Mohr–Coulomb, and Drucker–Prager criteria in the context of conventional incremental plasticity theory, the failure criteria is used as the yield criteria. Then the incremental stress–strain relations can be derived based on the particular yield criterion and the concepts of normality and flow described in Chapter 9. For example, in the case of the Drucker–Prager model, f in Eq. (10-6) can be used as the yield criterion together with the flow rule [Eq. (9-77)]. In the following section, we derive the incremental stress–strain relation for the Drucker–Prager model; here the results are expressed in matrix notation so that they can be implemented in numerical solution procedures such as the finite element method.

DERIVATION OF INCREMENTAL EQUATIONS

Details of steps in deriving the incremental law for the Drucker–Prager model are given below. The steps are self-explanatory, and hence minimum commentary is included.

When the stress point is on the yield surface, Eq. (10-6) is always satisfied, and hence the variation in f is zero. That is,

$$df = 0 \tag{10-9a}$$

By using Eq. (10-9a), it is possible to write

$$df = \frac{\partial f}{\partial \sigma_{ij}} d\sigma_{ij} = 0 \tag{10-9b}$$

Substitution of Eq. (10-6) into Eq. (10-9b) leads to

$$df = \left(\frac{S_{ij}}{2\sqrt{J_{2D}}} - \alpha\delta_{ij} \right) d\sigma_{ij} = 0 \tag{10-9c}$$

where S_{ij} is the deviatoric stress tensor and the other quantities are the same as those defined in previous chapters. Assuming that the total strain $d\epsilon_{ij}$ can be separated into elastic and plastic components, we can write

$$d\epsilon_{ij}^e = d\epsilon_{ij} - d\epsilon_{ij}^p \tag{10-10a}$$

Here the superscripts e and p denote elastic and plastic components, respectively. However, the incremental plastic strains can be expressed by using the flow rule given in Eq. (9-77). By substituting the flow rule in Eq. (10-10a), we can write

$$d\epsilon_{ij}^e = d\epsilon_{ij} - \lambda \frac{\partial f}{\partial \sigma_{ij}} \tag{10-10b}$$

Then combining Eq. (10-10b) with Eq. (10-9c), we have

$$d\epsilon_{ij}^e = d\epsilon_{ij} - \lambda \left(\frac{S_{ij}}{2\sqrt{J_{2D}}} - \alpha\delta_{ij} \right) \tag{10-11}$$

Now, let us consider incremental form of the stress–strain relationship [Eq. (6-15a)], which is based on the theory of elasticity:

$$d\sigma_{ij} = K\, d\epsilon_{mm}^e \delta_{ij} + 2G\left(d\epsilon_{ij}^e - \frac{d\epsilon_{mm}^e}{3}\delta_{ij}\right) \tag{10-12}$$

From Eq. (10-11), the incremental volumetric strain can be expressed as

$$d\epsilon_{mm}^e = d\epsilon_{mm} + 3\lambda\alpha \tag{10-13}$$

Substitution of this in Eq. (10-12) leads to

$$d\sigma_{ij} = K\left[d\epsilon_{mm} + 3\lambda\alpha\right]\delta_{ij}$$
$$+ 2G\left[d\epsilon_{ij} - \frac{\lambda S_{ij}}{2\sqrt{J_{2D}}} + \lambda\alpha\delta_{ij} - \frac{d\epsilon_{mm}}{3}\delta_{ij} - \lambda\alpha\delta_{ij}\right] \tag{10-14a}$$

Hence

$$d\sigma_{ij} = K\left[d\epsilon_{mm} + 3\lambda\alpha\right]\delta_{ij} + 2G\left[d\epsilon_{ij} - \frac{\lambda S_{ij}}{2\sqrt{J_{2D}}} - \frac{d\epsilon_{mm}}{3}\delta_{ij}\right] \tag{10-14b}$$

Substitute Eq. (10-14b) into Eq. (10-9c) to yield

$$\left(\frac{S_{ij}}{2\sqrt{J_{2D}}} - \alpha\delta_{ij}\right)(d\epsilon_{mm} + 3\lambda\alpha)K\delta_{ij}$$
$$+ 2G\left(\frac{S_{ij}}{2\sqrt{J_{2d}}} - \alpha\delta_{ij}\right)\left(d\epsilon_{ij} - \frac{\lambda S_{ij}}{2\sqrt{J_{2D}}} - \frac{d\epsilon_{mm}}{3}\delta_{ij}\right) = 0 \tag{10-15a}$$

By noting that $S_{ij} \cdot \delta_{ij} = S_{ii}$, $\delta_{ij} \cdot \delta_{ij} = 3$, and $S_{ii} = 0$, Eq. (10-15a) leads to

$$-3\alpha(d\epsilon_{mm} + 3\lambda\alpha)K + \frac{GS_{ij}d\epsilon_{ij}}{\sqrt{J_{2D}}} - \lambda G = 0 \tag{10-15b}$$

Collecting terms in Eq. (10-15b), we have

$$\lambda(9\alpha^2 K + G) = \frac{GS_{ij}d\epsilon_{ij}}{\sqrt{J_{2D}}} - 3K\alpha\, d\epsilon_{mm} \tag{10-16}$$

Substituting $S_{ij} = \sigma_{ij} - (J_1/3)\delta_{ij}$ in Eq. (10-16), we obtain

$$\lambda = \frac{\sigma_{ij}d\epsilon_{ij} - d\epsilon_{mm}\left(\dfrac{J_1}{3} + \dfrac{3K\alpha\sqrt{J_{2D}}}{G}\right)}{\sqrt{J_{2D}}\,(1 + 9\alpha^2 K/G)} \tag{10-17}$$

Now let us define

$$p' = \frac{\sqrt{J_{2D}}}{k}\left(1 + \frac{9\alpha^2 K}{G}\right) \tag{10-18}$$

where p' is a nondimensional quantity. From the criterion in equation (10-6), we obtain

$$J_1 = \frac{\sqrt{J_{2D}} - k}{\alpha} \tag{10-19}$$

Let us consider the factor $(J_1/3) + (3K\alpha\sqrt{J_{2D}}/G)$ in Eq. (10-17). Using Eq. (10-19), we can write

$$\frac{J_1}{3} + \frac{3K\alpha\sqrt{J_{2D}}}{G} = \frac{\sqrt{J_{2D}} - k}{3\alpha} + \frac{3K\alpha\sqrt{J_{2D}}}{G} \tag{10-20a}$$

By simplifying Eq. (10-20a) we can obtain

$$\frac{J_1}{3} + \frac{3K\alpha\sqrt{J_{2D}}}{G} = \frac{1}{3\alpha}\left[\sqrt{J_{2D}}\left(1 + \frac{9K\alpha^2}{G}\right) - k\right] \tag{10-20b}$$

Substituting the value of p' from Eq. (10-18) in Eq. (10-20b) leads to

$$\frac{J_1}{3} + \frac{3K\alpha\sqrt{J_{2D}}}{G} = \frac{k}{3\alpha}(p' - 1) \tag{10-20c}$$

Substitution of Eq. (10-20c) in Eq. (10-17) leads to

$$\lambda = \frac{\sigma_{ij}\,d\epsilon_{ij} - (k/3\alpha)\,d\epsilon_{mm}(p' - 1)}{p'k} \tag{10-21}$$

Now consider Eq. (10-14b) and substitute in it the value of λ from Eq. (10-21) to obtain

$$d\sigma_{ij} = d\epsilon_{mm}\left(K - \frac{2G}{3}\right)\delta_{ij} + 2G\,d\epsilon_{ij}$$
$$+ \left[\frac{\sigma_{mn}\,d\epsilon_{mn} - (k/3\alpha)(p' - 1)\,d\epsilon_{mm}}{p'k}\right]\left[3\alpha K\delta_{ij} - \frac{GS_{ij}}{\sqrt{J_{2D}}}\right] \tag{10-22a}$$

Simplifying Eq. (10-22a), we can write

$$d\sigma_{ij} = d\epsilon_{mm}\left(K - \frac{2G}{3}\right)\delta_{ij} + 2G\,d\epsilon_{ij} + \frac{3\alpha K}{p'k}\sigma_{mn}\,d\epsilon_{mn}\delta_{ij}$$
$$- \frac{G\sigma_{mn}\,d\epsilon_{mn}}{p'k\sqrt{J_{2D}}}S_{ij} - \frac{K}{p'}\,d\epsilon_{mm}(p' - 1)\delta_{ij} + \frac{G\,d\epsilon_{mm}(p' - 1)}{3\alpha p'\sqrt{J_{2D}}}S_{ij} \tag{10-22b}$$

Because $S_{ij} = \sigma_{ij} - (J_1/3)\delta_{ij}$, Eq. (10-22b) becomes

$$d\sigma_{ij} = d\epsilon_{mm}\left(K - \frac{2G}{3}\right)\delta_{ij} + 2G\,d\epsilon_{ij} + \frac{3\alpha K}{p'k}\sigma_{mn}\,d\epsilon_{mn}\,\delta_{ij}$$

$$- \frac{G\sigma_{mn}\,d\epsilon_{mn}}{p'k\sqrt{J_{2D}}}\sigma_{ij} + \frac{G\sigma_{mn}\,d\epsilon_{mn}\,J_1\delta_{ij}}{3p'k\sqrt{J_{2D}}} - \frac{K(p'-1)}{p'}d\epsilon_{mm}\,\delta_{ij}$$

$$+ \frac{G\,d\epsilon_{mm}(p'-1)}{3\alpha p'\sqrt{J_{2D}}}\sigma_{ij} - \frac{G\,d\epsilon_{mm}(p'-1)J_1\delta_{ij}}{9\alpha p'\sqrt{J_{2D}}} \qquad (10\text{-}23\text{a})$$

Collecting terms in Eq. (10-23a) yields

$$d\sigma_{ij} = 2G\,d\epsilon_{ij} - \left[\frac{2G}{3}\left(1 - \frac{3K}{2G}\right) + \frac{k(p'-1)}{p'} + \frac{G(p'-1)J_1}{9\alpha p'\sqrt{J_{2D}}}\right]d\epsilon_{mn}\,\delta_{mn}\delta_{ij}$$

$$+ \left[\frac{3\alpha K}{p'k} + \frac{GJ_1}{3p'k\sqrt{J_{2D}}}\right]\sigma_{mn}\,d\epsilon_{mn}\,\delta_{ij} + \left[\frac{G(p'-1)}{3\alpha p'\sqrt{J_{2D}}}\right]\delta_{mn}\,d\epsilon_{mn}\,\sigma_{ij}$$

$$- \frac{G\sigma_{mn}\,d\epsilon_{mn}}{p'k\sqrt{J_{2D}}}\sigma_{ij} \qquad (10\text{-}23\text{b})$$

Hence

$$d\sigma_{ij} = 2G\,d\epsilon_{ij} - 2G\left[\frac{1}{3}\left(1 - \frac{3K}{2G}\right) + \frac{k}{2Gp'}(p'-1) + \frac{(p'-1)J_1}{18p'\alpha\sqrt{J_{2D}}}\right]d\epsilon_{mn}\,\delta_{mn}\delta_{ij}$$

$$+ \frac{2G}{p'k}\left[\frac{3\alpha K}{2G} + \frac{J_1}{6\sqrt{J_{2D}}}\right]\sigma_{mn}\,d\epsilon_{mn}\,\delta_{ij}$$

$$+ \frac{2G(p'-1)}{6\alpha p'\sqrt{J_{2D}}}\delta_{mn}\,d\epsilon_{mn}\,\sigma_{ij} - 2G\frac{\sigma_{mn}\,d\epsilon_{mn}\sigma_{ij}}{3p'k\sqrt{J_{2D}}} \qquad (10\text{-}23\text{c})$$

Let us now define a nondimensional quantity, h, as

$$h = -\left(\frac{3\alpha K}{2G} + \frac{J_1}{6\sqrt{J_{2D}}}\right) \qquad (10\text{-}24)$$

We have seen that

$$\frac{k}{3\alpha}(p'-1) = \frac{J_1}{3} + \frac{3K\alpha\sqrt{J_{2D}}}{G} \qquad (10\text{-}20\text{c})$$

Hence the value of h in Eq. (10-24) can be expressed as

$$h = -\frac{k(p'-1)}{6\alpha\sqrt{J_{2D}}} \qquad (10\text{-}25)$$

Substitution of Eq. (10-25) into Eq. (10-23c) leads to

$$d\sigma_{ij} = 2G\,d\epsilon_{ij} - 2G\left[\frac{1}{3}\left(1-\frac{3K}{2G}\right)+\frac{K}{2Gp'}(p'-1)+\frac{(p'-1)J_1}{18\alpha\sqrt{J_{2D}}\,p'}\right]d\epsilon_{mn}\,\delta_{mn}\delta_{ij}$$

$$-\frac{2Gh}{p'k}\left(\sigma_{mn}\delta_{ij}+\sigma_{ij}\delta_{mn}\right)d\epsilon_{mn}-\frac{2G\sigma_{mn}\sigma_{ij}d\epsilon_{mn}}{2p'k\sqrt{J_{2D}}} \tag{10-26}$$

Hence Eq. (10-26) can be written as

$$d\sigma_{ij} = 2G\,d\epsilon_{ij} - 2G\left[A\left(\sigma_{mn}\delta_{ij}+\sigma_{ij}\delta_{mn}\right)+B\delta_{mn}\delta_{ij}+C\sigma_{mn}\sigma_{ij}\right]d\epsilon_{mn} \tag{10-27}$$

where

$$A = \frac{h}{p'k} \tag{10-28}$$

$$p' = \frac{\sqrt{J_{2D}}}{k}\left(1+9\alpha^2k/G\right) \tag{10-29}$$

$$h = \frac{-k(p'-1)}{6\alpha\sqrt{J_{2D}}} = -\left(\frac{3\alpha K}{2G}+\frac{J_1}{6\sqrt{J_{2D}}}\right) \tag{10-30}$$

$$B = \frac{2h^2\sqrt{J_{2D}}}{p'k}-\frac{3\nu K}{E} = \frac{2h^2}{1+9\alpha^2K/G}-\frac{3\nu K}{E} \tag{10-31}$$

$$C = \frac{1}{2kp'\sqrt{J_{2D}}} \tag{10-32}$$

The matrix form of Eq. (10-27) can be written as

$$\{d\sigma\} = [C^{ep}]\{d\epsilon\} \tag{10-33}$$

where $[C^{ep}]$ is the elastoplastic constitutive matrix which is expanded in Table 10-1.

Limitations of Drucker–Prager Model

The incremental plastic strains for the Drucker–Prager yield criterion can be seen from Eq. (10-11) as

$$d\epsilon_{ij}^p = \lambda\left(\frac{S_{ij}}{2\sqrt{J_{2D}}}-\alpha\delta_{ij}\right) \tag{10-34a}$$

The incremental plastic volumetric strains can be found from Eq. (10-34a) as

$$d\epsilon_{ii}^p = -3\alpha\lambda \tag{10-34b}$$

As shown in Fig. 10-4, the incremental plastic strain vector has a negative volumetric component, and its magnitude is given by Eq. (10-34b). This

TABLE 10-1 Stress–Strain Matrix for Ducker–Prager Material Model

$$
\begin{Bmatrix} d\sigma_{11} \\ d\sigma_{22} \\ d\sigma_{33} \\ d\sigma_{12} \\ d\sigma_{23} \\ d\sigma_{13} \end{Bmatrix} = 2G
\begin{bmatrix}
1 - T_1\sigma_{11} - R_1 & -(T_1\sigma_{22} + R_1) & -(T_1\sigma_{33} + R_1) & -(T_1\sigma_{12}) & -(T_1\sigma_{23}) & -(T_1\sigma_{13}) \\
-(T_2\sigma_{11} + R_2) & 1 - T_2\sigma_{22} - R_2 & -(T_2\sigma_{33} + R_2) & -(T_2\sigma_{12}) & -(T_2\sigma_{23}) & -(T_2\sigma_{13}) \\
-(T_3\sigma_{11} + R_3) & -(T_3\sigma_{22} + R_3) & 1 - T_3\sigma_{33} - R_3 & -(T_3\sigma_{12}) & -(T_3\sigma_{23}) & -(T_3\sigma_{13}) \\
-T_1\sigma_{12} & -T_2\sigma_{12} & -T_3\sigma_{12} & \tfrac{1}{2} - C\sigma_{12}^2 & -C\sigma_{12}\sigma_{23} & -C\sigma_{12}\sigma_{13} \\
-T_1\sigma_{23} & -T_2\sigma_{23} & -T_3\sigma_{23} & -C\sigma_{12}\sigma_{23} & \tfrac{1}{2} - C\sigma_{23}^2 & -C\sigma_{13}\sigma_{23} \\
-T_1\sigma_{13} & -T_2\sigma_{13} & -T_3\sigma_{13} & -C\sigma_{12}\sigma_{13} & -C\sigma_{13}\sigma_{23} & \tfrac{1}{2} - C\sigma_{13}^2
\end{bmatrix}
\begin{Bmatrix} d\epsilon_{11} \\ d\epsilon_{22} \\ d\epsilon_{33} \\ d\gamma_{12} \\ d\gamma_{23} \\ d\gamma_{13} \end{Bmatrix}
$$

where $T_a = A + C\sigma_{aa}$ and $R_a = A\sigma_{aa} + B$; $a = 1, 2, 3$. Here a repeated subscript *does not* mean summation.

indicates volume increase or dialation at failure. However, experimental data on normally consolidated clays and loose sands indicates only compressive deformations or decrease in volume during shear, which is in contradiction with predictions from the Drucker–Prager model. This discrepancy may be due to several reasons. It is possible that the normality rule may not be valid here and the Drucker–Prager model may not be applicable to these materials.

Some problems can be solved without using the normality rule in the Drucker–Prager criterion. Here the stresses satisfy the yield criterion, while the strain states are forced to satisfy certain conditions such as no volume changes. The major difficulty in discarding normality is that it implies that the material is unstable according to Drucker's postulates (7). The limitations of the Drucker–Prager model have also been discussed by Christian and Desai (4).

Hardening Behavior

It has been observed that many geologic materials experience plastic deformations almost from the very start of loading. In other words, if the material is unloaded from a given state during loading, it does not return to its original configuration, and experiences inelastic or plastic deformations. Thus during a loading path $L–L'$ (Fig. 10-4), the material undergoes a process of continuous yielding until it finally reaches the conventional failure or ultimate state defined by the function f (Fig. 10-4), which may be considered as the final yield surface.

Furthermore, even during hydrostatic loading, many materials experience plastic deformations. For instance, as indicated in Fig. 10-5(b), upon unloading from state A to state A', the material undergoes volumetric plastic strains. When reloaded from A', the material may behave elastically up to state A and then start experiencing plastic deformations.

It is possible to define the foregoing behavior by defining a series of yield surfaces prior to reaching the failure or the final yield surface. Since the behavior is plastic under hydrostatic loading, the successive yield surfaces should intersect the J_1 axis [Fig. 10-5(a)]. This idea was first proposed by Drucker, et al. (8) with respect to the behavior of a soil specimen under conventional triaxial test.

During the successive yielding, the material generally hardens, and exhibits strain or work hardening behavior. For such a hardening material, when the stress point moves beyond the current yield surface, a new yield surface is established. The yield surfaces that define the hardening behavior are often called *hardening caps* [Fig. 10-5(c)].

In the context of plasticity, the yield surfaces should be convex (see Chapter 9). For simplicity, Drucker et al. (8) assumed, as an approximation, the hardening cap to be a circular arc. As we shall see subsequently, the shapes of these surfaces were assumed to be different by different investigators. In

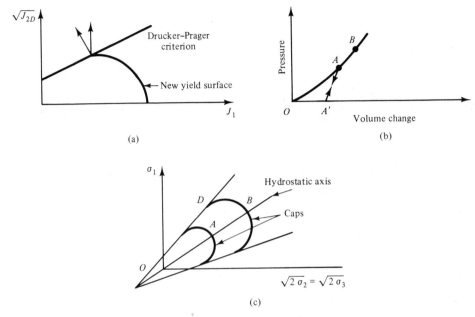

Figure 10-5 Hardening behavior of geologic media: (a) yield surface intersecting J_1-axis; (b) behavior under hydrostatic loading and unloading; (c) hardening caps.

actuality, the shape of the surface should be determined from appropriate laboratory tests for a given material. This idea of successive yield surfaces can explain the hardening behavior of geologic materials (under hydrostatic loading), and can permit the use of a conventional normality rule.

A new yield surface intersects the Drucker–Prager failure surface at an angle [Fig. 10-5(a)]. It is possible to explain the observed volumetric behavior of geologic media if we assume that the incremental plastic strain vector is normal to the hardening yield surface at the point of intersection, D. When the hydrostatic stress is increased to bring the specimen to point B as shown in Fig. 10-5, the yield curve must also expand to point B. Hence there will be a family of yield curves corresponding to points on the hydrostatic axes. In other words, as the material work hardens, the yield surface expands to a new position. However, when the state of stress reaches a point on the yield surface which is locally parallel to the hydrostatic axis, the plastic volumetric changes will be zero and hence no further hardening will take place. This leads to the fact that incremental plastic strain vector at point D [Fig. 10-5(c)], is normal to the curved yield surface as well as the hydrostatic axis.

Details of various approaches for the introduction of successive yield surfaces are given in Chapter 11. In the remainder of this chapter we present examples of determination of parameters for the Tresca, von Mises, Mohr–Columb, and Drucker–Prager models covered in this chapter and Chapter 9.

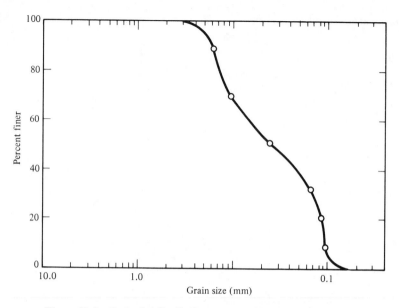

Figure 10-6 Grain size distribution curve for silty sand (Refs. 10, 11).

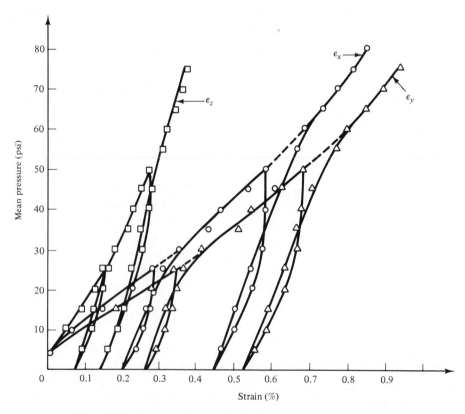

Figure 10-7 Stress–strain curves for hydrostatic compression stress path.

DETERMINATION OF PARAMETERS

Example 10-1. Parameters for a Silty Sand

The gradation curve for the soil considered here is shown in Fig. 10-6. The soil has a specific gravity of 2.59 and the water content in the field of about 9.0%. The truly triaxial tests were conducted under various stress paths with an initial density of about 2.00 g/cm³. Further details of the soil and the testing program are given in Refs. 9 to 12. The test results for hydrostatic tests are shown in Fig. 10-7, and those for CTC, RTE, CTE, RTC, TC, TE, and SS tests are shown in Figs. 10-8 to 10-14; they are adopted from Refs. 9 to 12.

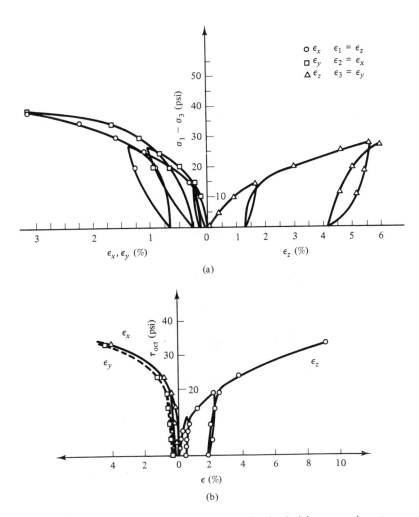

Figure 10-8 Stress–strain curves from conventional triaxial compression stress path. (a) $\sigma_0 = 10$ psi; (b) $\sigma_0 = 20$ psi.

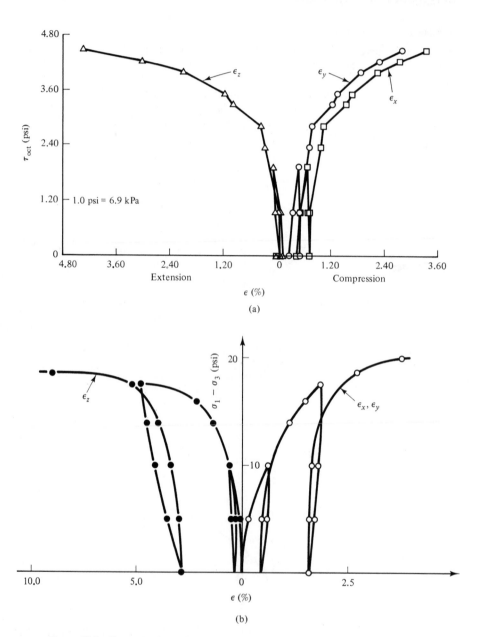

Figure 10-9 Stress–strain curves from RTE stress path: (a) $\sigma_0 = 10$ psi; (b) $\sigma_0 = 20$ psi.

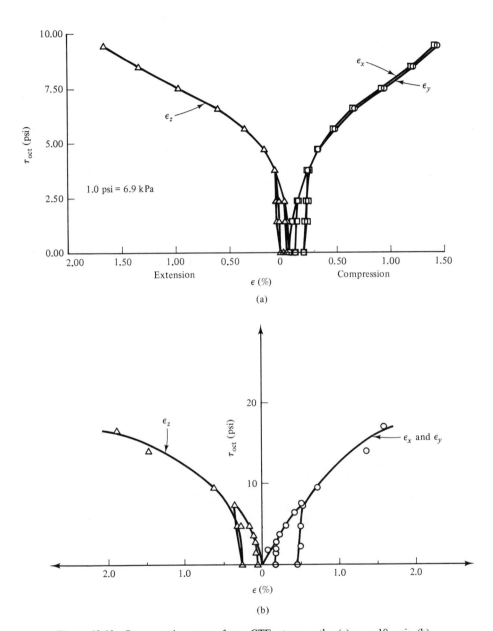

Figure 10-10   Stress–strain curves from CTE stress path: (a) $\sigma_0 = 10$ psi; (b) $\sigma_0 = 20$ psi.

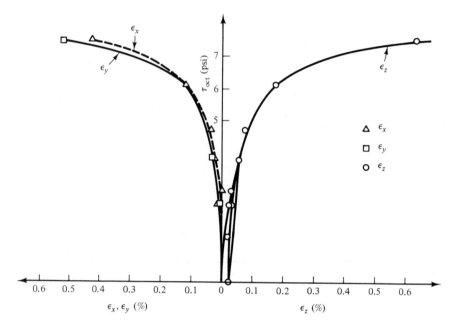

Figure 10-11   Stress–strain curves from RTC stress path: $\sigma_0 = 20$ psi.

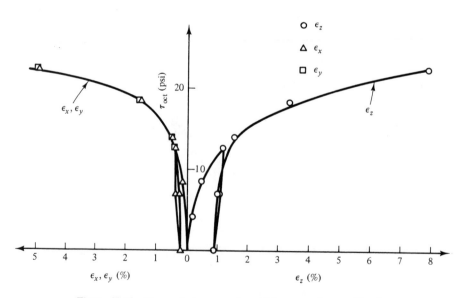

Figure 10-12   Stress–strain curves from TC stress path: $\sigma_0 = 25$ psi.

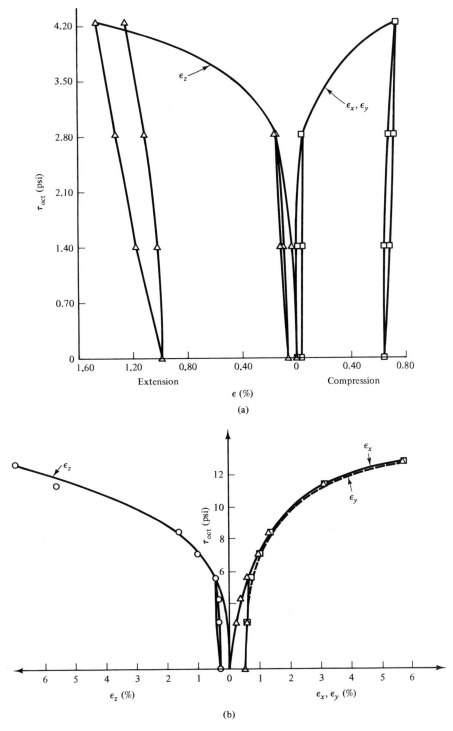

Figure 10-13 Stress–strain curves from TE stress path: (a) $\sigma_0 = 10$ psi; (b) $\sigma_0 = 20$ psi.

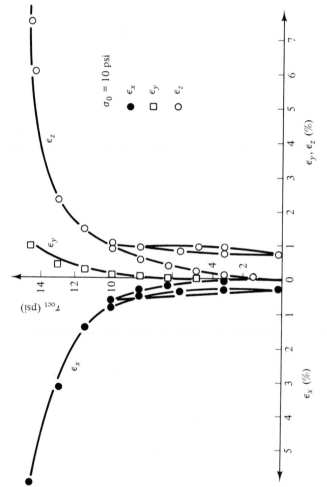

Figure 10-14 Stress–strain curves from SS path: $\sigma_0 = 10$ psi.

260

It is required to find the parameters for the Tresca, von Mises, Mohr–Coulomb, and Drucker–Prager criteria. The derivation of parameters will be performed first (procedure I) by using the two CTC tests with one of the RTE tests. Then in procedure II, the parameters are obtained by using a number of tests. Only a limited number of tests are used here, although all available data are included for additional use by the reader. The parameters determined from both the approaches are compared to assess the influence of using a greater number of test results for finding the parameters.

I. Computations using two CTC tests and one RTE test:

1. *CTC Test,* $\sigma_0 = \sigma_{oct} = 10$ *psi (8.69 kPa) (1 psi = 6.89 kPa).* From Fig. 10-8(a), we have

$$(\sigma_1 - \sigma_3) \text{ ultimate} \simeq 40 \text{ psi}(276 \text{ kPa})$$

$$\sigma_1 = 40 + \sigma_3 = 40 + 10 = 50$$

$$\sigma_1 + \sigma_3 = 50 + 10 = 60$$

$$\frac{\sigma_1 - \sigma_3}{2} = 20$$

$$\frac{\sigma_1 + \sigma_3}{2} = 30$$

$$J_{2D} = \tfrac{1}{6}\left[(\sigma_1 - \sigma_2)^2 + (\sigma_2 - \sigma_3)^2 + (\sigma_3 - \sigma_1)^2\right]$$

$$= \tfrac{1}{6}\left[(50 - 10)^2 + (10 - 50)^2\right] = 533$$

$$\sqrt{J_{2D}} = 23$$

$$J_1 = 50 + 10 + 10 = 70$$

2. *CTC Test,* $\sigma_0 = 20$ *psi (138 kPa) [Fig. 10-8(b)]:*

$$(\tau_{\text{oct}})_u = 33 \text{ psi } (228 \text{ kPa})$$

$$(\sigma_1 - \sigma_3) = 70$$

$$\sigma_1 = 70 + 20 = 90$$

$$\frac{\sigma_1 + \sigma_3}{2} = \frac{90 + 20}{2} = 55$$

$$\frac{\sigma_1 - \sigma_3}{2} = \frac{90 - 20}{2} = 35$$

$$\sqrt{J_{2D}} = 40.4$$

$$J_1 = 90 + 20 + 10 = 130$$

3. *RTE Test,* $\sigma_0 = 20$ *psi (138 kPa) [Fig. 10-9(b)]:*

$$(\tau_{oct})_u = 9 \text{ psi } (62 \text{ kPa})$$

$$\tau_{oct} = \tfrac{1}{3}\sqrt{(\sigma_1 - \sigma_2)^2 + (\sigma_2 - \sigma_3)^2 + (\sigma_3 - \sigma_1)^2}$$

$$= \frac{\sqrt{2}}{3}(\sigma_3 - \sigma_1)$$

$$(\tau_{oct})_u = 9 = \frac{\sqrt{2}}{3}(\sigma_3 - \sigma_1)$$

$$\sigma_1 = \sigma_3 - \frac{27}{\sqrt{2}} = 1$$

$$\frac{\sigma_1 + \sigma_3}{2} = 10.5$$

$$\frac{\sigma_1 - \sigma_3}{2} = 9.6$$

$$\sqrt{J_{2D}} = 11$$

$$J_1 = 41$$

Figures 10-15(a) and (b) show plots of $(\sigma_1 + \sigma_3)/2$ versus $(\sigma_1 - \sigma_3)/2$ and J_1 versus $\sqrt{J_{2D}}$, respectively, from procedure I above using two CTC tests and one RTE test. The parameters for various models are evaluated as follows by using these figures:

Tresca (Chapter 9):

$$\frac{\sigma_1 - \sigma_3}{2}\bigg|_{average} = k = 21.5 \text{ psi } (149 \text{ kPa})$$

Mohr–Coulomb:

$$\frac{\sigma_1 - \sigma_3}{2} = \frac{\sigma_1 + \sigma_3}{2}\sin\phi + c\cos\phi$$

From Fig. 10-15(a), the equation of the failure envelope is given by

$$\frac{\sigma_1 - \sigma_3}{2} = k + \frac{\sigma_1 + \sigma_3}{2}\tan\theta$$

Therefore, $k = c\cos\phi$; and

$$\frac{\sigma_1 + \sigma_3}{2}\tan\theta = \frac{\sigma_1 + \sigma_3}{2}\sin\phi$$

or

$$\tan\theta = \sin\phi$$

and

$$\phi = \sin^{-1}(\tan\theta)$$

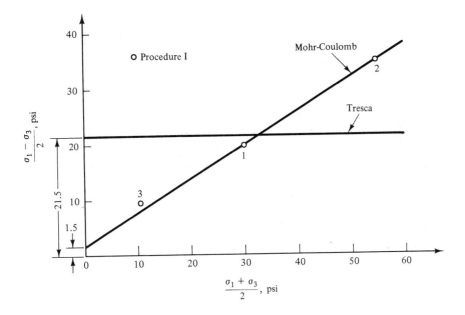

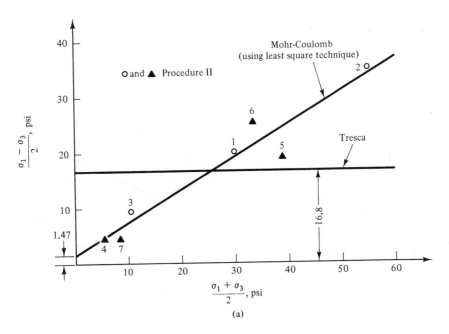

Figure 10-15 (a) Envelopes for Tresca and Mohr–Coulomb models; (b) envelopes for von Mises and Drucker–Prager models.

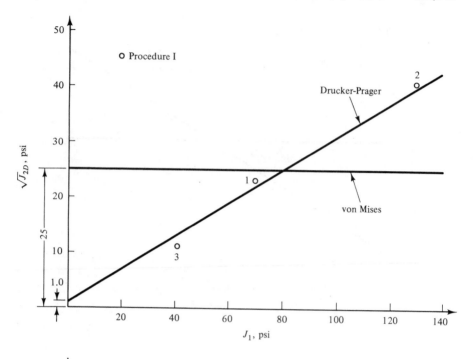

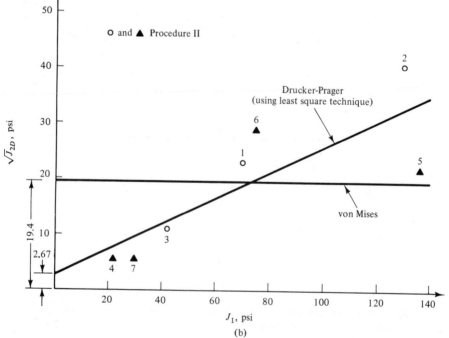

(b)

Figure 10-15 (*continued*)

From Fig. 10-15(a),

$$\tan \theta \simeq 0.608$$

$$\phi = 38°$$

$$k = 1.5$$

$$c = \frac{k}{\cos \phi} = \frac{1.5}{\cos 38} = 1.90 \text{ psi } (13.1 \text{ kPa})$$

The plot of J_1 versus $\sqrt{J_{2D}}$ [Fig. 10-15(b)] provides parameters for the von Mises and Drucker–Prager criteria.

von Mises:

$$\sqrt{J_{2D}}\Big|_{\text{average}} = k = 25.0 \text{ psi } (172 \text{ kPa})$$

Drucker–Prager:

$$\sqrt{J_{2D}} - \alpha J_1 - k = 0$$

$$k = 1.0 \text{ psi}$$

$$\alpha = \frac{d\sqrt{J_{2D}}}{dJ_1} = \frac{30.8}{100.00} = 0.308$$

Note: k in von Mises and Drucker–Prager models is different from k in the Tresca and Mohr–Coulomb models.

II. Computations using other available tests:

4. *RTE Test, $\sigma_0 = 10$ psi (69 kPa):*

$$(\tau_{\text{oct}})_u = 4.20 \text{ psi } (29 \text{ kPa})$$

$$= \frac{\sqrt{2}}{3}(\sigma_3 - \sigma_1)$$

Hence

$$\sigma_3 - \sigma_1 = 9.0$$

$$\sigma_1 = 10.0 - 9.0 = 1.0$$

$$\frac{\sigma_3 - \sigma_1}{2} = 4.5$$

$$\frac{\sigma_1 + \sigma_3}{2} = 5.5$$

$$\sqrt{J_{2D}} = 5.2$$

$$J_1 = 21.0$$

5. *CTE Test, $\sigma_0 = 20$ psi (138 kPa):*

$$(\tau_{\text{oct}})_u = 18 \text{ psi } (124 \text{ kPa})$$

$$= \frac{\sqrt{2}}{3}(\sigma_3 - \sigma_1)$$

Hence

$$\sigma_3 - \sigma_1 = 38.0$$

$$\sigma_2 = \sigma_3 = 38 + 20 = 58$$

$$\frac{\sigma_3 - \sigma_1}{2} = 19$$

$$\frac{\sigma_1 + \sigma_3}{2} = 39$$

$$\sqrt{J_{2D}} = \frac{\sigma_3 - \sigma_1}{\sqrt{3}} = 22$$

$$J_1 = 58 + 58 + 20 = 136$$

6. *TC Test, $\sigma_0 = 25$ psi (172 kPa):*

$$(\tau_{\text{oct}})_u = 24 \text{ psi}$$

$$= \frac{\sqrt{2}}{3}(\sigma_1 - \sigma_3)$$

Hence

$$\sigma_1 - \sigma_3 = \frac{72}{\sqrt{2}} = 51$$

For the TC test, we increase, say σ_1 by $\Delta\sigma$, and decrease σ_2 and σ_3 by $\Delta\sigma/2$:

$$\sigma_1 - \sigma_0 = \Delta\sigma$$

$$\tfrac{3}{2}\Delta\sigma = 51$$

$$\Delta\sigma = 34$$

$$\sigma_1 = 59$$

$$\sigma_2 = \sigma_3 = 25 - \frac{34}{2} = 8$$

$$\frac{\sigma_1 - \sigma_3}{2} = \frac{59 - 8}{2} = 25.5$$

$$\frac{\sigma_1 + \sigma_3}{2} = \frac{59 + 8}{2} = 33.5$$

$$\sqrt{J_{2D}} = 29$$

$$J_1 = 75$$

7. *TE Test*, $\sigma_0 = 10$ psi (69 kPa):

$$(\tau_{oct})_u = 4.30$$

$$= \frac{\sqrt{2}}{3}(\sigma_3 - \sigma_1)$$

$$\sigma_3 - \sigma_1 = 9.1$$

$$\Delta\sigma = 6.0$$

$$\sigma_1 = 10 - 6 = 4$$

$$\sigma_2 = \sigma_3 = 10 + \frac{6}{2} = 13$$

$$\frac{\sigma_3 - \sigma_1}{2} = 4.5$$

$$\frac{\sigma_1 + \sigma_3}{2} = 8.5$$

$$\sqrt{J_{2D}} = \frac{\sigma_3 - \sigma_1}{\sqrt{3}} = 5.3$$

$$J_1 = 30$$

Figure 10-15(a) shows a plot of $(\sigma_1 + \sigma_3)/2$ versus $(\sigma_1 - \sigma_3)/2$ for all the tests above, and Fig. 10-15(b) shows the corresponding plot of J_1 versus $\sqrt{J_{2D}}$. The parameters for various models from the figures are found as follows:

Tresca [Fig. 10-15(a)]:

$$\left.\frac{\sigma_1 - \sigma_3}{2}\right|_{average} = k = 16.8 \text{ psi } (115 \text{ kPa})$$

Mohr–Coulomb [Fig. 10-15(a)]:

$$\frac{\sigma_1 - \sigma_3}{2} = \frac{\sigma_1 + \sigma_3}{2}\sin\phi + c\cos\phi$$

$$\phi = \sin^{-1}(\tan\theta)$$

By using the least square technique, the values of $\tan\theta$ and k are calculated as

$$\tan\theta = 0.59$$

$$k = 1.47$$

Hence,

$$\phi = 36°$$

$$c = \frac{k}{\cos\phi} = \frac{1.47}{\cos 36} = 1.82 \text{ psi}$$

Von Mises [Fig. 10-15(b):

$$\left.\sqrt{J_{2D}}\right|_{average} = k = 19.4 \text{ psi}$$

Drucker–Prager [Fig. 10-15(b)]:

$$\sqrt{J_{2D}} - \alpha J_1 - k = 0$$

By using the least square technique, the values of k and α are found to be

$$k = 2.67 \text{ psi}$$
$$\alpha = 0.23$$

Summary:

	Procedure I	Procedure II
Tresca	$k = 21.5^*$	$k = 16.8^*$
von Mises	$k = 25.0$	$k = 19.4$
Mohr–Coulomb	$c = 1.9, \phi = 38°$	$c = 1.82, \phi = 36°$
Drucker–Prager	$k = 1.0; \alpha = 0.308$	$k = 2.67; \alpha = 0.23$

*psi

It can be seen that the parameters obtained by using a greater number of tests are different from those from a smaller number of tests.

APPLICATIONS

Models based on plasticity theory described in this chapter have been used in solving various boundary value problems in geomechanics. In this section a few examples are presented for the purpose of showing their applications.

Example 10-2   Strip Footing on Tresca Material

One early example of an analysis of a strip footing on a Tresca-type soil is given in Ref. 13. Here the Tresca plasticity model has been implemented in conjunction with a finite difference procedure. A typical example of a strip footing is shown in Fig. 10-16; the

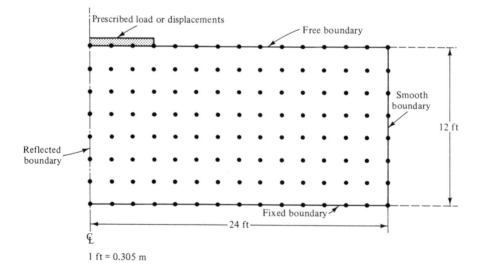

1 ft = 0.305 m

Figure 10-16   Footing on half-space (Ref. 13).

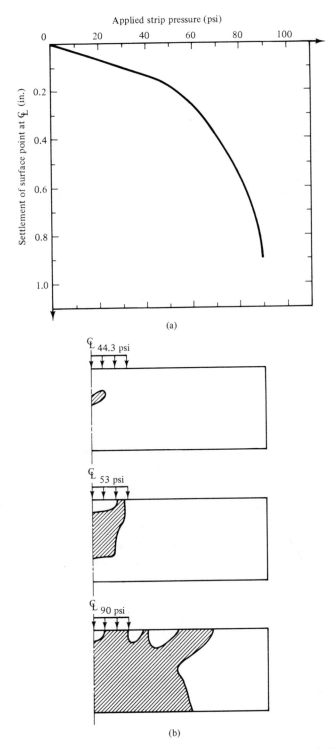

Figure 10-17 (a) Load–displacement curve and (b) plastic zones (Ref. 13).

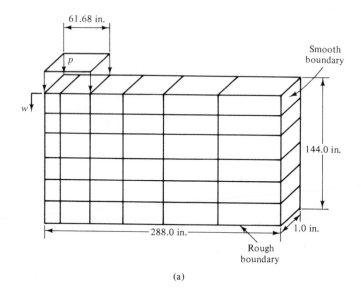

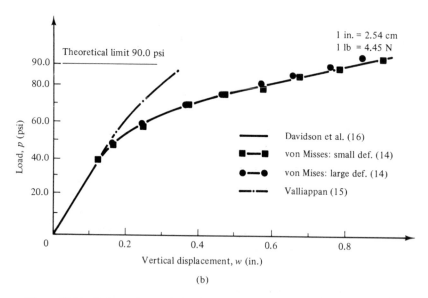

Figure 10-18 Deformation analysis of strip footing: (a) finite element mesh; (b) load versus displacement plot (Ref. 14).

geometry and boundary conditions are also shown in this figure. The properties used in the analysis are:

$$E = 30,000 \text{ psi} \left(207,000 \text{ kN/m}^2\right)$$

$$\nu = 0.3$$

$$c = 17.5 \text{ psi} \left(120.75 \text{ kN/m}^2\right)$$

The computed load–displacement curve and the spread of plastic zones with increasing load are shown in Figs. 10-17(a) and (b), respectively. Closed-form solutions for the failure pressure of a punch into a perfectly plastic material has been calculated as $(2 + \pi)c$, which is equal to 90 psi (621 kN/m²). The agreement between this value and the numerically computed ultimate load [Fig. 10-17(a)] is considered good.

Example 10-3   Strip Footing on von Mises Material

An example problem of a soil mass subjected to a strip footing load was analyzed by Phan (14) by assuming undrained conditions and by using the von Mises constitutive model. The model has been implemented in conjunction with a three-dimensional finite element procedure for large deformation analysis, and a typical finite element

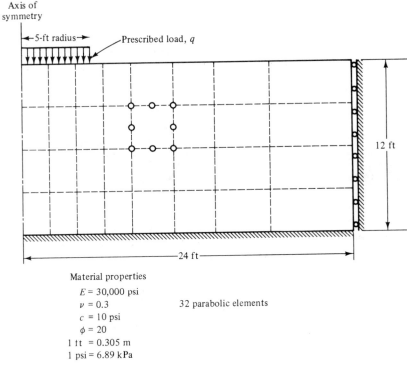

Figure 10-19   Circular footing, geometric data, and mesh (Ref. 17).

mesh is shown in Fig. 10-18(a). The parameters used for the soil are:

$$E = 30,000 \text{ psi } (207,000.0 \text{ kN/m}^2)$$
$$\nu = 0.3$$
$$c = 17.5 \text{ psi } (120.75 \text{ kN/m}^2)$$
$$\phi = 0°$$

The nonlinear analysis was carried out using the Newton–Raphson technique, and the computed load–displacement curve at the center of the footing is shown in Fig. 10-18(b). The predictions are compared with the results obtained previously (15, 16) by using a two-dimensional procedure with constant-strain triangular elements for the case of small deformations for this problem. The predictions with small- and large-strain assumptions appear to yield similar results. This could be due to the fact that for this problem, the deformations in the material are governed predominantly by the material nonlinearies rather than by geometric nonlinearity.

Example 10-4   Circular Footing on Mohr–Coulomb Material

An application of the Mohr–Coulomb criterion to an investigation of the stability of a circular footing is given by Zienkiewicz and Humpheson (17). This constitutive model has been implemented in a two-dimensional finite element procedure, and a typical mesh used in the study is shown in Fig. 10-19. The geometry, boundary conditions, and material properties are also shown on the figure. Details of the implementation are given in Ref. 17, and only representative results are presented here. Figure 10-20 shows

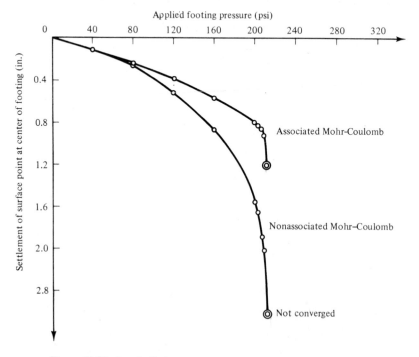

Figure 10-20   Load–displacement curves for circular footing (Ref. 17).

the predicted load–displacement curves for both the associated and nonassociated (these terms are defined elsewhere in the book) Mohr–Coulomb plasticity models. The computed collapse load for both cases was the same, although the displacements were significantly different.

Example 10-5   Lateral Earth Pressure in Mohr–Coulomb Material

An example of the application of the Mohr–Coulomb criterion for the analysis of lateral earth pressure is given by Christian et al. (18). A two-dimensional finite element procedure has been used in their investigation, and Fig. 10-21 shows a typical finite element mesh for the retaining wall problem. The computed load–displacement curves for a smooth wall are shown in Fig. 10-22. Here two procedures have been used for computing wall loads. The first uses the stress in the elements immediately behind the wall integrated over the wall height to obtain the load on the wall. The second uses nodal forces to obtain the load on the wall. The load computed based on element stresses compare well with the Rankine theory. The three curves for each of the foregoing procedures shown in Fig. 10-22 correspond to different schemes for correcting the stresses (18).

Example 10-6   Strip footing on Drucker–Prager Material

Details of computational algorithms for implementation of a Drucker–Prager model in conjunction with finite element procedures for two- and three-dimensional problems are given by Siriwardane and Desai (19). In one of the examples solved, the nonlinear behavior of a strip footing was analyzed using the three-dimensional finite element procedure. Plane strain conditions were simulated in the three-dimensional analysis by constraining the deformations to one plane. A simulated plane strain (constrained three-dimensional) section and the finite element mesh are shown in Fig. 10-23. Here constrained three-dimensional analysis was carried out by using 8- and 16-node brick elements. Furthermore, the analysis was also carried out using the two-dimensional procedure with the 8-node quadrilateral element. In the constrained analysis, 8-node bricks simulate 4-node plane elements, and 16-node bricks simulate 8-node plane elements. Computed nonlinear responses from three- and two-dimensional analysis are compared below.

Figure 10-24 shows the results obtained using the Drucker–Prager model with the following properties:

$$E = 10,000.0 \text{ psi } \left(690.0 \text{ kN/m}^2\right)$$

$$\nu = 0.3$$

$$c = 15.0 \text{ psi } \left(103.5 \text{ kN/m}^2\right)$$

$$\phi = 20.0°$$

Results obtained by using 8-node quadrilateral elements compare well with those obtained with 16-node constrained three-dimensional elements. This can be expected since the number of degrees of freedom in the plane of deformation is the same for both analyses. Use of 8-node constrained three-dimensional elements seems to give a stiffer response and a higher ultimate value. The ultimate values predicted by using 8-node plane and 16-node constrained three-dimensional elements compare well with that obtained from Terzaghi's (20) bearing capacity formula (Fig. 10-24); the bearing capacity is 265.0 psi (1828.5 kN/m^2).

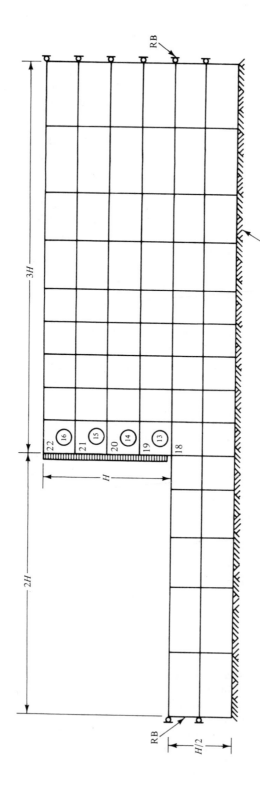

Figure 10-21 Finite element mesh for retaining wall: **RB**, reflected boundary; **FB**, fixed boundary; $H = 10$ m (Ref. 18).

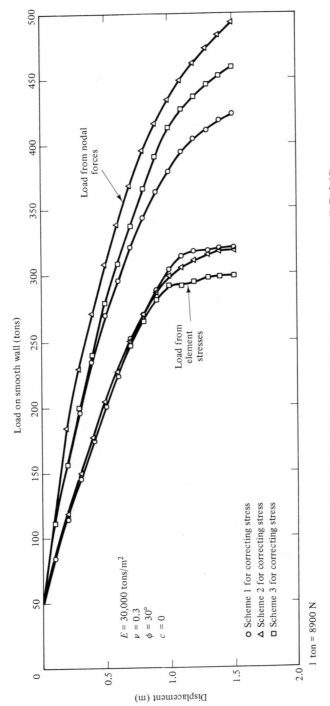

Figure 10-22 Load on displacement behavior of smooth retaining wall (Ref. 18).

$E = 30,000$ tons/m²
$\nu = 0.3$
$\phi = 30°$
$c = 0$

o Scheme 1 for correcting stress
△ Scheme 2 for correcting stress
□ Scheme 3 for correcting stress

1 ton = 8900 N

Load from nodal forces

Load from element stresses

Load on smooth wall (tons)

Displacement (m)

275

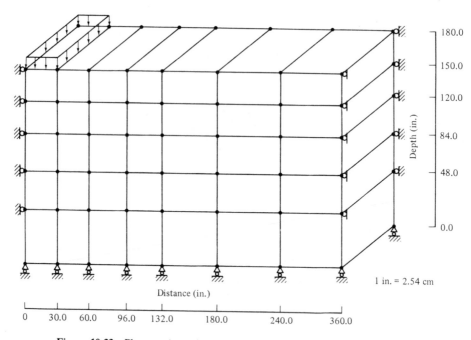

Figure 10-23 Plane strain analysis using a three-dimensional procedure (simulated plane strain) (Ref. 19).

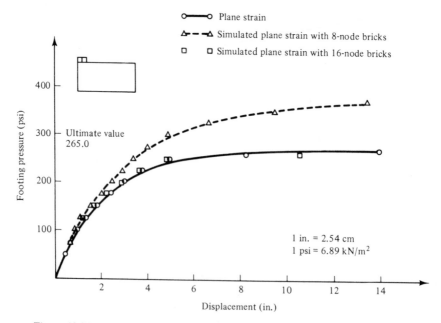

Figure 10-24 Load–displacement behavior of strip footing with Drucker–Prager model (Ref. 19).

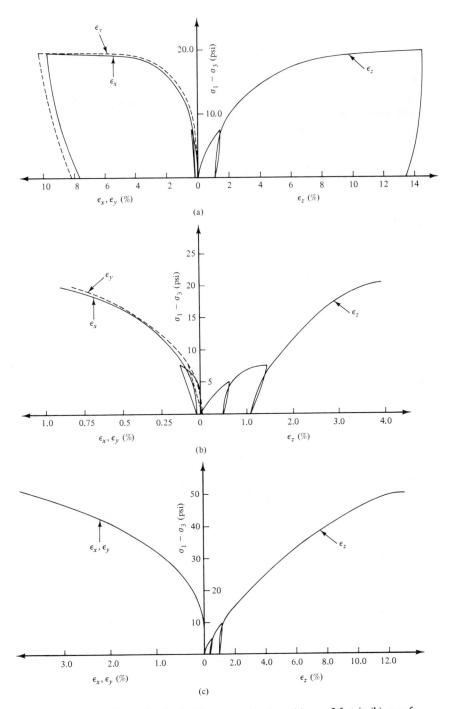

Figure 10-25 Conventional triaxial compression test: (a) $\sigma_0 = 2.5$ psi; (b) $\sigma_0 = 5$ psi; (c) $\sigma_0 = 10$ psi.

PROBLEM

10-1. Figures 10-25 to 10-28 show stress–strain results for three CTC tests and one TC, TE, and SS test each for a cohesive soil. Figure 10-29 (for a HC test) is included for the sake of completeness. Compute parameters for Tresca, von Mises, Mohr–Coulomb, and Drucker–Prager criteria first by using the three CTC tests, and then by adding a number of other tests. Compare the parameters obtained from the two procedures.

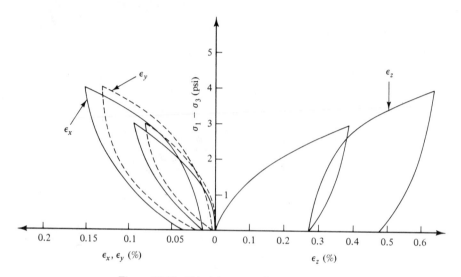

Figure 10-26   Triaxial compression test: $\sigma_0 = 5$ psi.

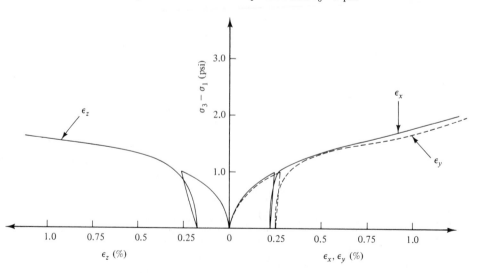

Figure 10-27   Triaxial extension test: $\sigma_0 = 5$ psi.

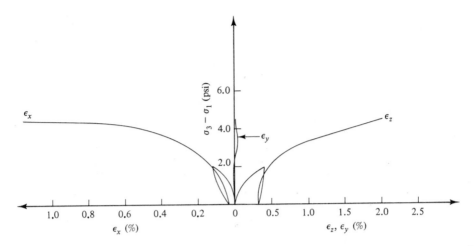

Figure 10-28   Simple shear test: $\sigma_0 = 5$ psi.

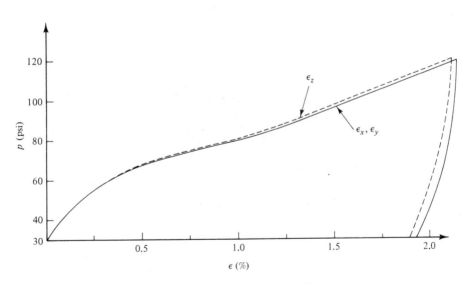

Figure 10-29   Hydrostatic compression test.

REFERENCES

1. Nayak, G. C., and Zienkiewicz, O. C., "A Convenient Form of Invariants and Its Application in Plasticity," *Proc. ASCE*, Vol. 98, No. ST4, 1972, pp. 949–954.
2. Lode, W., "Versuche über den Einfluss der mitt leren Hauptspannung auf das Fliessen der Metalle, Eisen, Kuper und Nickel," *Z. Phys.*, Vol. 36, 1926, pp. 913–939.
3. Drucker, D. C., and Prager, W., "Soil Mechanics and Plastic Analysis or Limit Design," *Quart Appl. Math.*, Vol. 10, No. 2, 1952, pp. 157–165.
4. Christian, J. T., and Desai, C. S., "Constitutive Laws for Geologic Media," in *Numerical Methods in Geotechnical Engineering*, C. S. Desai and J. T. Christian (Eds.), McGraw-Hill Book Company, New York, 1977, Chapter 2.
5. Reyes, S. F., "Elastic–Plastic Analysis of Underground Openings by the Finite Element Method," Ph.D. thesis, University of Illinois, Urbana-Champaign, 1966.
6. Shieh, W. J. J., and Sandhu, R. S., "Application of Elasto-plastic Analysis in Earth Structures," *Proc. Natl. Meet. Water Resources Eng.*, ASCE, Memphis, Tenn., Jan. 1970.
7. Drucker, D. C., "Some Implications of Work Hardening and Ideal Plasticity," *Quart. Appl. Math.*, Vol. 7, No. 4, 1950, pp. 411–418.
8. Drucker, D. C., Gibson, R. E., and Henkel, D. J., "Soil Mechanics and Work Hardening Theories of Plasticity," *Proc. ASCE*, Vol. 81, Paper 798, Sept. 1955.
9. Siriwardane, H. J., "Nonlinear Soil–Structure Interaction Analysis of One-, Two-, and Three-Dimensional Problems Using Finite Element Method," Ph.D. dissertation, Va. Polytech. Inst. and State Univ., Blacksburg, 1980.
10. Janardhanam, R., "Constitutive Laws of Materials in Track Support Structures," Ph.D. dissertation, Va. Polytech. Inst. and State Univ., Blacksburg, 1981.
11. Desai, C. S., Siriwardane, H. J., and Janardhanam, R., "Interaction and Load Transfer in Track-Guide-Way Systems," Nos. DOT/RSPA/DMA-50/83/11 & 12, Dept. of Transportation, Washington, D.C., 1982.
12. Munster, C. L., "Constitutive Modelling of a Granular Soil under Three-Dimensional States of Stress," M.S. thesis, Va. Polytech. Inst. and State Univ., Blacksburg, 1981.
13. Höeg, K., Christian, J. T., and Whitman, R. V., "Settlement of Strip Load on Elastic–Plastic Soil," *J. Soil Mech. Found. Div.*, *ASCE*, Vol. 94, No. SM2, 1968, pp. 431–445.
14. Phan, H. V., "Geometric and Material Nonlinear Analysis of Three-Dimensional Soil–Structure Interaction," Ph.D. dissertation, Va. Polytech. Inst. and State Univ., Blacksburg, 1979.
15. Valliappan, S., Discussion, *J. Soil Mech. Found. Div.*, *ASCE*, Vol. 95, No. SM2, 1969, pp. 676–678.
16. Davidson, H. L., and Chen, W. F., "Elastic–Plastic Large Deformation Response of Clay to Footing Loads," *Rep. 355-18*, Fritz Eng. Lab., Lehigh Univ., Bethlehem, Pa., 1974.

17. Zienkiewicz, O. C., and Humpheson, C., "Viscoplasticity: A Generalized Model for Soil Behavior," in *Numerical Methods in Geotechnical Engineering*, C. S. Desai and J. T. Christian (Eds.) McGraw-Hill Book Company, New York, 1977, Chapter 3.

18. Christian, J. T., Hagmann, A. J., and Marr, W. A., "Incremental Plasticity Analysis of Frictional Soils," *Int. J. Numer. Anal. Methods Geomech.*, Vol. 1, No. 4, 1977.

19. Siriwardane, H. J., and Desai, C. S., "Computational Procedures for Nonlinear Three-Dimensional Analysis with Some Advanced Constitutive Laws," *Int. J. Numer. Anal. Methods Geomech.*, Vol. 7, No. 2, 1983, pp. 143–171.

20. Terzaghi, K., and Peck, R. B., *Soil Mechanics in Engineering Practice*, 2nd ed., John Wiley & Sons, Inc., New York, 1967.

11

PLASTICITY MODELS: II

In this chapter we present models that allow for the continuous yielding and hardening behavior introduced at the end of Chapter 10. We consider primarily the Cam clay and cap models, based on the critical state concepts.

CAM CLAY MODELS BASED ON THE CRITICAL STATE CONCEPT

A scientific study of yielding of soils can be considered to have started with the works of Rendulic (1) and Hvorslev (2). Here it was observed that the constant void ratio contours on the triaxial plane (σ_1 versus $\sqrt{2}\,\sigma_2 = \sqrt{2}\,\sigma_3$) were the same under undrained and drained conditions. Subsequently, the findings of Rendulic and Hvorslev were investigated by Roscoe et al. (3–7), who proposed models for the yielding of soils based on the theory of plasticity. Some of the important parameters used in the development of critical state models were p, q, and e. With respect to the cylindrical triaxial configuration, these parameters take the forms described below.

For the axisymmetric triaxial conditions, $\sigma_2 = \sigma_3$ and $\epsilon_2 = \epsilon_3$; hence the work done on a test specimen per unit volume, is given by

$$dW = \sigma_1\, d\epsilon_1 + 2\sigma_3\, d\epsilon_3$$

$$= \left(\frac{\sigma_1 + 2\sigma_3}{3}\right)(d\epsilon_1 + 2\,d\epsilon_3) + (\sigma_1 - \sigma_3)\frac{2}{3}(d\epsilon_1 - d\epsilon_3) \qquad (11\text{-}1)$$

Now, let us define the quantities as follows:

$$p = \frac{\sigma_1 + 2\sigma_3}{3} = \frac{J_1}{3} = \text{mean pressure} \qquad (11\text{-}2a)$$

$$q = \sigma_1 - \sigma_3 = \sqrt{3J_{2D}} \qquad (11\text{-}2b)$$

$$d\epsilon_v = d\epsilon_1 + 2\,d\epsilon_3 = \text{volumetric strain} \qquad (11\text{-}2c)$$

$$d\epsilon_s = \tfrac{2}{3}(d\epsilon_1 - d\epsilon_3) = \text{deviatoric strain} \qquad (11\text{-}2d)$$

Therefore, the quantity dW can be written as

$$dW = p\,d\epsilon_v + q\,d\epsilon_s \qquad (11\text{-}3)$$

Concept of Critical Void Ratio (Critical Density)

When a loose soil sample is sheared, it passes through progressive states of yielding before reaching a state of collapse. That is, the stress path passes through several yield surfaces (hardening caps), causing plastic deformations. The yielding continues to occur until the material reaches a critical void ratio, after which the void ratio remains constant during subsequent deformations (Fig. 11-1). That is, the material will reach a state in which the arrangement of

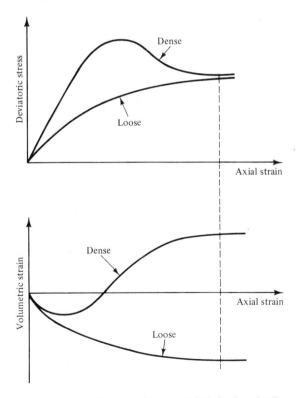

Figure 11-1 Schematic of stress–strain behavior of soils.

the particles is such that no volume change takes place during shearing. This particular void ratio is called the *critical void ratio* (8); this can be considered as the critical state of the material.

When a dense soil sample is sheared, it reaches a peak stress as shown in Fig. 11-1, and then reaches a residual stress. Initially the material compacts, and then it dilates until the volumetric strain reaches a constant value which corresponds to its critical value. This concept is similar to the idea of critical density proposed by Casagrande (8). It has been observed that a soil with a

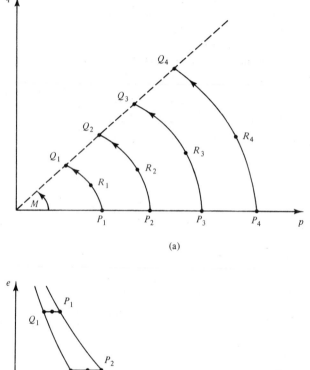

(a)

(b)

Figure 11-2 Drained and undrained behavior: (a) typical undrained stress paths; (b) undrained stress paths on e–p plot (Ref. 3).

void ratio lower than the critical value deforms in such a manner as to increase its volume, whereas at a void ratio higher than the critical value the deformations will decrease the volume.

In an attempt to study the yielding behavior of normally consolidated clays, Roscoe and coworkers (3-7) analyzed several tests on samples of a saturated clay. Typical results of several undrained tests are shown schematically in Fig. 11-2. As can be seen, the stress paths were geometrically similar, and their ultimate states were observed to lie on a straight line on the q–p plot. The ultimate states Q_1, Q_2, Q_3, and Q_4 were observed to lie on a curve which was similar to the isotropic consolidation line on the e–p plot [Fig. 11-2(b)]. It is interesting to note that on the e–$\ln p$ plot, these two curves were observed to be essentially parallel to each other (Fig. 11-3).

The results from a series of drained triaxial tests on the same soil have shown that the ultimate states (or failure points) lie on the same critical void ratio line observed in the undrained behavior. The slope of this line on the q–p plot is denoted by 'M', which is a material parameter. The stress paths followed in drained tests are shown in Fig. 11-4. This shows that failure takes place when the material reaches a critical void ratio, and as discussed previously, the critical state line (CSL) is parallel to the isotropic consolidation line on e–$\ln p$ plot (Fig. 11-3).

Figure 11-5 shows a schematic representation of the behavior of a soil as a function of the shear stress q, normal stress p, and void ratio e. The material can sustain and pass through any state (p, q, e) below the surface formed by two surfaces that intersect at the critical state. The surface $ABCD$ is referred to as the state boundary surface or Roscoe surface, while BB^1C^1C is referred to as the Hvorslev surface. States of wet, normally consolidated soil lie below the

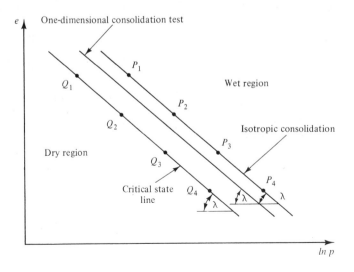

Figure 11-3 e–$\ln p$ plot (Refs. 3–7).

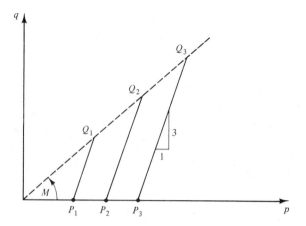

Figure 11-4   Drained stress paths.

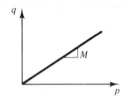

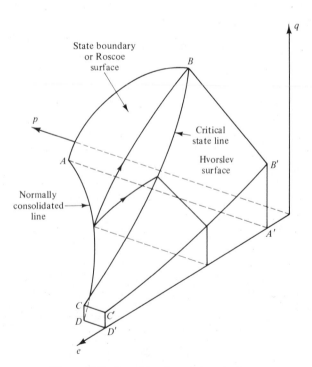

Figure 11-5   State boundary surface (Ref. 3).

state boundary surface. On the other hand, dry and overconsolidated soils exhibit behavior on the right side of the CSL.

Before the material reaches the critical state, it passes through successive states of yielding; during this yielding, the material hardens. This continuously yielding state can lead to the critical state or ultimate yield state, at which the material flows without changes in volume and in shear stress.

Figure 11-6 shows a projection of the critical state line on the q–p space together with projections of typical sections of the state boundary surface. The projection of the CSL is usually a straight line passing through the origin. The projection of the state boundary surface form continuous curves and are referred to as yield surfaces, yield loci, or yield caps.

The shape of the yield surfaces will depend on the material; shapes such as spherical, bullet, and elliptical have often been assumed (3–7, 9). A yield surface intersects the projection of CSL at the critical point (Fig. 11-6). For normally consolidated soils, yield surfaces exist only between the CSL and the p-axis; continuation of this surface up to the origin is shown in Fig. 11-6 to indicate its complete shape. The pressure corresponding to the intersection of the yield loci with the p-axis is denoted by p_0. Each surface has its unique p_0 which is used to define the hardening behavior.

The behavior in Fig. 11-5 can be represented in terms of three-dimensional states of stress. In terms of the stress paths discussed in Chapter 5, the hydrostatic path can describe the behavior in p–e space, and the other stress paths represent the general three-dimensional behavior. In the models developed by Roscoe and coworkers, the behavior in Fig. 11-5 has been interpreted in terms of states of stress simulated commonly in classical soil mechanics. The behavior in the p–q space is simulated from cylindrical or conventional triaxial tests and that in the p–e space from consolidation tests.

The projection of the critical state line on the p–q plot is similar to the idea of fixed failure envelopes in the conventional models such as von Mises and Mohr–Coulomb. In contrast to the conventional plasticity models, the critical state surface is treated as an open fixed yield surface, and additional yield surfaces are introduced to account for the continuous yielding of soils. Thus in the critical state concept, the preultimate states are represented by additional closed yield surfaces or caps that can expand during the deformation process.

One of the characteristics of the foregoing combined yield surface, composed of a part of the fixed surface and the associated caps, is that at their intersection, which represents the point on the critical state, the tangent to the cap is horizontal. That is, with the associated flow rule, any plastic flow at the critical state occurs with constant volume. In the following sections, we discuss how the critical state concept is used to formulate stress–strain models for soils.

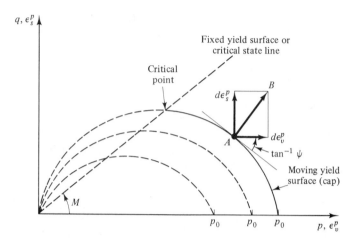

Figure 11-6 Yield locus in $q-p$ space (Ref. 6).

Stress–Strain Behavior

Let us consider the isotropic loading of a saturated clay as shown in Fig. 11-7 on the $e-\ln p$ plot. If the material is normally consolidated at A, the isotropic loading will follow path AB. Let us now unload the sample to the mean pressure p_A. Because of the elastoplastic nature, the unloading path will not follow loading path AB. Instead, the material will follow path BD upon unloading. When the material is reloaded from pressure p_A to p_B, it will usually follow the same path as unloading, that is, DB. Since the unloading and reloading follow the same path, they show elastic behavior. As shown in Fig. 11-7, the slope of the loading path is denoted by λ, and the slope of unloading–reloading path is denoted by κ. The vertical distance AD shows the plastic component in the change in volume, and DE shows the elastic component of the change in volume. Now we can write the total change in e during the loading–unloading cycle as

$$e = e_A - e_B = \lambda \ln\left(\frac{p_B}{p_A}\right) = \lambda(\ln p_B - \ln p_A) \qquad (11\text{-}4\text{a})$$

$$e^e = e_D - e_E = \kappa \ln\left(\frac{p_B}{p_A}\right) = \kappa(\ln p_B - \ln p_A) \qquad (11\text{-}4\text{b})$$

Differentiation of the two relationships above leads to

$$de = -\lambda \frac{dp}{p} \qquad (11\text{-}5\text{a})$$

$$de^e = -\kappa \frac{dp}{p} \qquad (11\text{-}5\text{b})$$

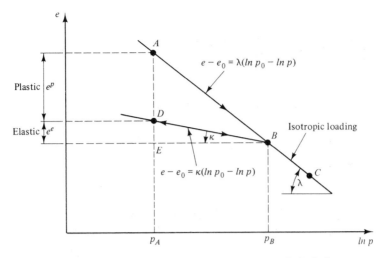

Figure 11-7 Consolidation behavior in e–ln p space (Refs. 3–6).

where the superscript e denotes the recoverable elastic component. Hence

$$de^p = de - de^e = -(\lambda - \kappa)\frac{dp}{p} \qquad (11\text{-}5c)$$

where de^p is the plastic component of the incremental void ratio.

With compressive volumetric strain positive, we have

$$d\epsilon_v = -\frac{de}{1 + e_0} = \frac{\lambda}{1 + e_0}\frac{dp}{p} \qquad (11\text{-}6a)$$

Hence the elastic volumetric strain dv^e can be written as

$$d\epsilon_v^e = -\frac{de^e}{1 + e} = \frac{\kappa}{1 + e}\frac{dp}{p} \qquad (11\text{-}6b)$$

In the stress–strain theory based on the critical state concept, it is assumed that there is no recoverable energy associated with shear distortion (i.e., $d\epsilon_s^e = 0$). Therefore, at all times

$$d\epsilon_s = d\epsilon_s^p \qquad (11\text{-}7)$$

where $d\epsilon_s$ is the shear strain defined in Eq. (11-2d).

According to the normality condition, the incremental plastic strain vector is normal to the yield surface at any point. With reference to Fig. 11-6, this can be expressed as

$$\frac{d\epsilon_s^p}{d\epsilon_v^p} = -\frac{dp}{dq} \qquad (11\text{-}8)$$

where $d\epsilon_v^p$ is the volumetric plastic strain. Note that $d\epsilon_v^p$ versus $d\epsilon_s^p$ plot is superimposed on the p–q plot.

Equation of the Yield Locus

Let us define stress ratio

$$\eta = \frac{q}{p} \tag{11-9a}$$

or

$$q = \eta p \tag{11-9b}$$

Therefore,

$$dq = p\, d\eta + \eta\, dp \tag{11-10}$$

Let us assume that the slope of the yield curve at any point (p, q) in Fig. 11-6 be ψ. Since q decreases with p, the sign of ψ is negative:

$$\frac{dq}{dp} = -\psi \tag{11-11a}$$

or

$$dq = -\psi\, dp \tag{11-11b}$$

Substitution of Eq. (11-11b) in Eq. (11-10) yields

$$p\, d\eta + \eta\, dp = -\psi\, dp \tag{11-12a}$$

By rearranging, this can be written as

$$\frac{dp}{p} + \frac{d\eta}{\eta + \psi} = 0 \tag{11-12b}$$

Equation (11-12b) defines a yield locus. Since for this model the successive yield locii (hardening caps) are geometrically similar, ψ is a function of η only. Therefore, any yield curve passing through a known point can be obtained by integrating Eq. (11-12b). That is,

$$\int_{p_0}^{p} \frac{dp}{p} + \int_{0}^{\eta} \frac{d\eta}{\eta + \psi} = 0 \tag{11-13a}$$

$$\ln p - \ln p_0 + \int_{0}^{\eta} \frac{d\eta}{\eta + \psi} = 0 \tag{11-13b}$$

Equation (11-13b) represents the yield curve passing through $(p_0, 0)$; here p_0 is treated as a variable which has a unique value for any yield surface. In fact, this can be considered as the hardening parameter.

Equations (11-13) can be expressed in the differential form as

$$\frac{dp_0}{p_0} - \frac{dp}{p} - \frac{d\eta}{\psi + \eta} = 0 \tag{11-14}$$

When a material changes its state from one yield locus to another, the change in the hardening parameter is the same irrespective of the stress path followed.

Using Eq. (11-5), we have

$$de = -\lambda \frac{dp_0}{p_0} \qquad (11\text{-}15a)$$

$$de^e = -\kappa \frac{dp_0}{p_0} \qquad (11\text{-}15b)$$

Hence

$$de^p = -(\lambda - \kappa) \frac{dp_0}{p_0} \qquad (11\text{-}15c)$$

Substituting Eq. (11-14) in Eq. (11-15c), we can write

$$de^p = -(\lambda - \kappa)\left(\frac{dp}{p} + \frac{d\eta}{\psi + \eta} \right) \qquad (11\text{-}15d)$$

Therefore, the plastic volumetric strain can be written as

$$d\epsilon_v^p = \frac{de^p}{1+e} = \frac{\lambda - \kappa}{1+e}\left(\frac{dp}{p} + \frac{d\eta}{\psi + \eta} \right) \qquad (11\text{-}16)$$

The ratio ψ of plastic components of shear and volumetric strain can be obtained by considering the dissipated energy while undergoing deformation on the state boundary surface. Here various assumptions can be made to define the magnitude of the dissipated energy dW, and depending on the assumption made, any number of hypotheses can be developed. In the Cam clay models (3) it was assumed that

$$dW = Mp\, d\epsilon_s \qquad (11\text{-}17)$$

Hence

$$p\, d\epsilon_v^p + q\, d\epsilon_s^p = Mp\, d\epsilon_s \qquad (11\text{-}18a)$$

However, by assuming that $d\epsilon_s = d\epsilon_s^p$ [Eq. (11-7)], we have

$$\frac{d\epsilon_s^p}{d\epsilon_v^p} = \frac{1}{M - \eta} \qquad (11\text{-}18b)$$

In the Cam clay models, the associated flow rule is assumed; hence the incremental strain vector AB (Fig. 11-6) is normal to the yield surface. Consequently, the ratio of $d\epsilon_s^p$ to $d\epsilon_v^p$ is equal to $1/\psi$ [Eq. (11-11)]. This leads to

$$\psi_c = M - \eta \qquad (11\text{-}19)$$

Here the subscript c denotes ψ for the Cam clay model.

In the *modified Cam clay* model (6), the dissipated energy dW is assumed as

$$dW = p\sqrt{\left(d\epsilon_v^p \right)^2 + M^2 \left(d\epsilon_s^p \right)^2} \qquad (11\text{-}20)$$

This leads to

$$\frac{d\epsilon_s^p}{d\epsilon_v^p} = \frac{2\eta}{M^2 - \eta^2} \qquad (11\text{-}21a)$$

Therefore,

$$\psi_{cm} = \frac{M^2 - \eta^2}{2\eta} \qquad (11\text{-}21b)$$

Here the subscript cm denotes modified Cam clay model.

Once ψ is known, the yield locus for the modified Cam clay model can be found by integrating Eq. (11-13a). That is,

$$\int_0^\eta \frac{2\eta \, d\eta}{M^2 + \eta^2} = -\int_{p_0}^p \frac{dp}{p} \qquad (11\text{-}22a)$$

$$\ln(M^2 + \eta^2) - \ln(M^2) = -\ln p + \ln p_0 \qquad (11\text{-}22b)$$

Simplification leads to

$$\frac{M^2 + \eta^2}{M^2} = \frac{p_0}{p} \qquad (11\text{-}22c)$$

or

$$M^2 p^2 - M^2 p_0 p + q^2 = 0 \qquad (11\text{-}22d)$$

This is the equation of an ellipse on q–p plot. Substituting the value of ψ_{cm} from Eq. (11-21) in Eq. (11-16), one can write the following expressions for incremental quantities based on the modified Cam clay model:

$$d\epsilon_v^p = \frac{\lambda - \kappa}{1 + e}\left(\frac{dp}{p} + \frac{2\eta \, d\eta}{M^2 + \eta^2}\right) \qquad (11\text{-}23a)$$

$$d\epsilon_v = \frac{\lambda}{1 + e}\left[\frac{dp}{p} + \left(1 - \frac{\kappa}{\lambda}\right)\frac{2\eta \, d\eta}{M^2 + \eta^2}\right] \qquad (11\text{-}23b)$$

$$d\epsilon_s^p = \frac{\lambda - \kappa}{1 + e}\left(\frac{dp}{p} + \frac{2\eta \, d\eta}{M^2 + \eta^2}\right)\frac{2\eta}{M^2 - \eta^2} \qquad (11\text{-}23c)$$

Derivation of Elastoplastic Constitutive Relation

Using Eqs. (11-6), the plastic volumetric strain can be written as

$$d\epsilon_v^p = d\epsilon_v - d\epsilon_v^e$$

$$= \left(\frac{\lambda - \kappa}{1 + e_0}\right)\frac{dp_0}{p_0} \qquad (11\text{-}24a)$$

where e_0 is the initial void ratio before the increment. As shown in Eq.

(11-22d), the yield locus takes the form

$$f = M^2 p^2 - M^2 p_0 p + q^2 = 0 \qquad (11\text{-}22\text{d})$$

where for the cylindrical triaxial state of stress

$$q = \sigma_1 - \sigma_3 \qquad (11\text{-}24\text{b})$$

However, in order to apply the yield surface to a three-dimensional incremental plasticity theory (for solving boundary value problems), it can be appropriate to introduce the stress invariants in the equation of yield surface [Eq. (11-22d)]. For a general state of stress

$$q = \frac{3}{\sqrt{2}} \tau_{\text{oct}}$$

$$= \frac{1}{\sqrt{2}} \left[(\sigma_{11} - \sigma_{22})^2 + (\sigma_{22} - \sigma_{33})^2 + (\sigma_{11} - \sigma_{33})^2 + 6\sigma_{12}^2 + 6\sigma_{23}^2 + 6\sigma_{13}^2 \right]^{\frac{1}{2}}$$

$$(11\text{-}25\text{a})$$

In terms of stress invariants, Eq. (11-22d) can be written as

$$f = M^2 J_1^2 - M^2 J_1 J_{01} + 27 J_{2D} = 0 \qquad (11\text{-}25\text{b})$$

Here, J_{01} is the value of J_1 at the intersection of the yield cap with the J_1 axis; this is analogous to p_0, which is defined in Eq. (11-13b). This is the hardening parameter which is assumed to depend on the plastic volumetric strain.

Now, at yield, as $f = f[q, p, p_0(\epsilon_v^p)]$, we have

$$df = \frac{\partial f}{\partial q} dq + \frac{\partial f}{\partial p} dp + \frac{\partial f}{\partial \epsilon_v^p} d\epsilon_v^p = 0 \qquad (11\text{-}26)$$

The normality rule for plastic deformations can be written as [Eq. (9-56)]

$$d\epsilon_{ij}^p = \bar{\lambda} \frac{\partial Q}{\partial \sigma_{ij}} \qquad (11\text{-}27)$$

Here Q is the plastic potential and $\bar{\lambda}$ is the scalar parameter of proportionality. Equation (11-27) can be written as

$$d\epsilon_{ij}^p = \bar{\lambda} A_{ij} \qquad (11\text{-}28\text{a})$$

where

$$A_{ij} = \frac{\partial Q}{\partial \sigma_{ij}} = \frac{\partial Q}{\partial p} \left(\frac{\partial p}{\partial \sigma_{ij}} \right) + \frac{\partial Q}{\partial q} \left(\frac{\partial q}{\partial \sigma_{ij}} \right) \qquad (11\text{-}28\text{b})$$

Let us define

$$B_{ij} = \frac{\partial F}{\partial \sigma_{ij}} = \frac{\partial f}{\partial p} \left(\frac{\partial p}{\partial \sigma_{ij}} \right) + \frac{\partial f}{\partial q} \left(\frac{\partial q}{\partial \sigma_{ij}} \right) \qquad (11\text{-}29)$$

Combining Eqs. (11-26) and (11-29), we obtain

$$df = B_{ij} d\sigma_{ij} + \frac{\partial f}{\partial \epsilon_v^p} d\epsilon_v^p = 0 \qquad (11\text{-}30)$$

The total strain increments have been assumed to be the vector sum of the elastic and plastic strain increments. That is,

$$d\epsilon_{ij}^e = d\epsilon_{ij} - d\epsilon_{ij}^p \tag{11-31}$$

Using the generalized Hooke's law, it is possible to write

$$d\sigma_{ij} = C_{ijkl}(d\epsilon_{kl} - d\epsilon_{kl}^p) \tag{11-32}$$

Substituting Eqs. (11-32) and (11-28) in Eq. (11-30) yields

$$df = B_{ij}C_{ijkl}(d\epsilon_{kl} - \bar{\lambda}A_{kl}) + \frac{\partial F}{\partial \epsilon_v^p}\bar{\lambda}A_{ii} = 0 \tag{11-33}$$

Simplifying Eq. (11-33), the value of $\bar{\lambda}$ is obtained as

$$\bar{\lambda} = \frac{B_{ij}C_{ijkl}\,d\epsilon_{kl}}{B_{ij}C_{ijkl}A_{kl} - \dfrac{\partial f}{\partial \epsilon_v^p}A_{ii}} \tag{11-34}$$

Substitution of Eq. (11-34) in Eq. (11-32) leads to

$$d\sigma_{ij} = C_{ijkl}\,d\epsilon_{kl} - \frac{C_{ijkl}A_{kl}B_{mn}C_{mnrs}\,d\epsilon_{rs}}{B_{mn}C_{mnrs}A_{rs} - \dfrac{\partial f}{\partial \epsilon_v^p}A_{ii}} \tag{11-35a}$$

or

$$d\sigma_{ij} = \left[C_{ijrs} - \frac{C_{ijkl}A_{kl}B_{mn}C_{mnrs}}{B_{mn}C_{mnrs}A_{rs} - \dfrac{\partial F}{\partial \epsilon_v^p}A_{ii}} \right] d\epsilon_{rs} \tag{11-35b}$$

Therefore, the plastic part of the constitutive relationship C_{ijrs}^p can be written as

$$C_{ijrs}^p = \frac{C_{ijkl}A_{kl}B_{mn}C_{mnrs}}{B_{mn}C_{mnrs}A_{rs} - \dfrac{\partial F}{\partial \epsilon_v^p}A_{ii}} \tag{11-36}$$

In order to write the constitutive relationship in matrix notation, let us denote

$$\left\{ \frac{\partial \mathbf{Q}}{\partial \sigma_{ij}} \right\}^{\mathrm{T}} = \left[\frac{\partial Q}{\partial \sigma_{11}} \quad \frac{\partial Q}{\partial \sigma_{22}} \quad \frac{\partial Q}{\partial \sigma_{33}} \quad \frac{\partial Q}{\partial \sigma_{12}} \quad \frac{\partial Q}{\partial \sigma_{23}} \quad \frac{\partial Q}{\partial \sigma_{13}} \right]$$

$$= \{\mathbf{A}\}^{\mathrm{T}} \tag{11-37}$$

$$\left\{ \frac{\partial \mathbf{f}}{\partial \sigma_{ij}} \right\}^{\mathrm{T}} = \left[\frac{\partial f}{\partial \sigma_{11}} \quad \frac{\partial f}{\partial \sigma_{22}} \quad \frac{\partial f}{\partial \sigma_{33}} \quad \frac{\partial f}{\partial \sigma_{12}} \quad \frac{\partial f}{\partial \sigma_{23}} \quad \frac{\partial f}{\partial \sigma_{13}} \right]$$

$$= \{\mathbf{B}\}^{\mathrm{T}} \tag{11-38}$$

Note that for associative flow rule $\{A\} = \{B\}$. Using Eq. (11-24), it is possible to write

$$\frac{\partial p_0}{\partial \epsilon_v^p} = \frac{p_0(1 + e_0)}{\lambda - \kappa} \qquad (11\text{-}39)$$

Using the chain rule of differentiation, we have

$$\frac{\partial f}{\partial \epsilon_v^p} = \frac{\partial f}{\partial p_0} \frac{\partial p_0}{\partial \epsilon_v^p} = -\frac{M^2 p p_0(1 + e_0)}{\lambda - \kappa} \qquad (11\text{-}40)$$

Substituting Eq. (11-40) into Eq. (11-36), the constitutive relationship can be expressed in the form

$$\{d\sigma\} = [C^{ep}]\{d\epsilon\} \qquad (11\text{-}41)$$

where $[C^{ep}]$ is the elastoplastic constitutive matrix given by

$$[C^{ep}] = [C^e] - [C^p] \qquad (11\text{-}42)$$

where $[C^e]$ is the elastic portion of $[C]$.

CAP MODELS

In developing models for various soils based on the critical state concept, the Cambridge group (3-7) expressed the behavior by using quantities relevant to the conventional (cylindrical) triaxial and conventional consolidation tests. As described previously, these parameters were $q = \sigma_1 - \sigma_3$ and $p = (\sigma_1 + 2\sigma_3)/3$, and the void ratio e.

The cap models (10-14) are also based on the concept of continuous yielding of soils, but they are expressed in terms of three-dimensional state of stress and are formulated on the basis of consistent mechanics principles. The

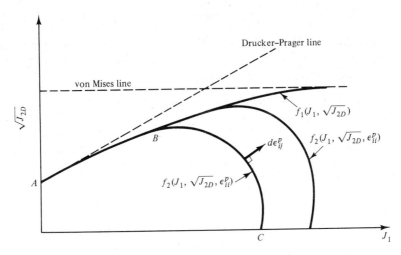

Figure 11-8 Cap model (Refs. 10–12).

name cap models derives from the shape of the elliptical yield surfaces which look like "caps."

There are certain other differences between the cap and Cam clay models. In the cap models, the portion of the cap above the fixed yield or failure surface (Fig. 11-8) is omitted, and only the yield surface marked ABC is considered. In the Cam clay models, the moving cap plays the main role in defining yielding, and the fixed yield surface is used essentially to define the critical state. On the other hand, in the cap model, both the fixed and moving surfaces are used to define the yielding process.

Fixed Yield Surface

The fixed yield surface, which can be considered to be an ultimate yield surface, is expressed as

$$f_1\left(J_1, \sqrt{J_{2D}}\right) = 0 \qquad (11\text{-}43\text{a})$$

In the initial Cap model (10), the fixed surface was assumed to be composed of an initial portion of the Drucker–Prager envelope joined smoothly to the subsequent von Mises surface (Fig. 11-8). The logic for adopting the von Mises surface at higher stresses was based on the observation that at higher stresses the material behaves like a "liquid." This was adopted particularly to simulate behavior of cohesionless materials subjected to high stresses caused by dynamic (blast) loads.

The expression for f_1 in Eq. (11-43a) adopted by DiMaggio and Sandler (10) is given by

$$f_1 = \sqrt{J_{2D}} + \gamma e^{-\beta J_1} - \alpha = 0 \qquad (11\text{-}43\text{b})$$

where α, β, and γ are material parameters.

Yield Caps

The yield surfaces are expressed as

$$f_2\left(J_1, \sqrt{J_{2D}}, k_1\right) = 0 \qquad (11\text{-}44\text{a})$$

where k_1 defines the deformation history, and usually is taken as the volumetric plastic strain $\epsilon_v^p = \epsilon_{ii}^p = I_1^p/3$. Consequently, a yield surface represents the locus of the points with the same volumetric plastic strain.

The fixed and moving yield surfaces are assumed to intersect such that the tangents to the yield surfaces at the intersection are parallel to the J_1-axis. As a result, with the associated plasticity, the increment of plastic strain vector is parallel to the $\sqrt{J_{2D}}$-axis implying no volume change once the fixed surface is reached. This is similar to the critical state concept in which the material does not change in volume at the critical state.

The yield surfaces intersect the J_1-axis at right angles, implying that under isotropic compression there are no shear deformations. This assumption removes the anomaly that can arise due to nonorthogonal yield surfaces to the J_1-axis.

The cap model has been applied successfully to simulate behavior of various geological materials such as McCormick Ranch Sand (11–13) and an artificial soil (15–19). Subsequent developments (14) have considered anistropic and kinematic hardening in the context of the cap model.

DiMaggio and Sandler (10) adopted an elliptic cap for representing yield surfaces for the cohesionless material considered by them. Hence the expression for f_2 in Eq. (11-44a) used was

$$f_2 = R^2 J_{2D} + (J_1 - C)^2 = R^2 b^2 \qquad (11\text{-}44b)$$

where $Rb = (X - C)$ (Fig. 11-9), R is the ratio of the major to minor axis of the ellipse, X the value of J_1 at the intersection of the cap with the J_1-axis, C the value of J_1 at the center of the ellipse, and b the value of $\sqrt{J_{2D}}$ when $J_1 = C$. The value of X, which is the hardening parameter similar to p_0 in the Cam clay models, depends on the plastic volumetric strain ϵ_v^p and is expressed as

$$X = -\frac{1}{D} \ln\left(1 - \frac{\epsilon_v^p}{W}\right) + Z \qquad (11\text{-}45)$$

where D, Z, and W are material parameters to be determined.

Comments

The Cam clay and cap models by no means can describe the behavior of materials under all significant conditions and stress paths. Usually, like other currently available models, they may be able to successfully simulate only a

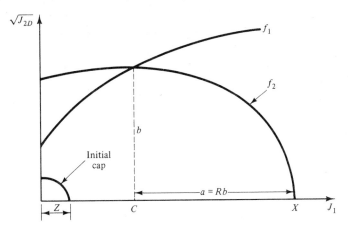

Figure 11-9 Yield surface for cap model (Refs. 10–12).

limited number of stress paths. Often, the gross load–displacement behavior of a system predicted by these models can be satisfactory. This may be due to the fact that the models simulate satisfactorily a part of the significant stress paths in boundary value problems for a given material (20).

EVALUATION OF PARAMETERS

Example 11-1   Cam Clay Model for a Clay

A hypothetical data set for a normally consolidated clay is presented in Figs. 11-10 to 11-18. Now we shall use these data to illustrate the procedure for determination of parameters for the critical state model. The data presented here pertain to conventional triaxial conditions.

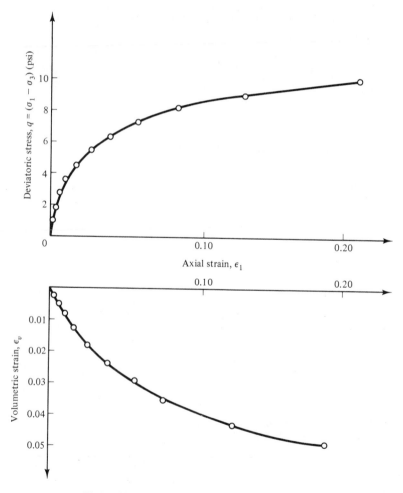

Figure 11-10   Constant pressure test: $p_0 = 10$ psi.

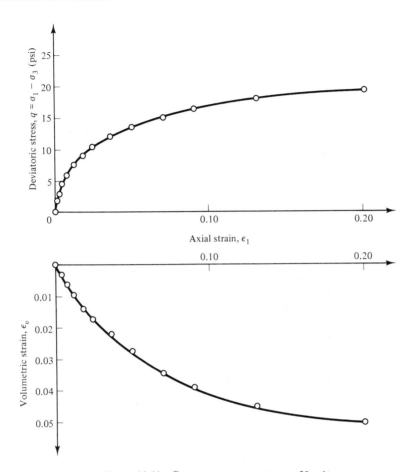

Figure 11-11   Constant pressure test: $p_0 = 20$ psi.

Figures 11-10 to 11-12 show the stress–strain relations for constant pressure tests under different initial conditions $p_0 = 10$ (69), 20 (139), and 30 (207) psi (kPa), respectively. Here the mean effective pressure of the soil sample is kept constant throughout the test. The last data point shown on the deviatoric stress (q) versus axial strain (ϵ_1) plot for any test is considered as the ultimate condition for that test. The void ratio values at the beginning and the end of each test are given in Table 11-1. Figures 11-13 to 11-15 show stress–strain plots for three fully drained tests performed at initial pressures 10, 20, and 30 psi. Here, there is no pore water pressure development, and the effective mean pressure increases during the test. The void ratio values at the beginning and end of each test are given in Table 11-1. Figures 11-16 to 11-18 show stress–strain behavior under undrained conditions. Here there is no change in volume of the sample. However, fluid pore pressures are developed during the test, thereby reducing the effective stresses until the ultimate (failure) state is reached.

Figure 11-19 shows the variation of void ratio with mean pressure which is obtained from a hydrostatic compression (HC) test with three load reversals (i.e.,

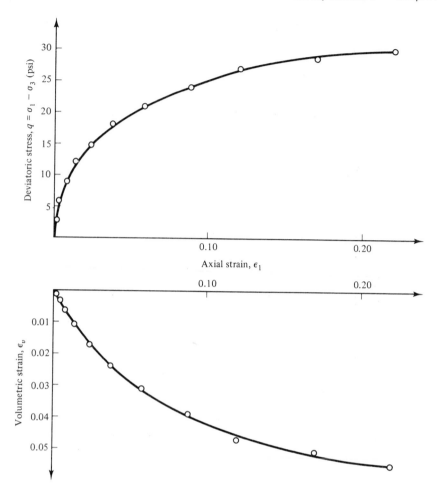

Figure 11-12   Constant pressure test: $p_0 = 30$ psi.

unloading–reloading cycles). In this figure, void ratio is plotted to a linear scale while the hydrostatic stress is plotted to a logarithmic (base 10) scale. In fact, this is the usual way of presenting one-dimensional consolidation or hydrostatic test results in geotechnical engineering practice.

Determination of the parameter, M: The parameter M is the slope of the critical state line on a q–p plot. To determine its value, the values of p and q at ultimate conditions for each test are plotted as shown in Fig. 11-20; the ultimate condition for each test is assumed to be the last data point plotted in Figs. 11-10 to 11-18. At the ultimate conditions, the sample undergoes excessive deformations under constant deviatoric stress, and hence it could be taken as the asymptotic stress to the curve. The slope of the critical state line (Fig. 11-20) is calculated as 1.0 for the soil above. That is,

$$M = 1.0 \qquad\qquad (11\text{-}46)$$

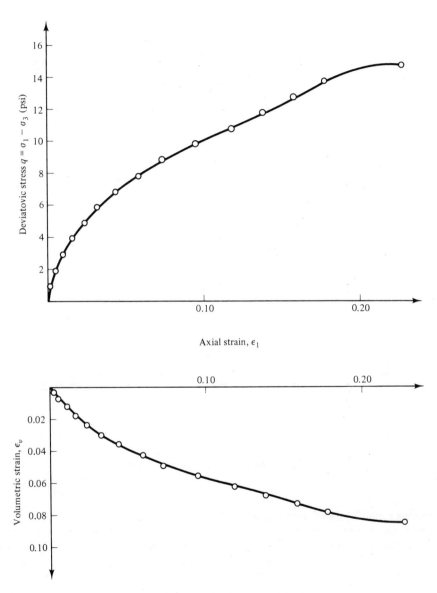

Figure 11-13   Drained test: $p_0 = 10$ psi.

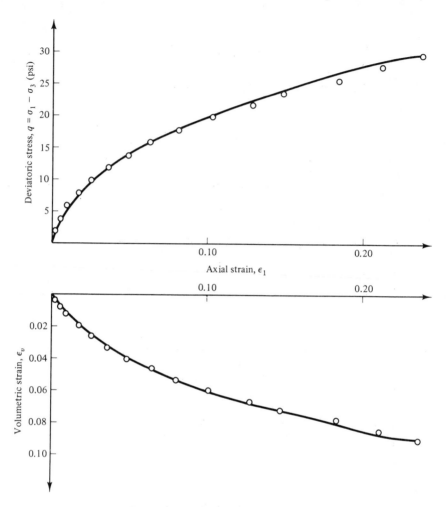

Figure 11-14   Drained test: $p_0 = 20$ psi.

Determination of parameters λ and κ: The values of λ and κ can be related to the commonly known quantities such as compression index (C_c) and swelling index (C_s). The compression index C_c is defined as the slope of virgin loading line on e–$\log_{10} p$ plot while the swelling index, C_s, is defined as the slope of unloading–reloading curves on the same plot. Usually, the compression index and swelling index are defined with respect to a one-dimensional consolidation test. However, it can be shown that the e–$\ln p$ curve for any constant stress ratio test, that is, for constant q/p ratio, is parallel to that obtained from a hydrostatic test (7), Fig. 11-21. In fact, one-dimensional consolidation is a special case of a constant q/p test. The e–$\ln p$ curve obtained from a hydrostatic test is parallel to that obtained under critical state conditions. The values of λ and κ can be related to C_c and C_s as follows. The virgin compression line can be

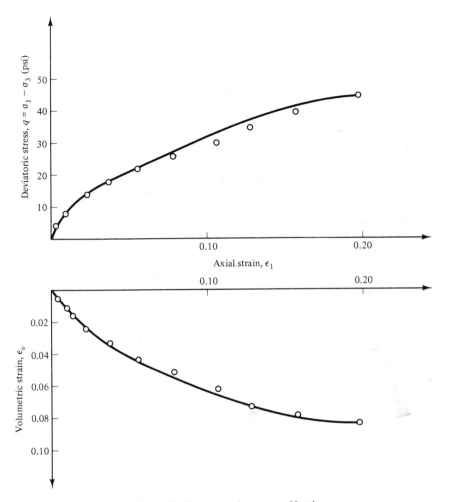

Figure 11-15 Drained test: $p_0 = 30$ psi.

expressed as

$$e - e_0 = C_c \log_{10}\left(\frac{p}{p_0}\right)$$ (11-47a)

or

$$e - e_0 = \lambda \ln\left(\frac{p}{p_0}\right)$$ (11-47b)

and the swelling line (unloading–reloading line) can be expressed as

$$e - e_0 = C_s \log_{10}\left(\frac{p}{p_0}\right)$$ (11-48a)

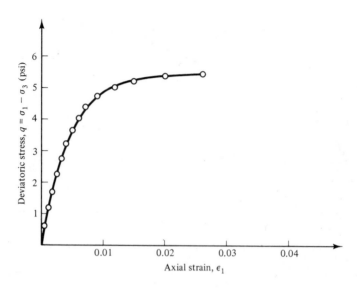

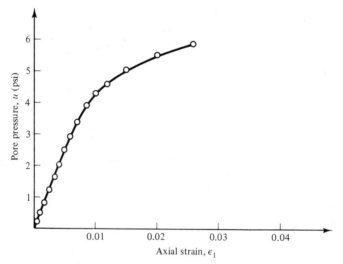

Figure 11-16   Undrained test: $p_0 = 10$ psi.

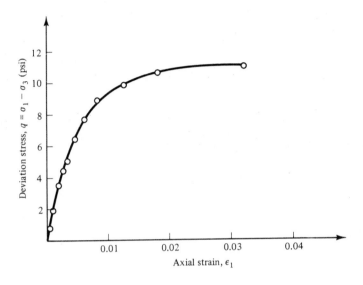

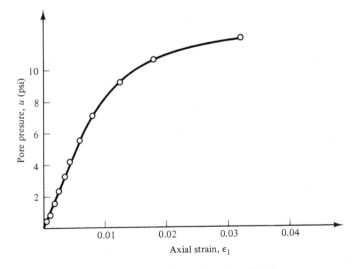

Figure 11-17 Undrained test: $p_0 = 20$ psi.

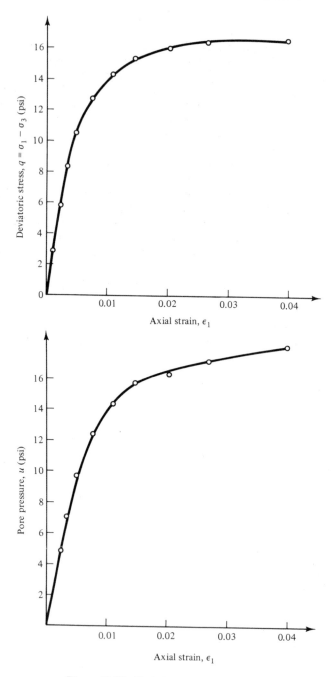

Figure 11-18   Undrained test: $p_0 = 30$ psi.

TABLE 11-1   Test Values for Cam Clay Model

Test	Initial Pressure, p_0 (effective) (psi)	Initial Void Ratio, e_0	Final Pressure, p_f (effective) (psi)	Final Void Ratio, e_f
p-constant	10	1.080	10	0.980
p-constant	20	0.959	20	0.860
p-constant	30	0.889	30	0.787
Drained	10	1.080	15	0.908
Drained	20	0.959	30	0.787
Drained	30	0.889	45	0.716
Undrained	10	1.080	5.55	1.080
Undrained	20	0.959	11.09	0.959
Undrained	30	0.889	16.64	0.889

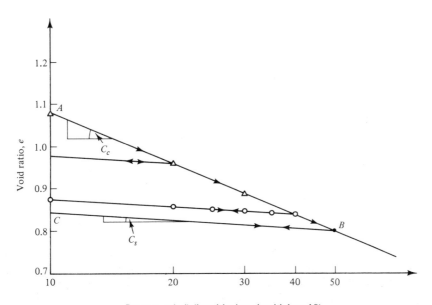

Pressure, p (psi) (logarithmic scale with base 10)

Figure 11-19   Hydrostatic compression test: $C_c = 0.40$, $C_s = 0.06$.

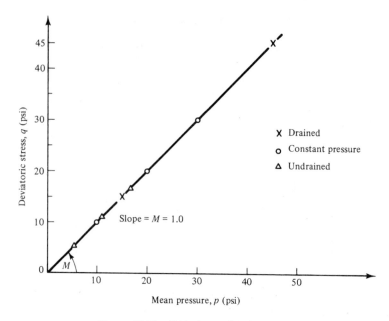

Figure 11-20   Critical state line in q–p space.

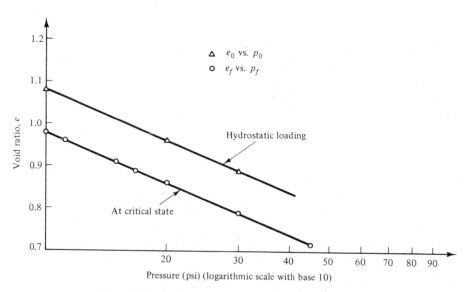

Figure 11-21   Isotropic consolidation. (Data from Table 11-1.)

or

$$e - e_0 = \kappa \ln\left(\frac{p}{p_0}\right) \qquad (11\text{-}48\text{b})$$

Therefore, comparing Eqs. (11-47a) and (11-47b), we have

$$\lambda = \frac{C_c}{\ln 10} = \frac{C_c}{2.303} \qquad (11\text{-}49)$$

and comparing Eqs. (11-48a) and (11-48b) yields

$$\kappa = \frac{C_s}{\ln 10} = \frac{C_s}{2.303} \qquad (11\text{-}50)$$

The value of C_c can be computed by considering two points, A and B, in Fig. 11-19 as

$$\begin{aligned}
C_c &= \frac{e_A - e_B}{\log p_B - \log p_A} \\
&= \frac{1.08 - 0.80}{\log 50 - \log 10} \\
&= 0.40 \qquad (11\text{-}51)
\end{aligned}$$

Here the subscript denotes the value at that point. The swelling index, C_s, can be computed by considering points B and C of the same figure as

$$\begin{aligned}
C_s &= \frac{e_C - e_B}{\log p_B - \log p_C} \\
&= \frac{0.842 - 0.80}{\log 50 - \log 10} \\
&= 0.06 \qquad (11\text{-}52)
\end{aligned}$$

Hence the values of λ and κ can be computed from Eqs. (11-49) and (11-50) as

$$\lambda = \frac{0.40}{2.303} = 0.174 \qquad (11\text{-}53)$$

$$\kappa = \frac{0.06}{2.303} = 0.026 \qquad (11\text{-}54)$$

Predictions of behavior: In a previous section, the following expressions for volumetric and shear strains for the modified Cam clay model were derived:

Volumetric strains:

$$d\epsilon_v^p = \frac{\lambda - \kappa}{1 + e}\left(\frac{dp}{p} + \frac{2\eta\, d\eta}{M^2 + \eta^2}\right) \qquad (11\text{-}55\text{a})$$

$$d\epsilon_v^e = \frac{\kappa}{1 + e}\frac{dp}{p} \qquad (11\text{-}55\text{b})$$

Adding Eqs. (11-55a) and (11-55b) we obtain the total volumetric strain as

$$d\epsilon_v = \frac{\lambda}{1 + e}\left[\frac{dp}{p} + \left(1 - \frac{\kappa}{\lambda}\right)\frac{2\eta\, d\eta}{M^2 + \eta^2}\right] \qquad (11\text{-}55\text{c})$$

Shear strains, ϵ_s:

$$de_s = de_s^p = \frac{\lambda - \kappa}{1 + e}\left(\frac{dp}{p} + \frac{2\eta\, d\eta}{M^2 + \eta^2}\right)\frac{2\eta}{M^2 - \eta^2} \qquad (11\text{-}56a)$$

$$= de_v^p\, \frac{2\eta}{M^2 - \eta^2} \qquad (11\text{-}56b)$$

The predictions of stress–strain behavior can be made using Eqs. (11-55) and (11-56). Here the state of stress along any given stress path is first adopted, and the corresponding shear and volumetric strains are computed from Eqs. (11-55) and (11-56). Since the equations above are expressed in an incremental form, total quantities are computed by accumulating incremental quantities along the particular stress path. This procedure is illustrated with respect to predictions of a fully drained test with an initial pressure of 30 psi (Table 11-2). The steps involved in the back prediction are as follows:

Step 1: Select a value of mean pressure p and compute the corresponding value of q along the stress path under consideration. (*Note*: It is also possible to select q, and compute the corresponding value of p.)

Step 2: Compute the value of stress ratio η as q/p.

Step 3: Compute increment in stress ratio η_{inc} as

$$\eta_{inc} = \left(\frac{q}{p}\right)_i - \left(\frac{q}{p}\right)_{i-1} \qquad (11\text{-}57)$$

where the subscript i denotes an incremental step.

Step 4: Compute the value of $d\epsilon_v$ during the increment of stress by using Eq. (11-55c).

TABLE 11-2   Predictions of Drained Test[a]

p	q	η	η_{inc}	$d\epsilon_v$	ϵ_v	de	e	$d\epsilon_s$	ϵ_s
30.0	0.0	0.000	0.00	0.0000	0.0000	0.0000	0.889	0.000	0.0000
31.0	3.0	0.097	0.097	0.0031	0.0031	0.0058	0.883	0.001	0.0010
32.0	6.0	0.188	0.091	0.0044	0.0075	0.0083	0.875	0.0018	0.0027
33.0	9.0	0.273	0.085	0.0054	0.0129	0.0101	0.865	0.0029	0.0056
34.0	12.0	0.353	0.080	0.0062	0.0190	0.0116	0.853	0.0043	0.0100
35.0	15.0	0.429	0.076	0.0067	0.0257	0.0124	0.840	0.0059	0.0158
36.0	18.0	0.500	0.071	0.0070	0.0327	0.0129	0.827	0.0078	0.0226
37.0	21.0	0.568	0.068	0.0071	0.0398	0.0130	0.814	0.0099	0.0335
38.0	24.0	0.632	0.064	0.0072	0.0470	0.0131	0.801	0.0122	0.0457
39.0	27.0	0.692	0.061	0.0072	0.0542	0.0130	0.788	0.0151	0.0608
40.0	30.0	0.750	0.058	0.0071	0.0613	0.0127	0.775	0.0186	0.0794
41.0	33.0	0.805	0.055	0.0070	0.0682	0.0124	0.762	0.0234	0.1028
42.0	36.0	0.857	0.052	0.0068	0.0750	0.0120	0.750	0.0302	0.1330
43.0	39.0	0.907	0.050	0.0066	0.0816	0.0116	0.738	0.0413	0.1743
44.0	42.0	0.955	0.048	0.0064	0.0881	0.0111	0.727	0.0631	0.2374
45.0	45.0	1.00	0.045	0.0063	0.0943	0.0109	0.716	0.128	0.3653

[a] Initial pressure, $p_0 = 30$ psi.

Step 5: Find the total volumetric strain by accumulating incremental quantity computed in step 4.

Step 6: Compute the change in void ratio from

$$\frac{de}{1+e} = d\epsilon_v \tag{11-58}$$

In Eq. (11-58), e is the value of the void ratio at the beginning of the current load increment. This leads to

$$de = (1+e)d\epsilon_v \tag{11-59}$$

Hence the void ratio after the present load increment can be computed as

$$e = e - de \tag{11-60}$$

Note that the initial void ratio for the next load increment is the updated value shown in Eq. (11-60).

Step 7: Compute incremental shear strain $d\epsilon_s$ from Eq. (11-56a). The total shear strain ϵ_s could be computed by accumulating the incremental quantities.

Figure 11-22 shows a comparison between the original data and the back predictions. The comparison is considered excellent; this can be expected since the original data set was found from back predictions with the following material parameters:

$$M = 1.0$$
$$\lambda = 0.174$$
$$\kappa = 0.026$$

The approach above (of using hypothetical data) was used since the objective in this example was just to illustrate the procedure for finding parameters for the Cam clay model.

Example 11-2   Cam Clay Model for a Silty Sand

The soil considered for this example is a silty sand, called Pueblo sand, found in the foundation of the Urban Mass Transportation Authority (UMTA) Test Section, Transportation Test Center (TTC), Pueblo, Colorado. Details of this soil and the test results used for this example are the same as those given in Chapter 10.

Figure 11-23 shows the ultimate stress conditions on the p–q plot. The ultimate stress was determined as the values of stresses asymptotic to the final portion of the stress–strain curves (see Chapter 10). The slope of the average line gives a value of M equal to about 1.24.

Figure 11-24 shows a graph of e versus $\ln p$ developed from the hydrostatic test (Fig. 10-7). The values of λ and κ found from this graph are

$$\lambda = 0.014$$
$$\kappa = 0.0024$$

The required elastic parameters are found from

$$K = \frac{p(1+e_0)}{\kappa}$$

$$G = \frac{3}{2}K\frac{1-2\nu}{1+\nu}$$

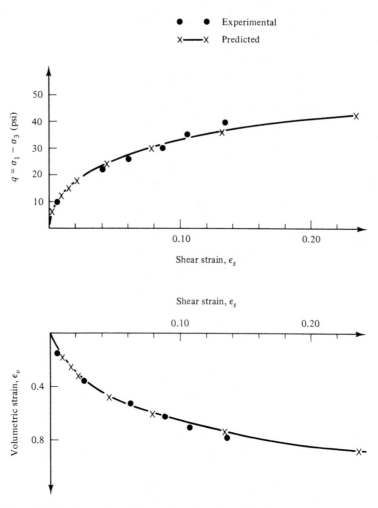

Figure 11-22 Comparison of experimental and predicted curves for drained test: $p_0 = 30$ psi.

where ν is the Poisson's ratio, and K and G are the bulk and shear moduli, respectively, which are functions of the current mean pressure. The Poisson's ratio can be found from the measurements of volumetric strains and the axial strains. For this soil it was found to be about 0.28.

The parameters for the Cam clay models for the Pueblo sand are

$$M = 1.24$$
$$\lambda = 0.014$$
$$\kappa = 0.0024$$
$$e_0 = 0.34$$
$$p_0 = 10 \text{ psi } (69 \text{ kPa})$$
$$\nu = 0.28$$

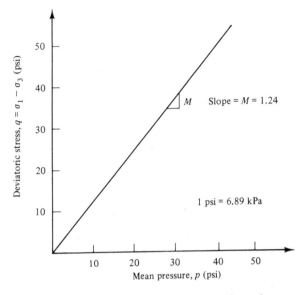

Figure 11-23 Critical state line for Pueblo sand.

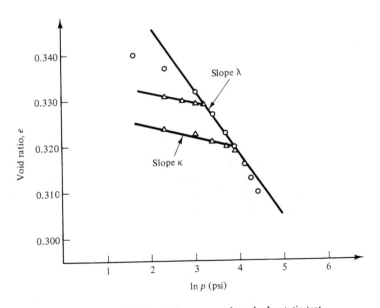

Figure 11-24 Void ratio–log p curve from hydrostatic test.

Back predictions: The foregoing parameters were used to back-predict the behavior of the CTC-20 test by using the procedure described in Example 11-1. The comparison between the predictions and the observations are shown in Fig. 11-25. It can be seen that the correlation between the two results for the CTC-20 test is satisfactory. It may be noted, however, that for most currently available models the correlation will usually be satisfactory for only some of the stress paths.

Example 11-3   Determination of Parameters for the Cap Model

In this example we consider the test results for the artificial soil described in Chapter 5. Figure 11-26 shows a plot of τ_{oct} versus γ_{oct} for four different tests. This and other calculations lead to the development of the fixed yield surface f_1 as shown in Fig. 11-27, where the range of ultimate stresses for different stress paths are marked on the $J_1 - \sqrt{J_{2D}}$ space. Ultimate stress for a given test is estimated as the value of stress asymptotic to the final stress–strain curves (see Chapter 5 and Fig. 11-26).

Within the range of (low) stress levels for the artificial soil, the shape of the failure surface was found to be somewhat different from that used in the original cap model (10). The entire surface can be expressed by using one polynomial in J_1 and $\sqrt{J_{2D}}$. However, for convenience, it was assumed to be composed of an initial Drucker–Prager surface which is connected by a smooth curve to another Drucker–Prager surface at higher stresses. A modified form of f_1 for the artificial soil was derived as (15, 16)

$$f_1 = \sqrt{J_{2D}} + \gamma e^{-\beta J_1} - \theta J_1 - \alpha = 0 \tag{11-61}$$

where θ is an additional material parameter. If $\theta = 0$, Eq. (11-61) reduces to Eq. (11-43b).

The form of the moving yield surfaces is assumed to be elliptical as in Eq. (11-44b), and the relation between the hardening parameter and volumetric plastic strain is adopted as in Eq. (11-45).

Moving yield surfaces or caps: In order to evaluate the parameters for the cap model it is necessary to obtain plots of the moving and fixed yield surfaces. The latter is obtained as shown in Fig. 11-27. The procedure for the determination of moving yield surfaces is described below.

The total volumetric plastic strains ϵ_v^p are found at various points along a given stress path, and are marked at those points (Fig. 11-28); the procedure for finding ϵ_v^p is described below. Then contours are drawn through the points of equal volumetric plastic strains. These contours are assumed to represent the yield caps.

Computation of volumetric strain: The volumetric plastic strains during shear loading can be found from laboratory test data. If during the test, all the three principal stresses change, the plots of $\sigma_1 - \epsilon_1$, $\sigma_2 - \epsilon_2$, and $\sigma_3 - \epsilon_3$ can be plotted as shown in Fig. 11-29(a). If only one (or two) of the stresses change during loading, the results can be plotted as shown in Fig. 11-29(b).

The unloading–reloading modulus, E_r, is usually adopted in order to delineate elastic and plastic strains. This modulus will vary in different regions of the stress–strain curves. It is often appropriate to adopt an average constant value of this modulus. For any increment of stress, i to $i + 1$, the increment of the elastic and plastic strains are found as shown in Fig. 11-29 by using the modulus E_r. This procedure permits computation of $d\epsilon_1^p$, $d\epsilon_2^p$, and $d\epsilon_3^p$. Then the increment of volumetric plastic strain is

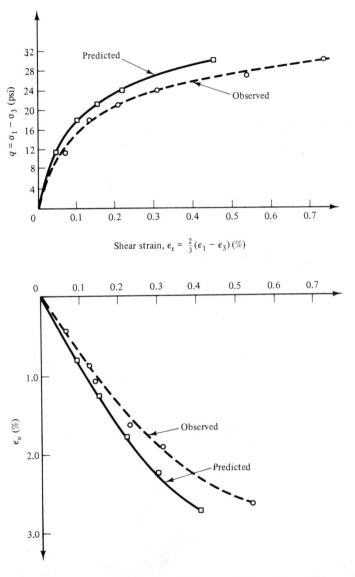

Figure 11-25 Comparisons between predictions and observations: CTC, $\sigma_0 = 20$ psi.

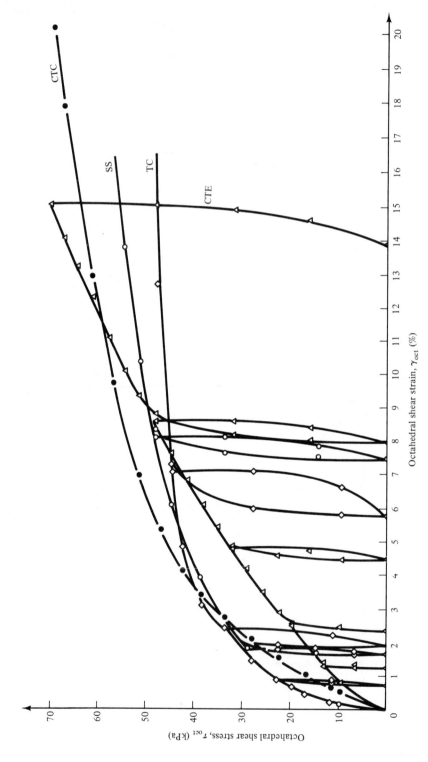

Figure 11-26 Octahedral shear stress–octahedral shear strain response (Refs. 15–17).

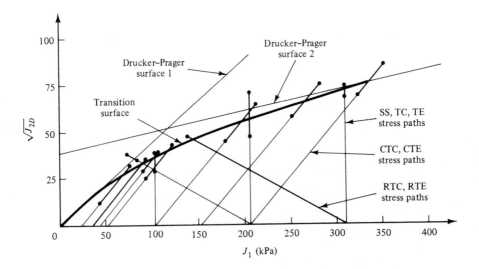

Figure 11-27 Ultimate (fixed) envelope for the artificial soil (Refs. 15–17).

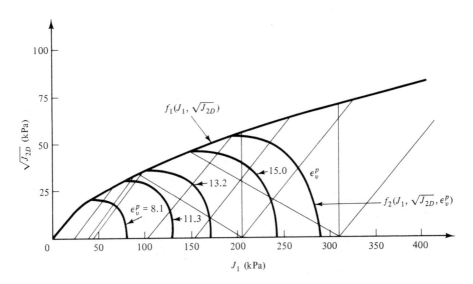

Figure 11-28 Contours of constant volumetric plastic strain for artificial soil (Refs. 15–17).

found as

$$d\epsilon_v^p = d\epsilon_1^p + d\epsilon_2^p + d\epsilon_3^p \qquad (11\text{-}62a)$$

and the total volumetric strain is evaluated as the sum of the increments up to $i + 1$:

$$\epsilon_v^p = \sum_1^{i+1} d\epsilon_v^p \qquad (11\text{-}62b)$$

The incremental value of the volumetric plastic strain can be found alternatively by using the elastic constitutive relation to first find the elastic strains as

$$\{d\epsilon^e\} = [C^e]^{-1}\{d\sigma\} \qquad (11\text{-}63a)$$

where $[C^e]$ is the elasticity matrix composed of elastic moduli such as E and ν. Then the plastic volumetric strains can be found from

$$d\epsilon_v^p = (d\epsilon_1 + d\epsilon_2 + d\epsilon_3) - (d\epsilon_1^e + d\epsilon_2^e + d\epsilon_3^e) \qquad (11\text{-}63b)$$

where $d\epsilon_1$, $d\epsilon_2$, and $d\epsilon_3$ are the total increments of strain measured from the stress–strain curves in Fig. 11-29.

Comment: The foregoing procedure for determination of yield surfaces based only on the volumetric plastic strain is approximate. In general, a measure based on total plastic strain may be more appropriate to define yield surfaces.

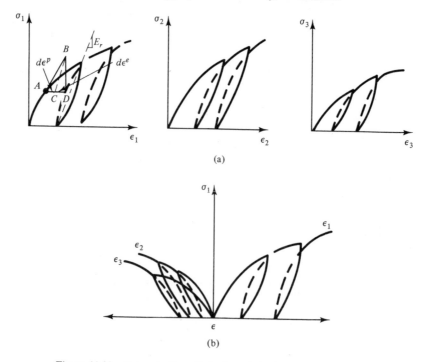

(a)

(b)

Figure 11-29 Determination of plastic and elastic strain increments.

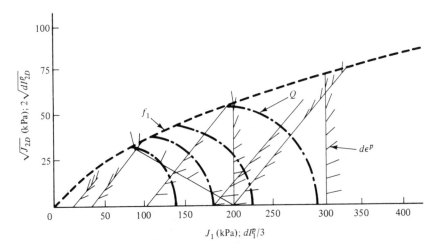

Figure 11-30 Plastic strain increment vectors and potential surfaces (Refs. 15–17).

Once the moving and the fixed yield surfaces are determined, we can proceed to curve-fitting schemes to find the required parameters. However, before proceeding toward that goal, we consider the question of plastic potential surfaces in order to verify the assumption of associative plasticity used in this model.

Plastic potential surface, Q: The incremental plastic strain vector $d\epsilon^p_{ij}$ is normal to the plastic potential surface. Hence the observed incremental plastic strain vectors can be used to find Q.

Figure 11-30 shows the directions of the incremental plastic strain vectors drawn at various points along different stress paths. Here the vectors are plotted in terms of $dI^p_1/3$ and $2\sqrt{I_{2D}}$ as coordinate axes which are superimposed on the $J_1-\sqrt{J_{2D}}$ stress space. The factors $\frac{1}{3}$ and 2 in $dI^p_1/3$ and $2\sqrt{I^p_{2D}}$ were selected from energy considerations, by expressing the (irrecoverable) work done due to the plastic deformation as

$$dW^p = J_1\left(\frac{dI^p_1}{3}\right) + \sqrt{J_{2D}}\left(2\sqrt{I^p_{2D}}\right) \tag{11-64}$$

The value of I^p_1 and $\sqrt{I^p_{2D}}$ can be obtained by using appropriate expressions from Chapter 3 once the values of $d\epsilon^p_1$, $d\epsilon^p_2$, and $d\epsilon^p_3$ are found from observed stress–strain curves (Fig. 11-29).

The plastic potential surfaces are plotted in such a way that the incremental plastic strain vectors are normal to them. A comparison of Figs. 11-30 and 11-28 shows that for the artificial soil, the yield surfaces f_2 based on volumetric plastic strains and the potential surfaces Q are essentially similar. Hence in the context of the Cap model it is considered appropriate to assume associative flow behavior for this soil.

Identification of parameters: We first state the parameters required for the cap model.

For elastic part of constitutive matrix, $[\mathbf{C}^e]$:

1. E, Young's modulus

2. ν, Poisson's ratio

For plastic part of constitutive matrix, $[\mathbf{C}^p]$:

3. α
4. γ
5. θ } For defining failure envelope f_1 [Eq. (11-61)]
6. β

7. D
8. W } For defining yield envelope f_2 [Eq. (11-44b)]
9. Z and hardening function [Eq. (11-45)]
10. R

Calculation of Elastic Parameters E and ν: The value of E can be calculated from a plot of τ_{oct} versus ϵ (Fig. 5-15), obtained from conventional triaxial test results. For the CTC test,

$$\tau_{oct} = \frac{\sqrt{2}}{3}(\sigma_1 - \sigma_3) \tag{11-65}$$

Since σ_3 is kept constant in a conventional triaxial test,

$$d\tau_{oct} = \frac{\sqrt{2}}{3}d\sigma_1 \tag{11-66}$$

Therefore, the initial slope of the τ_{oct} versus ϵ_1 plot can be related to the Young's modulus as

$$\frac{d\tau_{oct}}{d\epsilon_1} = \frac{\sqrt{2}}{3}\frac{d\sigma_1}{d\epsilon_1} = \text{slope}$$

$$= \frac{\sqrt{2}}{3}E \tag{11-67a}$$

Hence

$$E = \frac{3}{\sqrt{2}} \times (\text{slope}) \tag{11-67b}$$

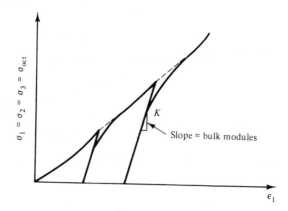

Figure 11-31   Schematic of hydrostatic compression test data.

From Fig. 5-15, the value of Young's modulus was determined as

$$E = 3904 \text{ psi } (27 \text{ MPa})$$

The bulk modulus K can be computed from the hydrostatic test data of the sample. The bulk modulus is defined as the slope of the unloading curve in a hydrostatic compression test [Figs. 5-14(b) and 11-31]. The bulk modulus was found as

$$K = 9000 \text{ psi } (62 \text{ MPa})$$

Since

$$K = \frac{E}{1 - 2\nu}$$

the value of Poisson's ratio can be found as

$$\nu = 0.283$$

Calculation of parameters α, γ, θ, and β for f_1: The equation of f_1 as given above is

$$\sqrt{J_{2D}} = \alpha + \theta J_1 + \gamma e^{-\beta J_1} \tag{11-61}$$

To determine the parameters, let us consider the state of stress when J_1 is equal to zero. Substituting zero value for J_1 in Eq. (11-61), we can write

$$\sqrt{J_{2D}} = \alpha - \gamma \tag{11-68}$$

Since β is assumed to be a positive quantity and compression is taken as positive, the quantity $e^{-\beta J_1}$ will be very small for large values of J_1. That is,

$$e^{-\beta J_1} \approx 0 \qquad \text{for large } J_1 \tag{11-69}$$

Then for large values of J_1, Eq. (11-61) reduces to

$$\sqrt{J_{2D}} = \alpha + \theta J_1 \tag{11-70a}$$

This equation represents Drucker–Prager yield criterion, and in the modified cap model, this is called the "Drucker–Prager II" surface (Fig. 11-32). The intercept of this

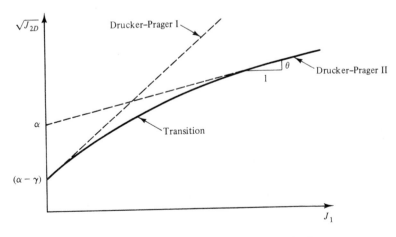

Figure 11-32 Interpretation of parameters for f_1.

second Drucker–Prager surface with the $\sqrt{J_{2D}}$ axis can be found by substituting a zero value for J_1 in Eq. (11-70a), that is,

$$\sqrt{J_{2D}} = \alpha \qquad (11\text{-}70\text{b})$$

The value of θ can be found from slope of the function in Eq. (11-70a). By differentiating Eq. (11-70a), we can write

$$\frac{d\sqrt{J_{2D}}}{dJ_1} = \theta \qquad (11\text{-}71)$$

In order to determine β, a point on the transition curve (Fig. 11-32) can be selected. By rearranging Eq. (11-61), we have

$$e^{-\beta J_1} = \frac{\alpha + \theta J_1 - \sqrt{J_{2D}}}{\gamma} \qquad (11\text{-}72\text{a})$$

Hence

$$\beta = -\frac{1}{J_1} \ln\left(\frac{\alpha + \theta J_1 - \sqrt{J_{2D}}}{\gamma} \right) \qquad (11\text{-}72\text{b})$$

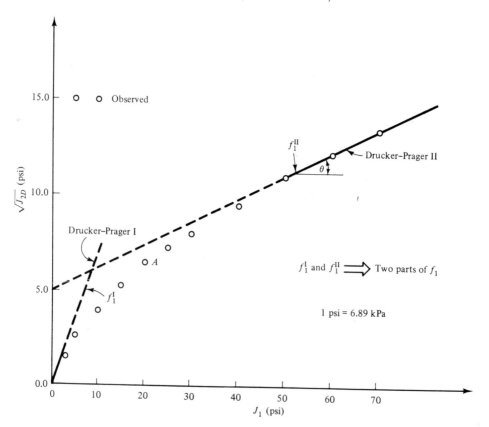

Figure 11-33 Ultimate (fixed) envelope for artificial soil.

TABLE 11-3   Points on Fixed Envelope f_1

J_1 (psi)	$\sqrt{J_{2D}}$ (psi)
0	0.00
5	2.60
10	4.00
15	5.25
20	6.50
25	7.25
30	8.00
40	9.50
50	11.00
60	12.25
70	13.50

Details of calculation: The plot of the ultimate envelope f_1 as shown in Fig. 11-27 can be used to tabulate (Table 11-3) values of J_1 and $\sqrt{J_{2D}}$. These values are replotted in Fig. 11-33 with the two Drucker–Prager surfaces indicated as f_1^I and f_1^{II}. From Fig. 11-33 the value of the intercept for f_1^I [Eq. (11-68)] is

$$\alpha - \gamma = 0.0$$

Now from Eq. (11-70a) the intercept for f_1^{II} is 5 psi (35 kPa); hence

$$\alpha = 5.0 \text{ psi } (35 \text{ kPa})$$

Substitution of the value of α in Eq. (11-68) gives

$$\gamma = 5.0 \text{ psi } (35 \text{ kPa})$$

From Fig. 11-33, the slope of Drucker–Prager II can be found as

$$\gamma = \frac{3.38}{28} = 0.12$$

In order to determine the value of β, let us select a point A on the transition curve (Fig. 11-33). The values of J_1 and $\sqrt{J_{2D}}$ at point A are

$$J_1 = 20.0 \text{ psi } (138 \text{ kPa})$$

$$\sqrt{J_{2D}} = 6.60 \text{ psi } (46 \text{ kPa})$$

Substituting these values in Eq. (11-72b), we obtain

$$\beta = 0.085$$

Calculation of parameters D, W, Z, and R for f_2 (Fig. 11-9):   The values of D, W, and Z have to be computed by using Eq. (11-45). The value of Z is zero when there is no significant initial yielding cap. For the artificial soil described in Chapter 5 there was no significant initial stress; hence

$$Z = 0$$

The values of D and W can be found from the hydrostatic test data [Fig. 5-14(b)]. Substituting $Z = 0$ in Eq. (11-45), we can write

$$X = -\frac{1}{D} \ln\left(1 - \frac{\epsilon_v^p}{W}\right) \tag{11-73a}$$

The value of X can be written in terms of mean pressure as

$$X = 3p \qquad (11\text{-}73\text{b})$$

where p is the mean pressure. Substitution of Eq. (11-73b) in Eq. (11-73a) leads to

$$3pD = -\ln\left(1 - \frac{\epsilon_v^p}{W}\right) \qquad (11\text{-}73\text{c})$$

Rearranging Eq. (11-73c), we have

$$\epsilon_v^p = W(1 - e^{-3pD}) \qquad (11\text{-}74)$$

However, the plastic volumetric strain can be expressed in terms of total and elastic components of strain as

$$\epsilon_v^p = \epsilon_v - \epsilon_v^e \qquad (11\text{-}75\text{a})$$

The elastic volumetric strains can be computed since the bulk modulus is known. Therefore,

$$\epsilon_v = W(1 - e^{-3pD}) + \frac{p}{K} \qquad (11\text{-}75\text{b})$$

The values of W and D are found so as to reproduce the observed loading path shown in Fig. 5-14(b). The test data for the hydrostatic compression will be used for this purpose. The observed data on mean pressure and volumetric strains [Fig. 5-14(b)] are shown in Table 11-4.

Details of calculation: Let us select two points on the loading portion of Fig. 5-14(b). In this calculation we have selected two points on which p is equal to 10 psi and 16 psi. Substituting these values in Eq. (11-75b), we can write

$$0.0742 - \frac{10}{9000} = -W(1 - e^{-3pD}) \qquad (11\text{-}76\text{a})$$

$$0.1138 - \frac{16}{9000} = -W(1 - e^{-3pD}) \qquad (11\text{-}76\text{b})$$

TABLE 11-4   Hydrostatic Compression Test Results

p (psi)	ϵ_v
0.	0.0
2	0.0139
4	0.0239
6	0.0471
8	0.0664
10	0.0742
12	0.0894
14	0.1117
16	0.1138
18	0.1170
20	0.1277

Equations (11-76) can be written as

$$0.0731 = -W(1 - e^{-30D}) \qquad (11\text{-}77\text{a})$$

$$0.1120 = -W(1 - e^{-48D}) \qquad (11\text{-}77\text{b})$$

Dividing Eq. (11-77a) by Eq. (11-77b), we obtain

$$\frac{1 - e^{-30D}}{1 - e^{-48D}} = 0.6524 \qquad (11\text{-}78)$$

The value of D can be found from a *trial-and-error* procedure to satisfy Eq. (11-78). The initial guess can be obtained by using the formula

$$\ln(0.6524) = (30 - 48)D \qquad (11\text{-}79)$$

Solution of this equation gives

$$D \approx 0.08 \qquad (11\text{-}80)$$

The value of D was then changed in steps of 0.01 until Eq. (11-78) was satisfied approximately. This can be done by hand calculation or by graphical methods. As an approximation D was found to be 0.0033. The accuracy of this value can be checked by substituting the value of D in Eq. (11-78):

$$\frac{1 - e^{-30(0.0033)}}{1 - e^{-48(0.0033)}} = 0.61 \ (\neq 0.6524)$$

The value of W was then calculated by using Eq. (11-76a). It should be noted that if Eq. (11-76b) were used, we will obtain a different value of W since the value of D was not correct; thus the approximate value of W may be calculated from Eq. (11-76a) or (11-76b). In this example the value of W was found from Eq. (11-76a) as

$$W = 0.695$$

By using the values of $D = 0.0033$ and $W = 0.695$, the volumetric strains were predicted for given values of p. A comparison of observed data and predictions from Eq. (11-75) are given in Table 11-5a. The comparison shows that the values of D and W approximately reproduce the observed behavior. However, to improve the accuracy, a search was made to obtain better values of D that will satisfy Eq. (11-78) more accurately.

In the next trial (trial 2), the value of D was taken as 0.00133. The values of W were found from both Eqs. (11-76a) and (11-76b). The computed values were

$$W_1 = 1.7914$$

$$W_2 = 1.6944$$

The average value of W is

$$W = 1.7429$$

Hence in the second trial the following values of W and D were used:

$$D = 0.00133$$

$$W = 1.7429$$

Observed data and predicted values are compared in Table 11-5b. As a further improvement, the value of D was selected as 0.0005 in the third trial. The corresponding value of W was found as 4.4 using the same procedure as described above.

TABLE 11-5a   Trial Prediction 1[a]

p (psi)	ϵ_v Observed	Predicted
0	0.0	0.0
2	0.0139	0.0143
4	0.0239	0.0288
6	0.0471	0.0436
8	0.0664	0.0588
10	0.0742	0.0742
12	0.0894	0.0898
14	0.1117	0.1060
16	0.1138	0.1224
18	0.1170	0.1391
20	0.1277	0.1561

[a] $D = 0.0033$, $W = 0.695$, $K = 9000$ psi.

TABLE 11-5b   Trial Predictions 2 and 3

Pressure, p (psi)	ϵ_v Observed	Trial 2[a]	Trial 3[b]
0	0.0	0.0	0.0
2	0.0139	0.0142	0.0134
4	0.0239	0.0286	0.0268
6	0.0471	0.0430	0.0404
8	0.0664	0.0576	0.0540
10	0.0742	0.0722	0.0676
12	0.0894	0.0870	0.0813
14	0.1117	0.1019	0.0948
16	0.1138	0.1170	0.1087
18	0.1170	0.1321	0.1224
20	0.1277	0.1474	0.1362

[a] $D = 0.00133$, $W = 1.7429$.
[b] $D = 0.0005$, $W = 4.4$.

Predicted values in the third trial are also given in Table 11-5b. It can be seen from this table that the third trial gave the best correlation between predictions and observed data. Hence the following values were used:

$$D = 0.0005 \, (\text{psi})^{-1} \left[0.000073 \, (\text{kPa})^{-1} \right]$$

$$W = 4.4$$

Alternative procedure for finding W and D: If the test was conducted up to high values of pressure, p, then the term e^{-3pD} in Eq. (11-75b) will approach to zero. Hence the value of the measured volumetric strain at the high pressure gives the value of W directly. Once W is known, D can be found from Eq. (11-75b).

It is possible to computerize the process of finding the parameters for a given (cap) model; this is described subsequently.

Determination of R: To determine the value of R it is required to know the size and shape of the yield caps, which can be represented as contours of constant plastic volumetric strains. For the artificial soil these contours were found to be approximately elliptical, and the ratio of major to minor axes of this ellipse was found to be approximately 2.0. In reality, the values of R were seen to lie in the range 1.67 to 2.0 in Fig. 11-28 in various zones of the $J_1-\sqrt{J_{2D}}$ space.

A summary of the parameters is given in Fig. 11-34. Hence the expressions for the cap model can be written in Foot-Pound–second (FPS) units/(psi) as

$$\sqrt{J_{2D}} = 5.00 + 0.12J_1 - 5e^{-0.085J_1} \tag{11-81a}$$

$$R^2J_{2D} + (J_1 - C)^2 = R^2b^2 \tag{11-81b}$$

$$X = \frac{1}{0.0005}\ln\left(1 - \frac{\epsilon_v^p}{4.4}\right) \tag{11-81c}$$

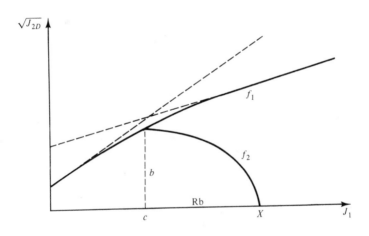

$$f_1 = \alpha + \theta J_1 - \gamma e^{-\beta J_1} = \sqrt{J_{2D}}$$

$$f_2 = R^2J_{2D} + (J_1 - C)^2 = R^2b^2$$

$$X = -\frac{1}{D}\ln\left(1 - \frac{\epsilon_v^p}{W}\right)$$

$$\epsilon_v = \frac{p}{K} + W(1 - e^{-3pD})$$

$E = 3904$ psi (27 MPa)

$v = 0.28$

$\alpha = 5.0$ psi (35 kPa)

$\theta = 0.12$

$\gamma = 5.0$ psi (35 kPa)

$\beta = 0.085$ (psi)$^{-1}$ [0.012 (kPa)$^{-1}$]

$D = 0.0005$ (psi)$^{-1}$ [0.000073 (kPa)$^{-1}$]

$W = 4.4$

$Z = 0.0$

$R = 1.67$ to 2.00

Figure 11-34 Parameters of modified cap model for artificial soil.

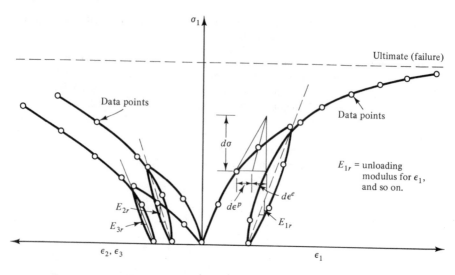

Figure 11-35 Symbolic stress–strain curves.

In the SI units (kPa), the expressions are

$$\sqrt{J_{2D}} = 35 + 0.12J_1 - 35e^{-0.085J_1} \qquad (11\text{-}82a)$$

$$R^2 J_{2D} - (J_1 - C)^2 = R^2 b^2 \qquad (11\text{-}82b)$$

$$X = -\frac{1}{0.000073} \ln\left(1 - \frac{\epsilon_v^p}{4.4}\right) \qquad (11\text{-}82c)$$

Back predictions: It is not as straightforward to back-predict stress–strain curves under various stress paths for the cap model as for the Cam clay models. However, the HC test can be predicted easily by using Eq. (11-75b); such predictions are presented in Ref. 20.

For other stress paths, the back predictions can be achieved by using the incremental constitutive relation

$$\{d\boldsymbol{\sigma}\} = [\mathbf{C}^{ep}]\{d\boldsymbol{\epsilon}\} \qquad (11\text{-}83)$$

The process of numerical integration of Eq. (11-83) can start from a given state of stress. Then depending on the stress path, we can choose values of the stress increments $\{d\boldsymbol{\sigma}\}$. The matrix $[\mathbf{C}^{ep}]$ can be found by using Eq. (11-35). The incremental stresses and strains can be accumulated after each increment. These predictions can be compared with the observed test data.

COMPUTER PROCEDURE FOR FINDING PARAMETERS

As can be seen from the calculations above, it is time consuming to evaluate manually parameters for the Cap model and other similar (complex) models. Hence it can be very useful to computerize the procedure for plotting raw laboratory data in the required form and evaluating the parameters by using the computer; this can also reduce the possibility of errors in the manual computations.

A computer procedure and code for evaluation of parameters for the Cap model have been developed by Zaman et al. (21). A description of the program is given below with brief details of an example problem; a listing and other features of the code are given in Appendix 4.

The procedure requires input of the following quantities:

1. A set of laboratory stress–strain curves under various stress paths. A number of points (Fig. 11-35) are chosen on each curve, and their coordinates (stress, strain) are input. An (average) value of the unloading modulus is required for each curve.
2. It is necessary to have data for at least one HC test, whereas data of one or more other stress paths are required.
3. For each curve, values of stresses at ultimate are input.

The program plots and determines the function f_1 [Eq. (11-61)] by an averaging scheme (Fig. 11-27). Based on specified increments for each curve, incremental stress and strain quantities are computed. Accumulated plastic volumetric strains are then plotted along each stress path (Fig. 11-28), and the computer plots yield caps f_2 [Eq. (11-44b)]. The vectors of total incremental strains are plotted as indicated in Fig. 11-30; at this time, the plastic potential functions are then manually drawn as curves orthogonal to the vectors.

All the required parameters (Fig. 11-34) for the Cap model are evaluated by the program and printed out. Appendix 4 gives details of raw data, plots of yield and potential surfaces, and values of parameters for the silty sand described in Chapter 10.

APPLICATIONS

A few examples of applications of the elastoplastic constitutive models in this chapter for solution of boundary value problems are presented below.

Example 11-4 Critical State Model for Laboratory Specimen

An example of a two-dimensional application of a critical state model is given in Ref. 22; here an axisymmetric triaxial specimen has been analyzed using the finite element method. Recently, Siriwardane and Desai (23) implemented this model in both two- and three-dimensional finite element procedures, and details of computational algorithms are given in Ref. 23. Since there is paucity of observed or previously solved data from two- and three-dimensional analysis with the critical state model, the accuracy has been verified by solving simple problems. Here behavior of a conventional triaxial specimen has been analyzed by using both two- (axisymmetric) and three-dimensional

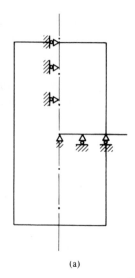

(a)

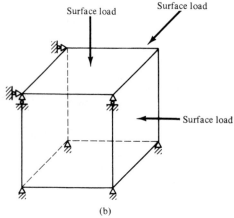

(b)

Figure 11-36 Simulation of triaxial test specimens: (a) cylindrical triaxial specimen; (b) truly triaxial specimen (Refs. 22, 23).

procedures, and the predictions are compared with those reported in Ref. 22. The finite element meshes used in the study are shown in Fig. 11-36. The following properties were used in the example:

$$\sigma_2 = \sigma_3 = 100 \text{ kN/m}^2$$
$$M = 1.0$$
$$\lambda = 0.14$$
$$\kappa = 0.026$$
$$\nu = 0.30$$
$$E_0 = 9900 \text{ kN/m}^2$$
$$e_0 = 1.08$$
$$p_0 = 114 \text{ kN/m}^2$$

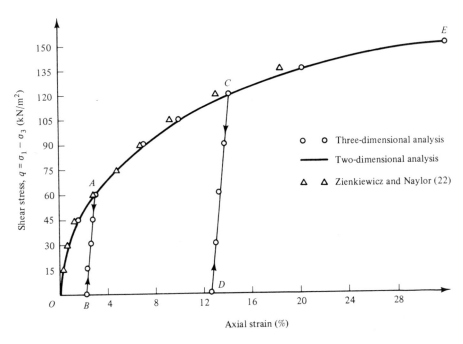

Figure 11-37 Triaxial test behavior from critical state model (Ref. 23).

The load was applied monotonically in increments of 15 kN/m² each, and one iteration was performed for each increment. The predictions from the analysis above are compared in Fig. 11-37 with those given by Zienkiewicz and Naylor (22). Both two- and three-dimensional procedures yield similar results, and the comparison is considered good. The computational algorithms have also been checked (23) with respect to stress paths involving load reversals; a typical computed path involving load reversal is given by $OABACDCE$ (Fig. 11-37). The unloading–reloading behavior is assumed to be elastic, and the virgin loading curve $OACE$ is the same as that obtained without load reversals; this shows the accuracy of the algorithm. An application of critical state model in a full-scale field experiment is given subsequently.

Example 11-5 Cap Model for Laboratory Specimen

Here the behavior of a conventional triaxial specimen is analyzed (23) using the Cap model. The finite element mesh involving one element was shown in Fig. 11-36(b). The following properties relevant to an artificial soil, reported in Refs. (15, 16, 20), are used:

$$\alpha = 5.6 \text{ psi } \left(38.58 \text{ kN/m}^2\right)$$

$$\gamma = 5.6 \text{ psi } \left(38.58 \text{ kN/m}^2\right)$$

$$\theta = 0.11$$

$$\beta = 0.062 \text{ (psi)}^{-1} \left[0.009 \text{ (kPa)}^{-1}\right]$$

$$R = 2.0$$

$$D = 0.05 \text{ (psi)}^{-1} \left[0.00725 \left(\text{kN/m}^2\right)^{-1}\right]$$

$$W = 0.18$$

$$E = 4000 \text{ psi } (27{,}560.0 \text{ kN/m}^2)$$

$$\nu = 0.35$$

$$p_0 = 10 \text{ psi } (68.90 \text{ kN/m}^2)$$

(*Note*: These parameters are somewhat different from those in Fig. 11-34 because a different number of tests were used for their determination.)

The analysis was carried out by using increments of 1.0 psi (6.89 kN/m²) with one iteration per increment. During loading, the cap surface expands continuously until an ultimate state is reached [Fig. 11-38(a)]. Predicted stress–strain curves are shown in Fig. 11-38(b). The results from the two-dimensional axisymmetric and

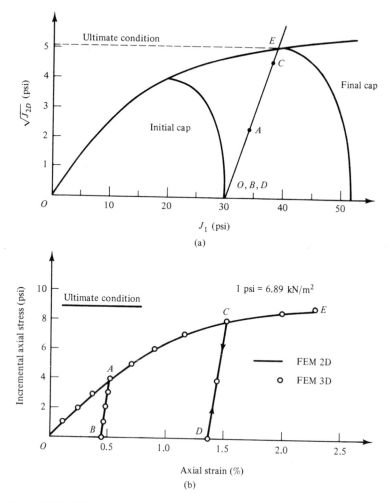

Figure 11-38 Triaxial test behavior from cap model: (a) experimental results and stress paths followed in CTC test; (b) stress–strain behavior (Ref. 23).

three-dimensional analysis are very close. The ultimate value of $\sqrt{J_{2D}}$ along the CTC stress path is 5.0 psi (34.45 kN/m^2); the corresponding σ_1 value is 8.66 psi (59.67 kN/m^2) [Fig. 11-38(a)]. This value compares very well with the finite element predictions of ultimate states [Fig. 11-38(b)].

The computational algorithm was also checked with respect to a path involving load reversals. Here the sample was loaded, unloaded, and reloaded along the path $OABACDCE$ (Fig. 11-38).

Example 11-6   Evaluation of Models for Footing in Soil

Elastoplastic models described in this chapter were used to predict the behavior of a plane strain laboratory footing test (20). The soil type used in the study was the artificial soil (15–18,20) described in Chapter 5. Parameters for different constitutive models determined from a comprehensive laboratory test on the artificial soil are given below.

$$E = 4000.0 \text{ psi} \qquad (1 \text{ psi} = 6.89 \text{ kPa})$$

$$\nu = 0.35$$

$$c = 0.0$$

$$\phi = 33°$$

$$\lambda = 0.11$$

$$\kappa = 0.001$$

$$\rho = 0.072 \text{ pci}$$

$$e_0 = 0.65$$

$$\alpha = 5.6 \text{ psi}$$

$$\beta = 0.062 \text{ (psi)}^{-1}$$

$$\gamma = 5.6 \text{ psi}$$

$$\theta = 0.11$$

$$R = 2.0$$

$$W = 0.18$$

$$D = 0.05 \text{ (psi)}^{-1}$$

where λ and κ are parameters for the critical state model; the critical state line has been assumed to be represented by the final yield surface, f_1, which is described by α, β, γ, and θ in the cap model [Eq. (11-61)]. The other parameters, c and ϕ, are related to Drucker–Prager model, and R, W, and D are additional parameters for the cap model [Eq. (11-45)].

(*Note*: The parameters for the Cap model are somewhat different from those given in Fig. 11-34 because a different number of tests were used for their determination).

The finite element mesh used in this study is shown in Fig. 11-39. The load application on the model footing was simulated in an incremental manner. Figure 11-40 shows the solutions obtained by using various constitutive models in comparison with the observations. From this figure it can be concluded that, for the artificial soil

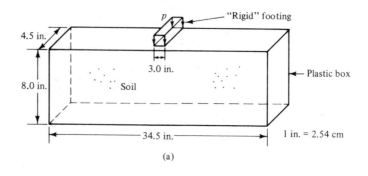

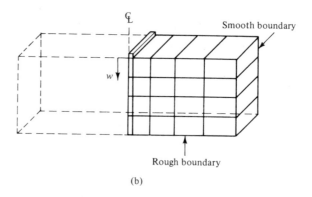

Figure 11-39 Finite element mesh used for the model footing test: (a) test model; (b) finite element mesh (Ref. 20).

used in the study, the Drucker–Prager model does not yield satisfactory results. The critical state model shows a better prediction for displacements. The cap models described in this chapter give excellent predictions compared with experimental observations.

For the artificial soil, hardening models that allow for the volume-change behavior are required because conventional models such as the Drucker–Prager model do not account for this aspect. This was evident in the laboratory observations, which indicated that the soil below the footing experienced the majority of deformations (20). Figure 11-41 shows displacement patterns in the soil mass near the ultimate load as computed using the Drucker–Prager and cap models. The results from the latter show trends that compare better with the laboratory observations in which the majority of downward deformations occurred under the footing.

Example 11-7   Various Models for Track Support Structure

In this example the deformation behavior of a track support structure (Fig. 11-42) is analyzed. Linear and nonlinear analyses were performed for the UMTA test section at the Transportation Test Center at Pueblo, Colorado (24, 25). Figure 11-43 shows the details of one of the sections along the tangent track with locations of the instruments (26).

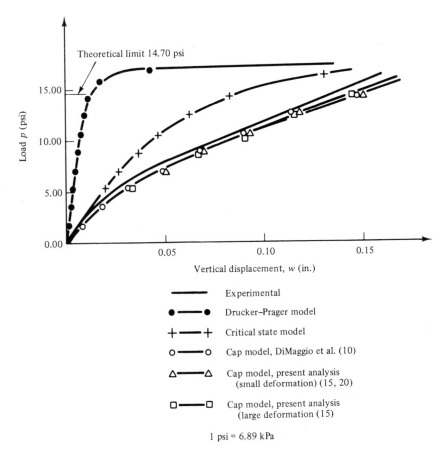

Figure 11-40 Comparisons between predictions from various models and observations (Refs. 15, 20).

Linear analysis: Material parameters for this analysis, adopted from the report by Stagliano et al. (26), are as follows:

	Rail	Tie	Ballast	Subballast	Subgrade (Sand)
E (psi)	30×10^6	5×10^6	30,000	20,000	5,000
(kN/m²)	(207×10^6)	(34.5×10^6)	(207×10^3)	(138×10^3)	(34.5×10^3)
ν	0.35	0.20	0.40	0.35	0.45

Nonlinear analysis: For the nonlinear analysis, the parameters were determined from a comprehensive series of truly triaxial tests for the subgrade silty sand, subballast, and wood in ties obtained from the test section (24, 25, 27). Based on the test result, it was found that the subgrade sand can be represented by a critical state model [Eq. (11-22d)], the subballast by a cap model [Eqs. (11-43) and (11-44)], and the ballast by a

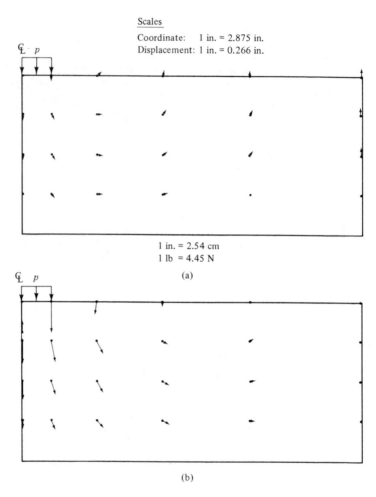

Figure 11-41 Comparison of displacement patterns: (a) Drucker–Prager model, $p = 16.0$ psi; (b) modified cap model, $p = 15.0$ psi (Ref. 20).

variable moduli model [Eq. (8-32)]. The behavior of the concrete tie and steel rail were assumed to be linear elastic. The geometric properties and constitutive parameters for various components used in the nonlinear analysis are given below (24).

Material properties:

Rail (steel)—linear elastic:

$E = 30 \times 10^6$ psi $(207.0 \times 10^6 \text{ kN/m}^2)$

$\nu = 0.3$

$I = 71.4 \text{ in}^4 \ (2971.9 \text{ cm}^4)$

$A = 11.65 \text{ in}^2 \ (75.16 \text{ cm}^2)$

Length $= 108$ in. (274.32 cm)

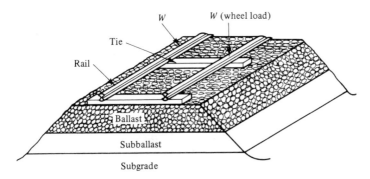

Figure 11-42 Schematic representation of track support structure (Ref. 24).

Subballast—cap model

$E = 20{,}000$ psi (138×10^3 kN/m^2)

$D = 0.308 \times 10^{-4}$ (psi)$^{-1}$ [0.446×10^{-5} (kN/m^2)$^{-1}$]

$\nu = 0.40$

$\alpha = 26.0$ psi (179.4 kN/m^2)

$\gamma_m = 0.0833$ lb/in^3 (0.00230 kg/cm^3)

$\gamma = 21.5$ psi (148.35 kN/m^2)

$R = 1.24$

$\beta = 0.02$

$W = 0.035$

Subgrade (sand)—modified Cam clay model

$M = 1.24$

$E_0 = 12{,}000$ psi (82.8×10^3 kN/m^2)

$\lambda = 0.014$

$\nu = 0.28$

$\kappa = 0.0024$

$\gamma_m = 0.081$ lb/in^3 (0.00224 kg/cm^3)

$e_0 = 0.340$

Ballast—variable moduli model

$K_0 = 4 \times 10^3$ psi (27.6×10^3 kN/m^2)

$G_0 = 1.714 \times 10^3$ psi (11.83×10^3 kN/m^2)

$K_1 = -5.4917 \times 10^3$ psi (-37.89×10^3 kN/m^2)

$\gamma_1 = 2.5593$

$K_2 = 2.536 \times 10^7$ psi (17.50×10^7 kN/m^2)

$\gamma_2 = -60.172$

$\gamma_m = 0.0613$ lb/in^3 (0.00169 kg/cm^3)

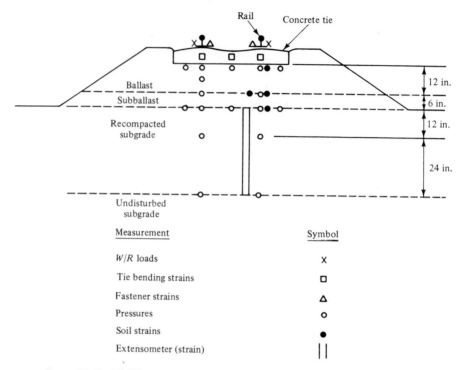

Figure 11-43   UMTA test section with detailed locations of instrumentation (Ref. 26).

Tie (concrete)—linear elastic

$E = 4.2 \times 10^6$ psi $(29 \times 10^6 \text{ kN/m}^2)$

$\nu = 0.20$

Width $= 10$ in. $(25.4$ cm$)$

Depth $= 6$ in. $(15.24$ cm$)$

The density γ_m of different materials is used to evaluate initial stresses. The value of the coefficient of earth pressure equal to 1 was adopted (24).

Interface behavior: The two- and three-dimensional models incorporate provision of interface elements that can permit relative motions and shear transfer between various components. In the present analysis, the concept of thin-layer element was used; details are given in Desai et al. (28). It was found that the thin-layer interface can provide results of the same or improved reliability as those of the conventional interface elements used previously. In this study interface elements were included only at the junction between the tie and ballast; its properties are as follows:

$$E = 30,000 \text{ psi } \left(208.5 \text{ kN/m}^2\right)$$

$$\nu = 0.30$$

$$G = 225.0 \text{ psi } \left(1553 \text{ kN/m}^2\right)$$

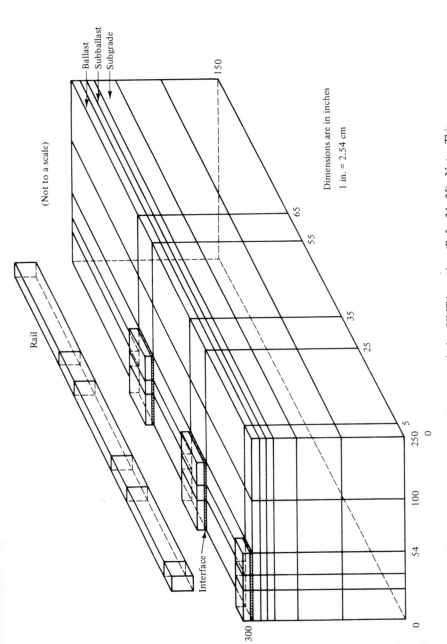

Figure 11-44 Typical finite element mesh for UMTA section (Refs. 24, 25). Note: This isometric view is not drawn to proportional scales; this is done to show clearly the longitudinal details. As a result, the elements in the transverse direction appear much smaller than they actually are.

(Not to a scale)

Rail

Interface

Ballast
Subballast
Subgrade

Dimensions are in inches
1 in. = 2.54 cm

339

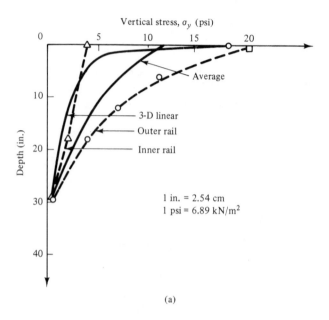

(a)

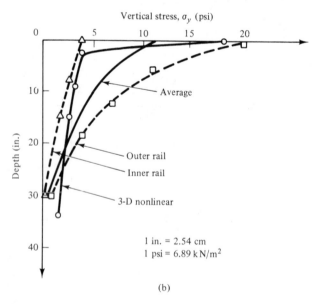

(b)

Figure 11-45 Comparisons of observed vertical stresses with (a) linear analysis and (b) nonlinear analysis (Ref. 24).

Mesh: Symmetry was assumed, and a quarter of the domain was discretized, as shown in Fig. 11-44.

Loading: A load of 16,000 lb (72.1 kN) for R-42 vehicle equivalent to a static wheel load of 13,100 lb (58.3 kN) as reported by Stagliano et al. (26) was applied on the rail above the central tie.

Results: Figure 11-45 shows comparison between the finite element predictions and observations below the inner and outer rails of the test section. It is surprising that the observations for the inner and outer rails at the tangent section are quite different. The data for the inner rail appears to show trends similar to those shown by closed-form solutions and other field observations (24). The numerical predictions and the observations together with an average of the two observations are shown in Fig. 11-45(a) and (b) for linear analyses, respectively. The predictions compare well with the observations at the inner rail.

It is reported in Ref. 26 the photographically observed values of rail displacement were not reliable due to unresolved inconsistencies. The numerical predictions for displacements are considered to be satisfactory because the predicted values of vertical displacement of 0.0202 in. (0.0513 cm) at the subgrade compare well with the observation of 0.021 in. (0.0533 cm) at the subgrade (26).

The seat load computed by integrating vertical stresses in the tie elements (Fig. 11-44) at the integration points just below the rail was found to be about 50% of the wheel load. This value compares well with that reported from the field data (26).

Comment: The results from the linear elastic analysis [Fig. 11-45(a)] show satisfactory trends with the observations at the inner rail; it may be noted here that the parameters for this analysis were adopted from Stagliano et al. (26). In the case of the nonlinear analysis [Fig. 11-45(b)], improved comparison is observed with respect to inner rail data; the parameters for the nonlinear analysis were derived from the special laboratory tests.

PROBLEMS

11-1. Find the parameters for the Cam clay model for the artificial soil described in Chapter 5.
Answers. $E = 4000$ *psi* (27,600 *kPa*), $\nu = 0.35$, $\alpha = 5.60$ *psi* (39 *kPa*), $\theta = 0.11$, $\gamma = 5.60$ *psi* (39 *kPa*), $\beta = 0.062$ $(psi)^{-1}$ [0.009 $(kPa)^{-1}$], $W = 0.18$, $D = 0.05$ $(psi)^{-1}$ [0.0072 $(kPa)^{-1}$], *and* $R = 2.00$.

11-2. Find the parameters for the cap model for the Pueblo soil given in Figs. 10-7 to 10-14.
Answers. E (*initial slope of CTC curves*) = 11,500 *psi* (79 *MPa*), $K = 19,450$ *psi* (134 *MPa*), $\nu = 0.40$, $\alpha = 13$ *psi* (90 *kPa*), $\theta = 0.10$, $\beta = 0.032$ $(psi)^{-1}$ [0.0046 $(kPa)^{-1}$], $\gamma = 13$ *psi* (90 *kPa*), $D = 0.0003$ $(psi)^{-1}$ [0.000044 $(kPa)^{-1}$], $W = 0.9$, *and* $Z = 0.0$.
(*Note*: Because of individual preferences in the curve-fitting procedure, the reader may find the values of the parameters somewhat different from the answers given in these problems.)

REFERENCES

1. Rendulic, L., "Pore-Index and Pore Water Pressure," *Bauingenieur*, Vol. 17, No. 559, 1936.

2. Hvorslev, M. J., "Physical Properties of Remolded Cohesive Soils," *Transl. 69-5*, U.S. Army Corps of Engineers Waterways Exp. Stn., Vicksburg, Miss., June 1969.

3. Roscoe, K. H., Schofield, A., and Worth, C. P., "On the Yielding of Soils," *Geotechnique*, Vol. 8, 1958, pp. 22–53.

4. Roscoe, K. H., and Pooroshasb, H. B., "A Theoretical and Experimental Study of Stress–Strain in Triaxial Compression Tests on Normally Consolidated Clays," *Geotechnique*, Vol. 13, 1963, pp. 12–28.

5. Roscoe, K. H., Schofield, A., and Thurairajah, A., "Yielding of Clays in State Wetter than Critical," *Geotechnique*, Vol. 13, No. 3, 1963, pp. 211–240.

6. Roscoe, K. H. and Burland, J. B., "On the Generalized Stress–Strain Behavior of 'Wet' Clay," in *Engineering Plasticity*, J. Heyman and F. A. Leckie (Eds.), Cambridge University Press, Cambridge, 1968, pp. 535–609.

7. Schofield, A. N., and Worth, C. P., *Critical State Soil Mechanics*, McGraw-Hill Book Company, London, 1968.

8. Casagrande, A., "Characteristics of Cohesionless Soils Affecting the Stability of Slopes and Earth Fills," *J. Boston Soc. Civil Eng.*, 1936, pp. 257–276.

9. Hagmann, A. J., Christian, J. T., and D'Appolonia, D. J., "Stress–Strain Models for Frictional Materials," *Rep. R70-18*, Massachusetts Institute of Technology, Cambridge, Mass., 1970.

10. DiMaggio, F. L., and Sandler, I. S., "Material Model for Granular Soils," *J. Eng. Mech. Div.*, *ASCE*, Vol. 97, No. EM3, 1971, pp. 935–950.

11. Baron, M. L., Nelson, I., and Sandler, I. S., "Influence of Constitutive Models on Ground Motion Predictions," *J. Eng. Mech. Div.*, *ASCE*, Vol. 99, No. EM6, 1973, pp. 1181–1200.

12. Sandler, I. S., DiMaggio, F. L., and Baladi, G. Y., "Generalized Cap Model for Geologic Materials," *J. Geotech. Eng. Div.*, *ASCE*, Vol. 102, No. GT7, July 1976, pp. 683–699.

13. Baron, M. L., Nelson, I., and Sandler, I. S., "Influence of Constitutive Models on Ground Motion Predictions," *Contract Rep. S-71-10*, U.S. Army Corps of Engineers Waterways Exp. Stn., Vicksburg, Miss., Nov. 1971.

14. Sandler, I. S., and Baron, M. L., "Recent Developments in the Constitutive Modeling of Geological Materials," *Proc. 3rd Int. Conf. Numer. Methods Geomech.*, Vol. 1, Aachen, W. Germany, W. Wittke (Ed.), Balkema Press, Rotterdam, 1979, pp. 363–376.

15. Desai, C. S., Perumpral, J. V., Sture, S., and Phan, H. V., "Geometric and Material Nonlinear Three-Dimensional Analysis of Structures Moving on Ground," *Report to NSF*, Dept. of Civil Eng., Va. Polytech. Inst. and State Univ., Blacksburg, 1979.

16. Sture, S., Desai, C. S., and Janardhanam, R., "Development of a Constitutive Law for an Artificial Soil," *Proc. 3rd Int. Conf. Numer. Methods Geomech.*, Vol. 1, Aachen, W. Germany, W. Wittke (Ed.), Balkema Press, Rotterdam, 1979, pp. 309–317.

17. Desai, C. S., "Some Aspects of Constitutive Laws for Geologic Media," *Proc. 3rd Int. Conf. Numer. Methods Geomech.*, Aachen, W. Germany, W. Wittke (Ed.), Balkema Press, Rotterdam, 1979.

18. Desai, C. S., Phan, H. V., and Perumpral, J. V., "Mechanics of Three-Dimensional Soil Structure Interaction," *J. Eng. Mech. Div., ASCE*, Vol. 108, No. EM5, Oct. 1982, pp. 731–747.

19. Siriwardane, H. J., and Desai, C. S., "Evaluation of Constitutive Parameters for Geologic Media: Modified Cam Clay and Cap Models," *Proc. Workshop Session, Symp. Implementation of Computer Procedures and Stress–Strain Laws in Geotech. Eng.*, Chicago, Aug. 1981.

20. Desai, C. S., Phan, H. V., and Sture, S., "Procedure, Selection and Application of Plasticity Models for a Soil," *Int. J. Numer. Anal. Methods Geomech.*, Vol. 5, 1981, pp. 295–311.

21. Zaman, M. M., Desai, C. S., and Faruque, M. O., "An Algorithm for Determining Parameters for Cap Model from Raw Laboratory Test Data," *Proc. 4th Int. Conf. Numer. Methods Geomech.*, Edmonton, Canada, 1982.

22. Zienkiewicz, O. C., and Naylor, D. J., "An Adoptation of Critical State Soil Mechanics Theory for Use in Finite Elements," *Proc. Roscoe Memorial Symp. on Stress–Strain Behavior of Soils*, Cambridge, Foulis, 1972.

23. Siriwardane, H. J., and Desai, C. S., "Computational Procedures for Nonlinear Three-Dimensional Analysis with Some Advanced Constitutive Laws," *Int. J. Numer. Anal. Methods Geomech.*, Vol. 7, No. 2, 1983, pp. 143–171.

24. Desai, C. S., and Siriwardane, H. J., "Analysis of Track Support Structures," *J. Geotech. Eng. Div., ASCE*, Vol. 108, No. GT3, March, 1982, pp. 461–480.

25. Siriwardane, H. J., "Nonlinear Soil–Structure Interaction Analysis for One-, Two- and Three-Dimensional Problems Using Finite Element Method," Ph.D. dissertation, Va. Polytech. Inst. and State Univ., Blacksburg, 1980.

26. Stagliano, T. R., Mente, L. J., Garden, E. C., Jr., Baxter, B. W., Hale, W. K., and Maccabe, R. W., Jr., "Pilot Study for Definitions of Track Component Load Environments," *Rep. UMTA-MA-06-0100-81-1*, Department of Transportation, Washington, D.C., 1981.

27. Janardhanam, R., "Constitutive Modelling of Materials in Track Support Structures," Ph.D. dissertation, Va. Polytech. Inst. and State Univ., Blacksburg, 1981.

28. Desai, C. S., Zaman, M. M., Lightner, J. G., and Siriwardane, H. J., "Thin-Layer Element for Interfaces and Joints," *Int. J. Numer. Anal. Methods Geomech.*, in press.

12

RECENT DEVELOPMENTS

The objective of this chapter is to present reviews of recent developments that go beyond the commonly used models covered in the previous chapters. The literature on the subject is vast and hence, the reviews have to be rather brief, and as a result, we found it difficult to make the explanations as simple as we would like them to be. Hence, the reader may wish to study this chapter in conjunction with relevant references cited for the specific topics. Also, some of the symbols used in this chapter have been adopted as they occur in the specific publications and may not have the same meanings as in the previous chapters.

ISOTROPIC AND ANISOTROPIC MATERIALS AND HARDENING

If the material behavior is not affected by orientation, it is called *isotropic*. A material that is initially isotropic may remain isotropic during the deformation process or may become anisotropic; the behavior of an anisotropic material is dependent on its orientation.

As discussed in Chapter 11, engineering materials often experience hardening during (plastic) straining. Hardening is described by using criteria that can define isotropic or anisotropic hardening. The former implies that an initially isotropic material remains isotropic during straining. However, an initially isotropic material can exhibit anisotropic behavior during anisotropic hardening. Indeed, an initially anisotropic material will usually exhibit aniso-tropic behavior due to isotropic or anisotropic hardening.

If a material is assumed to experience no hardening, then it is called *perfectly plastic.* Yield criteria for this type of material can be expressed as

or

$$f = f(J_i) \atop f_1(J_i) - k = 0 \Bigg\} i = 1, 2, 3$$

$$(12\text{-}1a)$$

$$(12\text{-}1b)$$

where f_1 is a function of the stress invariants J_1, J_2, and J_3, and k is a constant that defines yielding. Classical Tresca, von Mises, Mohr–Coulomb, and Drucker–Prager yield criteria belong to this category. Here the *shape, size,* and *orientation* of the yield surface remain unchanged during plastic straining [Fig. 12-1(a)].

If an initially isotropic material hardens isotropically (1, 2), the yield condition can be expressed as

or

$$f_1(J_i) - f_2(I_i^p) = 0 \atop f_1(J_i) - k(I_i^p) = 0 \Bigg\} i = 1, 2, 3$$

$$(12\text{-}2a)$$

$$(12\text{-}2b)$$

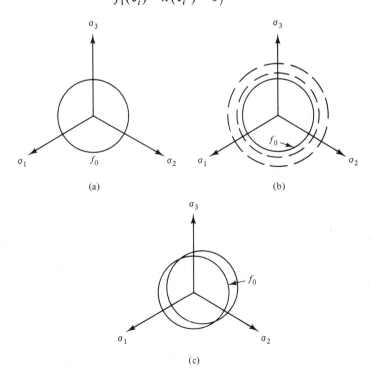

Figure 12-1 Plasticity models: (a) perfectly plastic; (b) isotropic hardening; (c) kinematic hardening.

where k is expressed as a function of parameters such as invariants of plastic strains, I_i^p. In other words, as hardening occurs the value of k changes and the size of the yield surface changes [Fig. 12-1(b)]. However, its shape and orientation do not change. Thus *isotropic hardening* takes place when the initial yield surface in the stress space expands uniformly without changing its shape or form during plastic flow and the center of the yield surface remains the same during the expansion. The initial yield surface f_0 is often defined as the locus of stress states when the first yielding occurs (Fig. 12-1); the states of stress within f_0 are considered elastic. These models cannot account for the Bauchinger effect involving different behavior in tension and compression.

The critical state and cap models described in Chapter 11 combine both Eqs. (12-1) and (12-2). The function in Eq. (12-1) is used to define an ultimate fixed surface, whereas that in Eq. (12-2) is used to define a series of expanding (or contracting) yield surfaces. Thus the critical state and cap models are isotropic hardening models for isotropic materials in which the size of the yield surface changes but the orientation and shape remain invariant (Figs. 11-6 and 11-8).

The idea of *kinematic hardening* was introduced by Ishlinski (3) and Prager (4). Here the shape and size of the yield surface remain the same but the yield surface can translate in the direction of the plastic strain increment in the stress space without rotation. This concept is often referred to as *Prager's kinematic hardening rule*. Here the size of the yield surface does not change.

Shield and Ziegler (5,6) modified and improved the Prager's rule by stipulating the translation of the yield surface to occur in the direction of radius vector joining the center of the yield surface to the stress point. Baltov and Sawczuk (7) introduced a further modification to permit rotation of the yield surface in the stress space.

Phillips et al. (8–10) have proposed hardening rules for isotropic material through a concept of loading surfaces that enclose the yield surface and describe reduction in yield strength during unloading.

The foregoing rules involve definition of the change (translation) of the yield or loading surfaces. However, they did not include a provision for hardening during (plastic) straining. In other words, the change in the size of the yield surface was not included; we now review works that permit nonlinear (kinematic) hardening that include modification in the size of the yield surface.

A general expression for elastic–plastic material was proposed by Green and Naghdi (11):

$$d\epsilon_{ij}^p = \lambda a_{ij} \frac{\partial f}{\partial \sigma_{kl}} d\sigma_{kl} \qquad (12\text{-}3)$$

where a_{ij} and σ_{kl} are second-order tensors and are functions of the state variables. This expression can be considered to include a number of rules discussed herein.

Kadashevitch and Novozhilov (12) proposed the following function for nonlinear hardening:

$$f\left[\sigma_{ij} - C\left[\left(\epsilon_{ij}^{p}\right)\left(\epsilon_{ij}^{p}\right)\right]\right] - k = 0 \qquad (12\text{-}4a)$$

where k is a constant and C a function of plastic strains which is included to define *plastic modulus*. An improvement of the rule in Eq. (12-4a) was proposed by Eisenberg and Phillips (13):

$$f\left(\sigma_{ij} - C(K_1)\epsilon_{ij}^{p}\right) - k = 0 \qquad (12\text{-}4b)$$

where $K_1 = d\epsilon_{ij}^{p}d\epsilon_{ij}^{p}$.

In the next developments, a series of (yield) surfaces were defined in the stress space, and the hardening was considered by defining constant or variable plastic work hardening moduli. This can be traced to the work of Koiter (14), who showed that the slip theory of Batdford and Budiansky (15) can be considered to represent a collection or nest of yield surfaces in the stress space. Mroz (16) and Iwan (17) proposed a family of *nested* yield surfaces in which the initial yield surface translates, as in kinematic hardening, and also expands with isotropic hardening. The hardening behavior is defined through a field of work hardening moduli which are assumed constant for each yield surface. Prevost (18–22) has proposed significant extensions and has formalized the nested yield surface model for application to soil behavior.

Dafalias and Popov (23–26) and Krieg (27) proposed the modification in the nested surface model of Mroz by considering two surfaces, a loading and a bounding surface. The intermediate surfaces are defined on the basis of the two surfaces, and the plastic work hardening moduli are expressed so as to vary smoothly with hardening.

Now we consider in detail some of the foregoing models that involve change in size and translation of the yield surface. Note that these models are special cases of general *anisotropic* hardening that can involve change in size, shape, and orientation of yield surface. A brief description of such general models will be given subsequently.

KINEMATIC HARDENING RULES

Prager (4) assumed that during plastic deformations the yield surface translates, which is expressed as

$$f(\sigma_{ij} - \alpha_{ij}) = k^2 \qquad (12\text{-}5a)$$

or

$$f_1(\sigma_{ij}, \alpha_{ij}) - k^2 = 0 \qquad (12\text{-}5b)$$

where α_{ij} represents total translation of the yield surface. Since α_{ij} is not necessarily isotropic, the material can experience (induced) anisotropy due to the hardening process. As discussed in Chapter 9, the increments of plastic strain are given by the following flow rule:

$$de^p_{ij} = d\lambda \frac{\partial f}{\partial \sigma_{ij}}, \qquad d\lambda > 0 \tag{12-6}$$

Prager further assumed that the surface in Eq. (12-5a) translates in the direction of de^p_{ij} [Fig. 12-2(a)]; this is expressed by

$$d\alpha_{ij} = c \, de^p_{ij} \tag{12-7}$$

where $d\alpha_{ij}$ is the increment (change) in α_{ij} and c the material constant. As described in Chapters 10 and 11, the value of $d\lambda$ is obtained from the consistency condition that the stress point remains on the yield surface during the plastic flow, hence

$$d\lambda = \frac{1}{c} \frac{(\partial f / \partial \sigma_{ij}) \, d\sigma_{ij}}{(\partial f / \partial \sigma_{km})(\partial f / \partial \sigma_{km})} \tag{12-8}$$

It has been shown that Prager's hardening rule is not invariant with respect to possible reductions in dimensions; that is, if the yield surface in the nine-dimensional space of σ_{ij} moves in the direction of the normal to the yield surface at the stress point, the yield locus in the two-dimensional space such as plane stress (σ_x, σ_y) does not move in the same direction (5, 6).

It was shown by Shield and Ziegler (5) that with the von Mises yield criterion, the yield surface moves in the direction of the radius connecting the

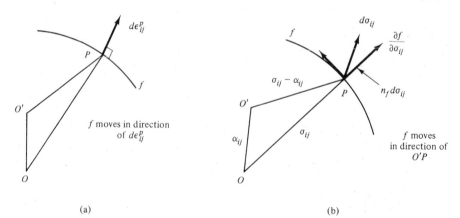

(a)   (b)

Figure 12-2 Kinematic hardening models: (a) Prager's kinematic hardening rule; (b) modification by Shield and Ziegler (Refs. 4, 5).

center (O') of the yield surface with the stress point (P) [Fig. 12-2(b)]. Hence the hardening rule was modified as

$$d\alpha_{ij} = (\sigma_{ij} - \alpha_{ij})\,d\mu, \qquad d\mu > 0 \qquad (12\text{-}9)$$

As before, the scalar $d\mu$ is determined from the consistency condition, hence

$$d\mu = \frac{(\partial f/\partial \sigma_{ij})\,d\sigma_{ij}}{(\sigma_{km} - \alpha_{km})(\partial f/\partial \sigma_{km})} \qquad (12\text{-}10)$$

The material behavior is completely defined by the flow rule [Eqs. (12-6) and (12-8)] and the hardening rule [Eqs. (12-9) and (12-10)]. If c is constant, we have a *linear hardening rule*; however, c can be assumed to be a function of the history of stress or plastic work, and then *nonlinear hardening* rules (in simple tension and compression) can be defined.

In the foregoing developments, kinematic hardening was considered in which the yield surface translates rigidly but does not change its shape and size. However, many materials may be better represented such that the yield surface changes (expands) and translates rigidly. Mroz (16) observed that the foregoing kinematic hardening models, although they are more realistic in accounting for the Bauschinger effects, and for anisotropy induced during plastic deformations, are not suitable for more complex behavior involving unloading and reloading along different stress paths. For instance, they may not be able to describe adequately the behavior of a material (metal) subjected to compression after it has undergone prestraining in tension. Consequently, Mroz proposed a modification that considers combination of both kinematic and isotropic hardening, expressed as

$$f_1(\sigma_{ij} - \alpha_{ij}) - f_2(\lambda) = 0 \qquad (12\text{-}11)$$

where $f_2(\lambda)$ is a measure of the expansion of the yield surface, α_{ij}, as before, represents the total translation, and λ is a monotonically increasing parameter which can be expressed in terms of rate or increments of plastic strains. For constant λ, Eq. (12-11) results in kinematic hardening in Eq. (12-5). For no translation and for $f_2(\lambda)$ to be a monotonically increasing function, we have isotropic hardening [Eq. (12-2)].

NESTED SURFACES MODELS

The rule proposed by Mroz (16) and Iwan (17) involving both kinematic and isotropic hardening is often referred to as the *nested* or *nesting surfaces* model. We now present the explanation given by Mroz (16) using the diagrams in Fig. 12-3.

Figure 12-3(a) shows a uniaxial stress–strain response discretized in n linear segments or pieces. A *constant* tangent modulus E_i ($i = 1, 2, \ldots, n$) or *constant* plastic modulus E_i^p ($i = 1, 2, \ldots, n$) is associated with each yield surface. The relation between E_i and E_i^p is obtained by assuming that the increment of total strain $d\epsilon$ can be expressed as the sum of the increment of elastic, $d\epsilon^e$, and plastic, $d\epsilon^p$, components as

$$d\epsilon = d\epsilon^e + d\epsilon^p \tag{12-12}$$

Now, since

$$d\epsilon = \frac{d\sigma}{E}, \qquad d\epsilon^e = \frac{d\sigma}{E^e}, \qquad d\epsilon^p = \frac{d\sigma}{E^p} \tag{12-13}$$

we have

$$\frac{1}{E} = \frac{1}{E^e} + \frac{1}{E^p} \tag{12-14a}$$

where $d\sigma$ is the increment of stress and E^e the elastic modulus. As we shall consider later, in generalization of Eq. (12-14a) for multiaxial cases, corresponding moduli H, H^e, and H^p are related as

$$\frac{1}{H} = \frac{1}{H^e} + \frac{1}{H^p} \tag{12-14b}$$

A representation of the curve in Fig. 12-3(a) in the stress space is shown as a series of surfaces f_i ($i = 0, 1, 2, \ldots, n$) [Fig. 12-3(b)], where f_0 is the initial yield surface. The surfaces are associated with each segment and are situated concentrically before the deformation process takes place.

Physical Interpretation

Surfaces $f_1, f_2, \ldots, f_n$ represent specific regions of constant work hardening moduli E_i^p. In other words, it is assumed that there exists, in the domain of interest from the initial state, f_0, to the limit, ultimate, or bounding state, f_n, a series of yield surfaces, each defining a specific part or region of the domain in the stress space. With deformation, the surfaces translate in the stress space and as soon as surface f_i touches the next surface, f_{i+1}, they both move together until they touch f_{i+2}, and so on. The surfaces touch each other tangentially and are not permitted to intersect each other.

For a simple explanation, consider a circular (two-dimensional space) or a spherical (three-dimensional space) vessel filled with a series of finite concentric rings or spheres in the neighborhood of each other (Fig. 12-4). A finite-size ring, which can be made very small, at the center of the vessel represents the initial surface f_0, and the boundary of the vessel represents the limit or bounding surface. Now assume that an external agency produces a

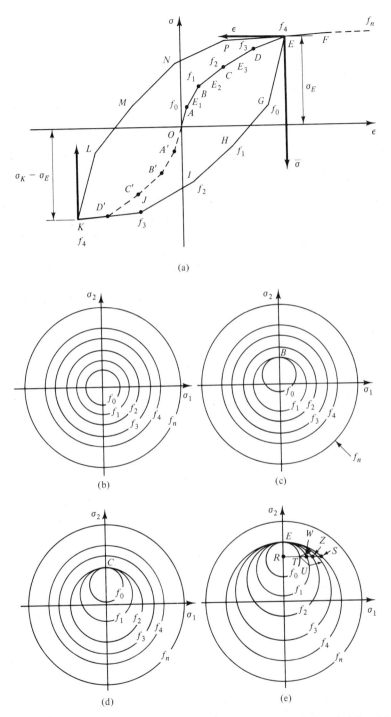

Figure 12-3 Nested surface model: (a) stress–strain curve; (b) before straining; (c) loading A to B; (d) loading B to C; (e) loading C to D to E and nonproportional loading (Ref. 16).

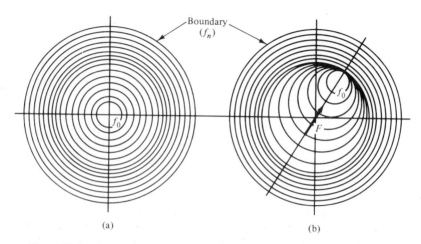

(a)   (b)

Figure 12-4 Physical interpretation: (a) vessel with finite number of concentric rings, initial condition; (b) movement of rings due to external force F.

disturbance at the central ring such that it is transmitted symmetrically in all directions. Then the rings will move uniformly and transmit the disturbance symmetrically (concentrically), which can be considered analogous to *isotropic hardening*. Now assume that the external agency causes expansion of the rings as well as movement or translation of the rings in the vessel; that is, the rings expand (uniformly), each inner ring pushing against the outer ring in its vicinity together with translation of their centers. This situation can be considered analogous to combined kinematic and isotropic hardening. This explanation can provide a simple understanding of the *nested* and *bounding* surface models.

Mathematical Interpretation

We now consider the mathematical interpretation of the concepts discussed above. First we give details of the nested surface model as proposed by Mroz (16). To explain the model, Mroz first considered proportional loading, specifically along the σ_z-axis [Fig. 12-3(a)] for the uniaxial case. For an increment of stress, the stress point moves from O to A [Fig. 12-3(a)], where it reaches the yield stress defined by f_0. The initial yield surface f_0 moves along the σ_z-axis, and at B, during plastic straining, it touches yield surface f_1 associated with point B; this is depicted in Fig. 12-3(c). The plastic strain induced during the movement from A to B is defined by the tangent modulus E_1. During this movement, all other surfaces remain fixed.

When the stress point moves from B to C, the surfaces f_0 and f_1 translate together until at C they touch f_2 associated with C [Fig. 12-3(d)]. The plastic strain during this movement is defined by modulus E_2. Similarly, during subsequent loadings the yield surfaces translate, taking with them the previous surfaces and touching the next ones during changes or increments in the state of stress. For the case in Fig. 12-3(e), the end or final state of the particular loading history up to E is represented by the surface, f_4. Because of the characteristic that the yield surfaces translate as a nest, this model is referred to as *nested* surface model.

Work Hardening Moduli.   The expression for the tangent moduli defining work hardening is obtained by assuming associated flow rule. The plastic strain increment is given in vectorial form as

$$d\epsilon^p = \frac{1}{H^p} n_f (d\sigma\, n_f) = \frac{1}{H^p} n_f d\sigma_f \qquad (12\text{-}15a)$$

where $d\sigma_f = d\sigma\, n_f$ is the projection of the stress change or increment $d\sigma$ $(d\sigma_{ij})$ on the unit normal vector n_f to the yield surface f, and H^p, the work hardening modulus, is given by

$$H^p = \frac{d\sigma_f}{d\epsilon^p} = \frac{d\sigma\, d\epsilon^p}{(d\epsilon^p)^2} = \frac{d\sigma_{ij} d\epsilon_{ij}^p}{d\epsilon_{ij}^p d\epsilon_{ij}^p} \qquad (12\text{-}15b)$$

H^p represents three-dimensional generalization of E^p in the uniaxial state of stress.

Unloading and Reloading.   Let us consider unloading from E along path $EGHIJK$ [Fig. 12-3(a)]. At G, (inverse) plastic flow occurs, and then the surface f_0 translates downward until it touches f_1 corresponding to point H [Fig. 12-5(a)]. Since this movement of f_0 is twice that for its movement from A to B during the loading, the stress difference between H and G is twice that between A and B. With subsequent unloading along HI, IJ, and JK corresponding to loadings BC, CD, and DE, respectively, the yield surfaces move and finally touch f_4 on the opposite side [Fig. 12-5(b)].

The curve of inverse loading or unloading $EGHIJK$ is defined uniquely by the primary loading curve $OABCDE$. The curve of reverse loading $OB'D'K$ is obtained as symmetrical with respect to loading curve $OABCDE$.

The magnitude of stress at K, σ_k, is equal to the stress at E, σ_E, and hence the stress difference between E and K is $2\sigma_E$. During cyclic loading between E and K, an identical curve $KLMNPE$ will be traced. For the given initial f_0 and the "final" state at E for the particular (loading), the yield surfaces f_0, f_1, f_2, and f_3 will translate, whereas f_4 will remain fixed. If a final (loading) state F were defined, its yield surface f_5 will remain stationary. The most final state for a given loading can reach the bounding or ultimate states on f_n.

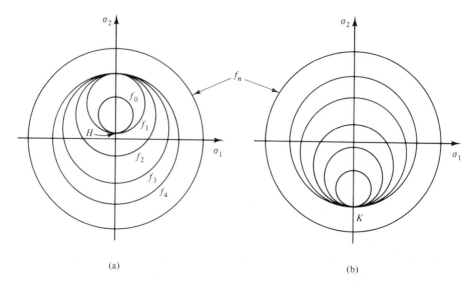

Figure 12-5 Unloading behavior: (a) unloading G to H; (b) unloading H to I, I to J, and J to K (Ref. 16).

The foregoing ideas can be generalized for paths other than the proportional. For such paths Mroz (16) adopted a criterion to control the manner of translation of the yield surfaces. Here it is assumed that the surfaces will not intersect each other but only touch or contact each other [Fig. 12-3(e)], and move. Consider a partial unloading (after E is reached) to a point R inside f_0. Then the stress point moves along RS, reaching f_0 at T. If f_0 now moves along the normal to itself at T, it will intersect f_1, which is not permitted. Hence f_0 must move such that point T contacts f_1 without intersecting it. This is achieved if T moves to a point U on f_1 because the outward normals at both T and U are parallel to each other. Thus instantaneous translation of f_0 will take place along TU, while the stress point moves along TS to point W on f_1, then to Z on f_2, and so on. The work-hardening moduli along the loading path reduce in magnitude with the translation of f_i, reaching a value equal to that for the final surface, such as f_4 in Fig. 12-3. It may be mentioned that this procedure of defining movements of the surfaces was adopted by Mroz (16); however, it is possible to evolve other and alternative schemes for a given material.

Movements of Yield Surfaces.   In the foregoing section, we described the model for movement of yield surfaces during loading, unloading, and reloading. Now we write the expression for the motion of the yield surfaces; here we follow the description given by Mroz (16).

Figure 12-6 shows locations of two typical yield surfaces f_m and f_{m+1} with centers O_m and O_{m+1}, respectively, defined by the position vectors α_{ij}^m and α_{ij}^{m+1} from the origin O; the superscript m denotes mth surface. The surfaces are expressed as

$$f\left(\sigma_{ij} - \alpha_{ij}^m\right) - \left(\sigma_0^m\right)^n = 0 \tag{12-16a}$$

$$f\left(\sigma_{ij} - \alpha_{ij}^{m+1}\right) - \left(\sigma_0^{m+1}\right)^n = 0 \tag{12-16b}$$

Here f is a homogeneous function of order n of its arguments, and σ_0^m and σ_0^{m+1} denote the size of the yield surfaces with respect to their origins. Their values can be constant or can vary during plastic straining.

Suppose that the stress point P (σ_{ij}^m) on f_m experiences a change in stress, $d\sigma_{ij}$. Then f_m will translate instantaneously along PR, where R is a stress point (σ_{ij}^{m+1}) on f_{m+1} corresponding to the same direction of the outward normal. The location of R is obtained by drawing a vector $O_{m+1}R$ parallel to $O_m P$. From Eq. (12-16) and the homogeneity of f, we have, by geometric proportion [Fig. 12-6(b)],

$$\sigma_{ij}^{m+1} - \alpha_{ij}^{m+1} = \frac{\sigma_0^{m+1}}{\sigma_0^m}\left(\sigma_{ij}^m - \alpha_{ij}^m\right) \tag{12-17}$$

The translation of f_m is given by

$$d\alpha_{ij}^m = \frac{d\mu}{\sigma_0^m}\left[\left(\sigma_0^{m+1} - \sigma_0^m\right)\sigma_{ij}^m - \left(\alpha_{ij}^m\sigma_0^{m+1} - \alpha_{ij}^{m+1}\sigma_0^m\right)\right] \tag{12-18a}$$

A special case of Eq. (12-18a) is obtained when the centers of the surfaces coincide, that is, $\alpha_{ij}^m = \alpha_{ij}^{m+1}$:

$$d\alpha_{ij}^m = d\mu\frac{\sigma_0^{m+1} - \sigma_0^m}{\sigma_0^m}\left(\sigma_{ij}^m - \alpha_{ij}^m\right) \tag{12-18b}$$

In this case, the instantaneous translation of f_m takes places along the radius vector, and we have the model proposed by Shield and Ziegler (5), Eq. (12-9).

The value of $d\mu$ is obtained from the condition that the stress point remains on the yield surface; then

$$\left(d\alpha_{ij}^m - d\sigma_{ij}^m\right)\frac{\partial f}{\partial \sigma_{ij}} = 0 \tag{12-19a}$$

and

$$d\mu = \frac{\left(\partial f/\partial \sigma_{ij}\right)d\sigma_{ij}}{\left(\partial f/\partial \sigma_{kl}\right)\left(\sigma_{kl}^{m+1} - \sigma_{kl}^m\right)} = \frac{d\sigma_f}{\left(\sigma^{m+1} - \sigma^m\right)n_f} \tag{12-19b}$$

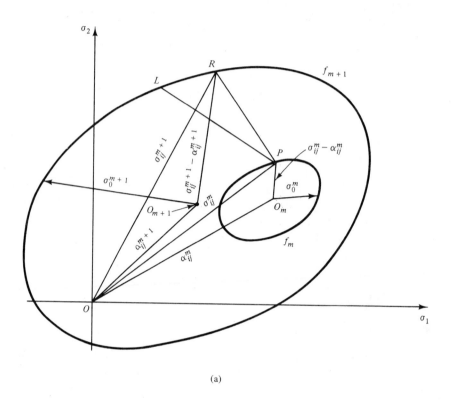

(a)

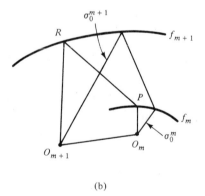

(b)

Figure 12-6 Movement of yield surfaces (Ref. 16).

Equation (12-17) can be used as a generic relation for defining location of the centers of the yield surfaces. For instance,

$$\sigma_{ij}^1 - \alpha_{ij}^1 = \frac{\sigma_0^1}{\sigma_0^0}\left(\sigma_{ij}^0 - \alpha_{ij}^0\right)$$

$$\sigma_{ij}^2 - \alpha_{ij}^2 = \frac{\sigma_0^2}{\sigma_0^1}\left(\sigma_{ij}^1 - \alpha_{ij}^1\right) \qquad (12\text{-}20)$$

$$\dotfill$$

$$\sigma_{ij}^n - \alpha_{ij}^n = \frac{\sigma_0^n}{\sigma_0^{n-1}}\left(\sigma_{ij}^{n-1} - \alpha_{ij}^{n-1}\right)$$

where n denotes final surface.

The foregoing model can be generalized to include expansion and contraction of the yield surfaces of constant work hardening moduli. Then, instead of being constant, σ_0^m, Eq. (12-16) can be assumed to be functions of a monotonically increasing scalar parameter λ as

$$\sigma_0^m = \sigma_0^m(\lambda), \qquad m = 0, 1, 2, \ldots, n \qquad (12\text{-}21\text{a})$$

Parameter λ can be expressed as a function of the length of the plastic strain trajectory:

$$\lambda = \int \left(d\epsilon_{ij}^p \, d\epsilon_{ij}^p\right)^{1/2} \qquad (12\text{-}21\text{b})$$

This will imply that the stress difference for a segment during unloading may not be twice that for the corresponding segment during loading (Fig. 12-3). The relation in Eq. (12-19b) will now be generalized to

$$d\mu = \frac{d\sigma_f - d\sigma_0^m}{(\sigma^{m+1} - \sigma^m)n_f} \qquad (12\text{-}22)$$

Prevost (18–22) has contributed useful modifications for application of this model to drained and undrained behavior of soils subjected to static and cyclic loading. These developments include undrained or pressure-independent models and drained or pressure-sensitive models. The former, which is expressed with respect to deviatoric behavior, is independent of mean (effective) normal stress, and is based on associative plasticity with the von Mises criterion. The pressure-sensitive model allows for nonassociative behavior on all yield surfaces except for the ultimate surface, for which associative behavior is assumed.

Undrained model (18).   Figure 12-7 includes plots similar to Fig. 12-3 showing schematic of stress–strain behavior in triaxial compression and extension. The nested surfaces $f_0, f_1, \ldots, f_m, f_n$ translate as discussed before, and

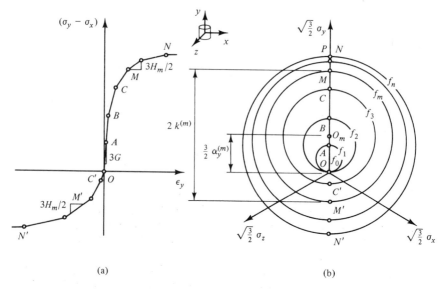

Figure 12-7 (a) Triaxial compression and extension stress–strain curves; (b) representation in stress space (Refs. 18, 20).

their sizes are defined by $k^0, k^1, \ldots, k^m, k^n$, which, in turn, define regions of constant plastic moduli H^p associated with each surface.

With the von Mises criterion, the yield surfaces have circular shapes and hence k represents radius of a given surface. Then the expression for mth yield surface can be written as

$$f_m = \left[\tfrac{3}{2}\left(S_{ij} - \alpha_{ij}^m\right)\left(S_{ij} - \alpha_{ij}^m\right)\right]^{1/2} - k^m = 0 \qquad (12\text{-}23)$$

where S_{ij} is the deviatoric stress tensor (Eq. 4-5d) and α_{ij}^m are the coordinates of the center of the yield surface f_m. The initial positions α_{ij}^m and the size k^m of the surfaces define the past history of the material element; α_{ij} is associated with the direction of loading and relates to the material's memory. Note that Eq. (12-23) is a special case of Eq. (12-16) with k^m replacing σ_0^m.

The physical reference coordinates x, y, and z are fixed with respect to the soil element, and if they coincide (initially) with the principal axes of the material anisotropy, then $\alpha_{xy}^m = \alpha_{yz}^m = \alpha_{xz}^m = 0$. In addition, if $\alpha_{xx}^m = \alpha_{zz}^m$ for all m, the soil will have rotational or transverse symmetry or cross-anisotropy. If $\alpha_{xx}^m = \alpha_{yy}^m = \alpha_{zz}^m = 0$, the material is (initially) isotropic. Thus, in general, for $\alpha_{ij}^m \neq 0$, the model allows for kinematic hardening and for stress-induced anisotropy.

The increments of plastic strains, $d\epsilon_{ij}^p$, for associated behavior are given by

$$d\epsilon_{ij}^p = \frac{1}{H^p} N_{ij}\langle L \rangle \qquad (12\text{-}24)$$

This equation is identical to Eq. (12-15), with the symbols N_{ij} and $\langle L \rangle$ replacing n_f and $d\sigma_f$, respectively. H^p is the plastic modulus associated with yield surfaces and N_{ij} are the components of the unit vector normal to the yield surface f:

$$N_{ij} = \frac{1}{|N|} \frac{\partial f}{\partial \sigma_{ij}}, \qquad |N| = \left(\frac{\partial f}{\partial \sigma_{kl}} \frac{\partial f}{\partial \sigma_{kl}} \right)^{1/2} \qquad (12\text{-}25a)$$

and the symbol $\langle L \rangle$ is defined as

$$\langle L \rangle = \begin{cases} L & \text{if } \dfrac{L}{H^p} \geqslant 0 \\ 0 & \text{otherwise} \end{cases} \qquad (12\text{-}25b)$$

Here $L = N_{kl} d\sigma_{kl}$.

For the undrained model the plastic modulus H^p_m plays the role of plastic shear modulus and is related to the tangent or slope of the stress–strain curve, H^p_m, which can be expressed as

$$\frac{1}{H^p_m} = \frac{1}{H_m} - \frac{1}{2G} \qquad (12\text{-}26)$$

where G is the (elastic) shear modulus.

In this undrained model, only deviatoric plastic strains occur, whereas the volumetric strains $d\epsilon^p_{ii} = 0$. Hence the increments of total deviatoric strains as the sum of the elastic and plastic components are given by

$$dE_{ij} = \frac{dS_{ij}}{2G} + \frac{3}{2H^p_m} \frac{S_{ij} - \alpha^m_{ij}}{(k^m)^2} \left[S_{kl} - \alpha^m_{kl} \right] dS_{kl} \qquad (12\text{-}27a)$$

and its inverse gives the deviatoric stress tensors as

$$dS_{ij} = 2G\, dE_{ij} - (2G - H_m) \frac{3}{2} \frac{S_{ij} - \alpha^m_{ij}}{(k^m)^2} \left[S_{kl} - \alpha^m_{kl} \right] dE_{kl} \qquad (12\text{-}27b)$$

Magnitudes of H^p_m. As explained earlier, the initial yield surface f_0 encloses a purely elastic region. It is not necessary that this surface has a finite size; that is, k^0 can be equal to zero. If the material experiences smooth transition from the elastic to plastic range, H^p_0 is infinite in Eq. (12-26) and can be set to a very high value.

As the loading progresses, the yield surfaces translate until the final or failure surface, f_n, is reached. Before f_n is reached, H^p_m is positive and any small increment of stress $d\sigma_{ij}$ can produce deformations such that if $d\sigma_{ij} d\epsilon_{ij} > 0$, the material is stable and $H^p_n = 0$. If H^p_n is negative, $d\sigma_{ij} d\epsilon_{ij} < 0$ and the material is unstable.

Loading and Unloading. Figure 12-8 shows a general schematic of the translation of yield surfaces during loading and unloading. For an infinitesimal stress increment, if

$$\left(S_{ij} - \alpha^m_{ij} \right) dS_{ij} > 0 \qquad (12\text{-}28)$$

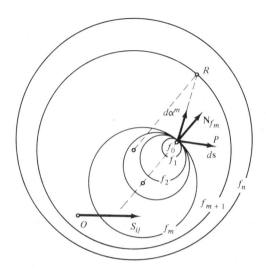

Figure 12-8 Field of yield surfaces in stress space (Refs. 18, 20).

loading occurs, and the incremental deviatoric stress tensor dS_{ij} denoted by the vector $d\mathbf{S}$ points in an outward direction with respect to f_m. The yield surfaces $f_0, f_1, \ldots, f_m$ which remain tangential to each other at the common point M then translate together toward f_{m+1}. As explained earlier (Fig. 12-6) the instantaneous translation is allowed to occur along PR, where R is a point on f_{m+1}. The instantaneous translation denoted by vector $d\boldsymbol{\alpha}^m$ is given by

$$d\alpha_{ij}^m = d\mu\,\mu_{ij} \tag{12-29}$$

where $\mu_{ij} = (k^{m+1}/k^m)(S_{ij} - \alpha_{ij}^m) - (S_{ij} - \alpha_{ij}^{m+1})$. From the condition that the stress points remain on the yield surface f_m during plastic deformations, the expression of $d\mu$ is obtained as [Eq. (12-22)]:

$$d\mu = \frac{1.5\left(S_{ij} - \alpha_{ij}^m\right)dS_{ij} - k^m\,dk^m}{k^{m+1}k^m - 1.5\left(S_{ij} - \alpha_{ij}^{m+1}\right)\left(S_{ij} - \alpha_{ij}^m\right)} \tag{12-30}$$

Since the size k^m of the yield surface can change, it can be expressed in terms of trajectory of plastic strain λ [Eq. (12-21b)].

If the stress point retreats from a yield surface f_m, unloading occurs which is assumed to be elastic and the value of $H_0^p = \infty$, that is, a high value can be adopted. Then

$$\left(S_{ij} - \alpha_{ij}^m\right)dS_{ij} < 0 \qquad \text{and} \qquad H_m^p = H_0^p \tag{12-31}$$

in Eq. (12-27a).

Specialization.   Prevost (18–22) has specialized the model for various common testing configurations, such as CTC, CTE, and SS stress paths. Here we include brief descriptions of the interpretation for static CTC and CTE, and cyclic CTC tests involving load reversals.

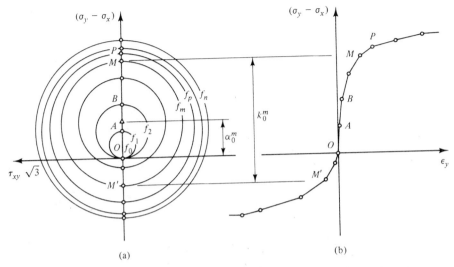

Figure 12-9 Representation of monotonic triaxial compression and extension tests: (a) on Π-plane; (b) stress–strain curves (Ref. 18).

Figure 12-9 shows the yield surfaces in the Π-plane and the corresponding stress–strain curves for compression and extension tests. For both test conditions $\sigma_x = \sigma_z$ and $\tau_{xy} = 0$ and the stress point moves along the $(\sigma_y - \sigma_x)$-axis. When it touches f_m, for the CTC condition:

$$\sigma_y - \sigma_x = \alpha_0^m + k_0^m \qquad (12\text{-}32a)$$

and for the CTE condition,

$$\sigma_y - \sigma_x = \alpha_0^m - k_0^m \qquad (12\text{-}32b)$$

where α_0^m and k_0^m = initial position and size of the yield surfaces, respectively [Fig. 12-9(a)]. In both cases, the incremental (vertical) strain is given by

$$d\epsilon_y = \left(\frac{1}{3G} + \frac{2}{3H_m^p} \right) d(\sigma_y - \sigma_x) = \frac{2}{3H_m} d(\sigma_y - \sigma_x) \qquad (12\text{-}33)$$

The values of α_0^m, k_0^m, and H_m are determined as shown in Fig. 12-9; α_0^m is the radius of the initial circle f_0, k_0^m is the radius of the circle f_m, and H_m is related to the chord or tangent slope of the stress–strain curve at f_m.

Figure 12-10 illustrates the conditions during cyclic triaxial test. Assume that during compression, the stress point P is reached on f_p. Then if reverse loading and plastic flow occur the surfaces translate downward [Fig. 12-10(b)]. The variation of k^m occurs during the first load reversal and is determined by comparing the stress difference corresponding to segments PA_1' and OA, $A_1'B_1'$ and AB, and so on. When $k^m = k_0^m$, as in Fig. 12-3, $PA_1' = 2k_0^1,\ldots, PM_1' = 2k_0^m$, and so on. As stated earlier, if k^m is a function of λ, the ratio of 2 for k_0^1/PA_1',

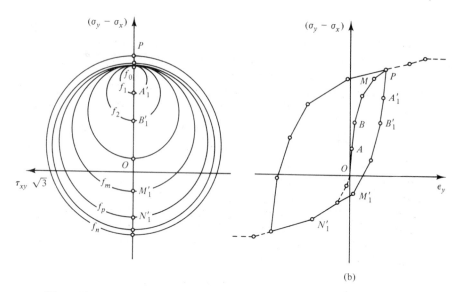

(b)

Figure 12-10 Cyclic triaxial test: (a) field of yield surfaces; (b) load-reverse load stress–strain curves (Ref. 18).

k_0^m/PM_1' may not hold. Then, it is necessary to find the value of k^m for each hysteresis loop separately.

Drained or Pressure-Sensitive Model (20).

The expression for yield surfaces for this case is an extension of Eq. (12-23) and is given by

$$f_m = \left[\tfrac{3}{2}\left(S_{ij} - \alpha_{ij}^m \right)\left(S_{ij} - \alpha_{ij}^m \right) + C^2 \left(p' - \beta^m \right)^2 \right]^{1/2} - k^m = 0$$

$$(12\text{-}34)$$

where p' is the mean effective normal stress, β^m the center of the yield surface along the hydrostatic axis, and C the material parameter referred to as yield surface axis ratio. Value of C is often adopted as $3/\sqrt{2}$ (Fig. 12-11).

For the pressure sensitive model, H_m^p in Eq. (12-33) becomes plastic bulk modulus, and for volumetric stress–strain response, we can write

$$d\epsilon_v = \left(\frac{1}{K} + \frac{3}{H^p} \right) dp'$$

$$(12\text{-}35)$$

where $d\epsilon_v = d\epsilon_x + d\epsilon_y + d\epsilon_z$, $dp' = \tfrac{1}{3}(d\sigma_x' + d\sigma_y' + d\sigma_z')$, and K is the elastic bulk modulus.

Prevost (18–22) has derived material parameters for a number of geological materials such as Drammen and Boston blue clays based on CTC and RTE tests. Sture (28) has described derivation of parameters for Drammen clay and Ottowa sand. The model can be considered to be in continuing development stage and additional implementation and verifications are desirable for its

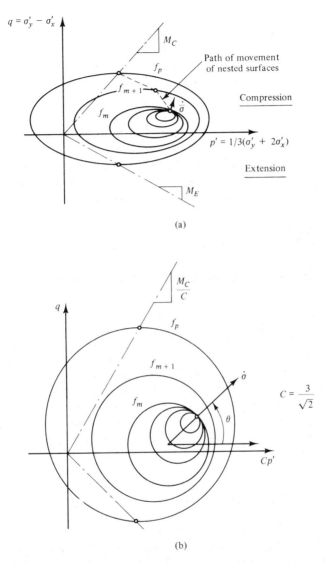

Figure 12-11 Drained nested surface model: (a) in $p'-q$ plane; (b) in transformed plane (Refs. 18–22).

detailed evaluation. Also, it will be appropriate to evaluate the model with respect to other stress paths (Chapter 5).

Mroz et al. (29, 30) have extended the foregoing models to include isotropic hardening due to changes in porosity and anisotropy due to initial consolidation. Incremental relations are derived for the case of conventional triaxial configuration for drained and undrained behavior after isotropic and anisotropic K_0 consolidation of clays.

BOUNDING SURFACE MODELS

Dafalias and Popov (23–25) and Krieg (27) have proposed an alternative, but
conceptually similar model to that by Mroz (16), for kinematic and isotropic
hardening. Instead of defining a series of surfaces $f_0, f_1, f_2, \ldots, f_n$ (Fig. 12-3),
individually, two surfaces, a bounding or limiting surface such as f_n and a
loading surface such as f_0 are defined. The parameters required to describe the
elastic–plastic deformation process are obtained by defining expressions that
among other factors depend on the two surfaces. As a result, the model is
considerably simplified as compared to the nested surface model. The plastic
moduli vary smoothly during straining and are not piecewise constant as in the
case of the nested surface model. We give below a description of the bounding
surface model based on the works of Dafalias and Popov (23–25).

Uniaxial Case

Figure 12-12(a) shows a schematic of uniaxial stress–strain response.
The material exhibits linear elastic response up to point A. Within this elastic
range, according to Eq. (12-14), the plastic modulus E^p is infinity. After
passing A, plastic strains occur and E^p starts assuming finite values which
change gradually and approach an ultimate value associated with the bounding
lines XX' and YY'.

In Fig. 12-12(a) OA represents the elastic part, and ABD the elastic-
plastic part. Elastic unloading is assumed to occur at point D along DD', and
plastic reloading occurs along $D'D''F$. The region FF' represents new elastic
unloading, which is followed by plastic reloading along $F'F''X$. After point B,
the curve followed during plastic reloading is the same as that during the first
plastic loading for purely kinematic hardening. In general, the bounds can
change with loading history.

In an elastic region such as FF', the plastic modulus E^p is infinite. The
value of E^p varies during the plastic behavior in regions such as $F'F''$. Along
FF', E^p was infinite, but right after F', it assumes a finite value which changes
as the process goes to F''. The zone $F''X$ represents plastic behavior during
which E^p is assumed to remain constant. The stress–strain response is thus
bounded by two lines XX' and YY'. In the multiaxial cases, the projections of
the points on the bounds are generalized into a bounding surface, hence the
name *bounding surface* model.

Dafalias and Popov (23–25) have presented explanation of the model by
projecting the stress–plastic strain space onto the stress space for the uniaxial
case (Fig. 12-13). Consider two points A and A' on the stress–plastic strain
curve. Unloading will occur from point A in the $\sigma-\epsilon^p$ space, along $A\bar{A}$, and the
projections of A and $\bar{A}$ on σ-axis are given by a and $\bar{a}$, respectively; these two
points play the role of loading surface associated with state A. Extension of A,
$\bar{A}$ to XX' and YY' leads to $B, \bar{B}$ with $b, \bar{b}$ the corresponding projections on the

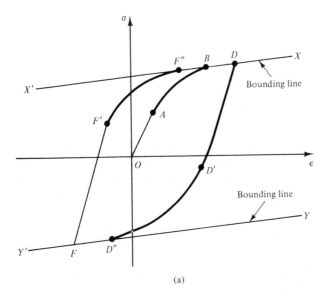

(a)

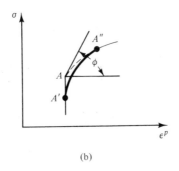

(b)

Figure 12-12 Schematic of bounding surface (line) model in (a) σ–ϵ; (b) σ–ϵ^P spaces (Ref. 23).

σ-axis. Points k and r are midpoints of $a\bar{a}$ and $b\bar{b}$ on the σ-axis and represent projections of points K and R along $B\bar{B}$. Corresponding nomenclature apply to the state of stress A'.

Under a stress increment $d\sigma$, the state of stress moves from point A to A'. Then the elastic zone $a\bar{a}$ moves to $a'\bar{a}'$ and $b\bar{b}$ to $b'\bar{b}'$. Now we define as $d\alpha$ the incremental displacement kk' of the center k of the inner line segment or elastic zone $a\bar{a}$, and $d\beta$ as the incremental movement rr' of the center r of the outer line segment $b\bar{b}$. Let us denote by S and dS the length and increment of length of the inner line segment, respectively, and by E_n^p the plastic modulus associated with the bounding line (surface). Then from the geometrical consid-

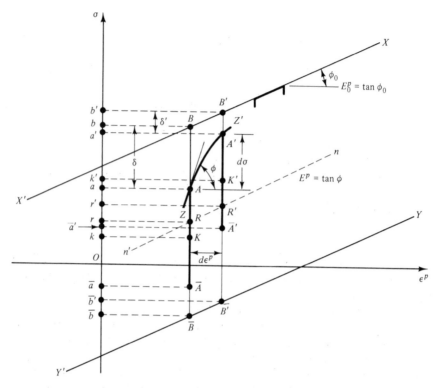

Figure 12-13 Bounding surface (line) model: uniaxial case (Ref. 23).

erations (Fig. 12-13), the following expressions can be written:

$$d\sigma = E^p \, d\epsilon^p \tag{12-36a}$$

$$d\alpha = d\sigma - \frac{dS}{2} \tag{12-36b}$$

$$d\beta = E_n^p \, d\epsilon^p \tag{12-36c}$$

$$\delta' - \delta = d\delta = \left(\frac{E_n^p}{E^p} - 1\right) d\sigma \tag{12-36d}$$

$$E^p = \bar{E}^p(\delta, W^p) \tag{12-36e}$$

where δ is the distance of A from B on XX' and δ' is the distance of A' and B' or XX'. The above applies when the two bounds are assumed to be straight lines. For the general case (26), $d\beta$ will be expressed in terms of a modulus that is different from E_n^p. This also contains the main hypothesis that the plastic modulus E^p associated with a point A can be expressed as a function of the modulus related to the bounding line (surface) and the distance of A from the bounding surface.

Note that $E_n^p = \bar{E}^p(0, W^p)$; that is, when $\delta = 0$, Eq. (12-36e) yields a plastic modulus relevant to the bounding line. Thus $\bar{E}^p$ is an increasing function of δ. It is a decreasing function of W^p if the material softens and increasing function of W^p if it hardens. Furthermore, the plastic modulus E^p varies during hardening or softening as a continuous function and need not be constant for different segments.

The rule of translation for the bounding surface is obtained from the movement $d\beta$ of the bounding surface as

$$d\beta = E_n^p \, d\epsilon^p$$

$$= E^p \, d\epsilon^p - \left(E^p - E_n^p \right) d\epsilon^p$$

$$= d\sigma - \left(E^p - E_n^p \right) d\epsilon^p \qquad (12\text{-}37)$$

Here $d\sigma$ is the incremental movement of the point on the loading surface that represents the current state of stress.

The quantities $d\epsilon^p$, $d\alpha$, and $d\beta$ are determined by using the equations above from the known values of $d\sigma$, W^p, and δ. For a subsequent stress increment, δ' is found from Eq. (12-36d) and a new value of E^p is found from Eq. (12-36e), and the process is repeated. If elastic unloading occurs, then for the subsequent plastic loading, a new value of W^p is used in Eq. (12-36).

The two line segments on the σ-axis move during the stress increment. The inner segment moves due to the stress increment $d\sigma$. The outer segment moves in a similar manner but at a slower rate. Hence the inner segment can eventually reach the outer one and then both move together. During the motion, E^p changes with δ and approaches E_n^p when $\delta = 0$. It is possible that E^p becomes E_n^p for $\delta \neq 0$. Then both segments move at the same rate without being in contact.

The foregoing description for uniaxial case contains the basic ingredients of the bounding surface model. It can easily be generalized for multiaxial loading.

Multiaxial Case

Figure 12-14 shows a schematic representation of bounding and loading surfaces. The expressions for δ, $d\epsilon_{ij}^p$, and H^p corresponding to δ, $d\epsilon^p$, and E^p for the uniaxial case are generalized as

$$\delta = \left[\left(\bar{\sigma}_{ij} - \sigma_{ij} \right) \left(\bar{\sigma}_{ij} - \sigma_{ij} \right) \right]^{1/2} \qquad (12\text{-}38\text{a})$$

$$d\epsilon_{ij}^p = \frac{1}{H^p} d\sigma \, n_{ij} \qquad (12\text{-}38\text{b})$$

$$H^p = \hat{H}_p \left(\delta, W^p \right) \qquad (12\text{-}38\text{c})$$

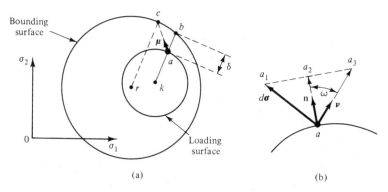

Figure 12-14 Bounding surface model: multiaxial case (Ref. 23).

where

$$W^p = \int \sigma_{ij}\, d\epsilon_{ij}^p \qquad (12\text{-}38d)$$

and σ_{ij} and $\bar{\sigma}_{ij}$ are stresses associated with points a and b, respectively (Fig. 12-14).

For generalizing the movement of the bounding surface, we need to obtain appropriate expressions for generalized $d\sigma$ in Eq. (12-37). Referring to Fig. 12-14(b), such expression is given by

$$aa_3 = \frac{d\sigma}{\cos\omega}\nu \qquad (12\text{-}39)$$

where aa_3 corresponds to $d\sigma$ in Eq. (12-37), ω is the generalized angle between unit normal n along aa_2 and unit normal ν along aa_3. It may be noted that the actual stress increment $d\sigma$ [Fig. 12-14(b)] is not the same as aa_3 in Eq. (12-39), which represents the incremental movement during hardening of a point a with stress σ.

For generalizing the part $(E^p - E_n^p)\, d\epsilon^p$ in Eq. (12-37), the idea proposed by Mroz (16) that the two surfaces must approach each other along the unit vector μ is used; this vector lies on the line segment ac (Fig. 12-14), with the same unit normal direction on the corresponding surfaces. Since it is assumed, for simplicity, that the bounding surface hardens only kinematically, that is, moves as a rigid body, the motion of point c is identical to that of the center. Thus Mroz's condition is satisfied if the translation $d\beta_{ij}$ is given by

$$\begin{aligned}d\beta_{ij} &= \frac{d\sigma}{\cos\omega}\nu_{ij} - \left(H^p - H_n^p\right)\left(d\epsilon_{mn}^p\, d\epsilon_{mn}^p\right)^{1/2}\mu_{ij} \\ &= d\sigma\left[\frac{1}{\cos\omega}\nu_{ij} - \left(1 - \frac{H_n^p}{H^p}\right)\mu_{ij}\right]\end{aligned} \qquad (12\text{-}40a)$$

Note that Eq. (12-40a) permits both surfaces to translate and harden isotropically or otherwise (26). This is in contrast to the nested surface model (16), in which only the inner surface translates and the outer remains fixed (except for isotropic hardening) until it is touched and pushed by the inner one(s).

If the loading surface moves as a rigid body, the incremental movement $d\alpha_{ij}$ of the center of the loading surface is given by the right-hand side of Eq. (12-39). Then, when the loading and the bounding surfaces come in contact, $\delta = 0$ for the common point and

$$d\alpha_{ij} = d\beta_{ij} \tag{12-40b}$$

This implies that both the surfaces move together at the same rate.

Dafalias and Popov (23) have specialized the concept above for Prager's and other models. Dafalias, Popov, and Herrmann have also extended the model and applied it for cyclic loading and plasticity with internal variable formalism (24, 26) and for behavior of soils under monotonic and cyclic loading (31, 32).

Physical Interpretation

The interpretation related to Fig. 12-4 can be used for the bounding surface model. The boundary of the vessel denotes the bounding surface f_n, whereas the loading surface f_0 is represented by the initial ring. Then under loading–unloading the initial ring expands or contracts, generating intermediate surfaces, and in the limit can reach the boundary.

Another simple interpretation can be given by considering a bounding surface f_n as a local peak or plateau on a mountain (Fig. 12-15). The initial or

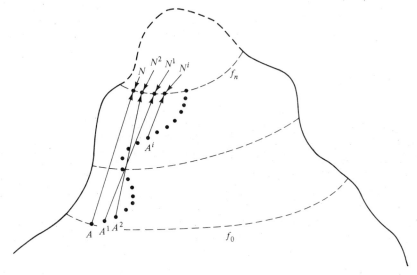

Figure 12-15   Physical interpretation of bounding surface models.

loading surface is given by the height or coordinates of the starting point A. Assume that a person starts to climb the mountain from A and aims to reach a point N on f_n. The dots show the path of movement which is guided by a beacon, light, or image point N (26, 31–32) on f_n. The person climbs in such a manner that the image point always remains in his view. Hence if his path is such that he loses the sight of the image point, the image point also moves. The characteristics of the movement are guided by the distance covered and remaining, and also the amount of work (W^p) expended during the climb. In other words, depending on the difficulties of the climb and the work expended, the rate of climb is affected.

Similar explanations apply for descent and subsequent climb, that is, a series of climb, descent, and climb, representing loading, unloading, and reloading.

Comparison of Nested, Bounding, and Other Models

In a general sense, all the models discussed so far involve essentially the same concept. For a given material there exists a limit, ultimate, or bounding state that can be reached during application of change in external loading starting from an initial state. They differ only in the manner in which the process of travel toward the limit state. Conceptually, there is no difference between the nested and bounding surface models. Further research will be needed to determine the trade-offs for these models or for evolution of other improved models. A comparison of the nested and bounding surface models is given by Prevost (33).

SOME GENERALIZATIONS AND SPECIAL FORMS

Kinematic Hardening, including Change in Orientation

Baltov and Sawczuk (7) have extended the kinematic and isotropic hardening model to include rotation of the yield surfaces in the stress space, and distortion and translation are also permitted. A generalization form of the von Mises criterion is defined as

$$2f = A_{ijkl}S_{ij}S_{kl} - 2(k)^2 = 0 \tag{12-41a}$$

where A_{ijkl} ($i, j, k, l = 1, 2, 3$) denotes the tensor of plastic moduli, which is assumed to be known, and k, as before, denotes the size of the yield surface. Combining the kinematic hardening rule given by

$$2f = \left(S_{ij} - \alpha_{ij}\right)\left(S_{ij} - \alpha_{ij}\right) - 2k^2 = 0 \tag{12-41b}$$

with the yield condition, [Eq. (12-41a)], for an initially anisotropic material leads to nondimensional form as

$$2\Phi = N_{ijkl}(S_{ij} - \alpha_{ij})(S_{kl} - \alpha_{kl}) - 1 = 0 \qquad (12\text{-}41c)$$

Baltov and Sawczuk have analyzed various specializations of Eq. (12-41c) and have applied the model for simulating experimental behavior of aluminum alloy 195 under combined tension and torsion. A special case of this model is described subsequently.

Some General Forms for Anisotropic Hardening

We can write a general form for yield surfaces, f, as

$$f = f(\sigma_{ij}, \epsilon_{ij}^p, W^p) \qquad (12\text{-}42a)$$

For initially isotropic materials, we have

$$f = f(J_i, I_i^p, K_j, W^p) \qquad (12\text{-}42b)$$

where W^p = plastic work and K_j ($j = 1, 2, 3, 4$) are the joint invariants of the stress and plastic strain tensors $(11, 34, 35)$

$$
\begin{aligned}
K_1 &= \text{tr}(\sigma \epsilon^p) = \sigma_{ij}\epsilon_{ji}^p \\[4pt]
K_2 &= \text{tr}(\sigma^2 \epsilon^p) = \sigma_{ik}\sigma_{kj}\epsilon_{ji}^p \\[4pt]
K_3 &= \text{tr}\left[\sigma(\epsilon^p)^2\right] = \sigma_{ik}\epsilon_{kj}^p\epsilon_{ji}^p \\[4pt]
K_4 &= \text{tr}\left[\sigma^2(\epsilon^p)^2\right] = \sigma_{ik}\sigma_{kj}\epsilon_{jm}^p\epsilon_{mi}^p
\end{aligned}
\qquad (12\text{-}42c)
$$

The inclusion of K_j can allow for induced anisotropy. The perfectly plastic, isotropic hardening, kinematic hardening, and nested surfaces models described above can be considered as special cases of Eq. (12-42).

A number of special forms of Eq. (12-42) have also been studied by Desai (36), Desai and Siriwardane (37), and Baker and Desai (35). For instance, the kinematic hardening model can be expressed as

$$f = f\left[(\sigma_{ij} - a\epsilon_{ij}^p), W^p\right] \qquad (12\text{-}43)$$

where a is a parameter with dimensions of stress.

The form invariance principle (38) states that a simultaneous rotation of the material element with elastic and plastic strains embedded in it and a system of stress is equivalent to the rotation of the reference coordinate system (Fig. 12-16), and the constitutive equation must be valid in the rotated and the original coordinate systems.

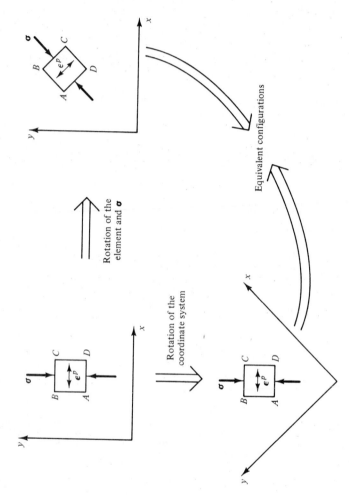

Figure 12-16 Equivalence of simultaneous rotation of $\boldsymbol{\epsilon}^p$ and $\boldsymbol{\sigma}$ to coordinate transformation (Ref. 35).

By applying the form invariance principle to Eq. (12-43), it can be shown that f may depend on $(\sigma_{ij} - a\epsilon_{ij}^p)$ through the following set of invariants:

$$L_1 = J_1 - aI_1^p$$

$$L_2^2 = J_2^2 - 2aK_1 + a^2(I_2^p)^2 \tag{12-44}$$

$$L_3^3 = J_3^3 - 3a^2K_2 - 3aK_3^2 - a^3(I_3^p)^3$$

which are functions of the joint invariants. It can be thus seen that the previous kinematic hardening models depend on joint invariants K_1, K_2, and K_3, and that the function in Eq. (12-42) contains Eq. (12-43) as a special case.

A General Basis for Deriving Yield and Plastic Potential Functions

A simple and special case of Eq. (12-42a) in which f is expressed as a complete polynomial in J_1, $J_2^{1/2}$, and $J_3^{1/3}$ was examined by Desai (36). As an example, a polynomial of degree three was expressed as

$$
\begin{aligned}
f(J_i) = {}& \alpha_0 + \alpha_1 J_1 + \alpha_2 J_2^{1/2} + \alpha_3 J_3^{1/3} + \alpha_4 J_1^2 + \alpha_5 J_1 J_2^{1/2} \\
& + \alpha_6 (J_2^{1/2})^2 + \alpha_7 J_2^{1/2} J_3^{1/3} + \alpha_8 (J_3^{1/3})^2 + \alpha_9 J_1 J_3^{1/3} \\
& + \alpha_{10} J_1^3 + \alpha_{11} J_1^2 J_2^{1/2} + \alpha_{12} J_1 (J_2^{1/2})^2 + \alpha_{13} (J_2^{1/2})^3 \\
& + \alpha_{14} (J_2^{1/2})^2 J_3^{1/3} + \alpha_{15} J_2^{1/2} (J_3^{1/3})^2 + \alpha_{16} (J_3^{1/3})^3 \\
& + \alpha_{17} J_1 (J_3^{1/3})^2 + \alpha_{18} J_1^2 J_3^{1/3} + \alpha_{19} J_1 J_2^{1/2} J_3^{1/3} \tag{12-45}
\end{aligned}
$$

Here α_i are material parameters which can be expressed as functions of quantities such as strains and density.

It is possible to express f as a function of J_1, $J_{2D}^{1/2}$, and $J_{3D}^{1/3}$ also. It was shown that the classical yield conditions such as von Mises, and Drucker–Prager appear as special cases of f in Eq. (12-45). Also, other recent criteria derived on the basis of experimental results for geological materials such as sands can also be shown to be embedded in Eq. (12-45). For instance, the criterion derived by Lade and Duncan (39–42),

$$\frac{J_1^3}{J_3} = \text{constant} \tag{12-46a}$$

is obtained from the eleventh and the seventeenth terms in Eq. (12-45). The criterion derived by Matsuoka and Nakai (43),

$$\frac{J_1 J_2}{J_3} = \text{constant} \tag{12-46b}$$

represents the thirteenth and the seventeenth terms in Eq. (12-45).

As illustrations, Desai (36) considered laboratory test data of two materials, artificial soil (Chapter 5) and Ottawa sand. Six functions were chosen (arbitrarily) from Eq. (12-45):

$$f_1 = \alpha_1 J_1 + \alpha_2 J_2^{1/2} = 0 \qquad \text{or} \qquad \frac{J_1}{J_2^{1/2}} = -\frac{\alpha_2}{\alpha_1} \qquad (12\text{-}47a)$$

$$f_2 = \alpha_{10} J_1^3 + \alpha_{16} J_3 = 0 \qquad \text{or} \qquad \frac{J_1^3}{J_3} = -\frac{\alpha_{16}}{\alpha_{10}} \qquad (12\text{-}47b)$$

$$f_3 = \alpha_{11} J_1^2 J_2^{1/2} + \alpha_{16} J_3 = 0 \qquad \text{or} \qquad \frac{J_1^2 J_2^{1/2}}{J_3} = -\frac{\alpha_{16}}{\alpha_{11}} \qquad (12\text{-}47c)$$

$$f_4 = \alpha_{13} \left(J_2^{1/2} \right)^3 + \alpha_{16} J_3 = 0 \qquad \text{or} \qquad \frac{\left(J_2^{1/2} \right)^3}{J_3} = -\frac{\alpha_{16}}{\alpha_{13}} \qquad (12\text{-}47d)$$

$$f_5 = \alpha_{13} \left(J_2^{1/2} \right)^3 + \alpha_{18} J_1^2 J_3^{1/3} \qquad \text{or} \qquad \frac{\left(J_2^{1/2} \right)^3}{J_1^2 J_3^{1/3}} = -\frac{\alpha_{13}}{\alpha_{18}} \qquad (12\text{-}47e)$$

$$f_6 = \alpha_6 \left(J_2^{1/2} \right)^2 + \alpha_9 J_1 J_3^{1/3} \qquad \text{or} \qquad \frac{\left(J_2^{1/2} \right)^2}{J_1 J_3^{1/3}} = -\frac{\alpha_6}{\alpha_9} \qquad (12\text{-}47f)$$

Tables 12-1 and 12-2 show values of f_1 to f_6 at ultimate for test results under various stress paths in the truly triaxial device (Chapter 5). Ultimate was defined as the value of stress corresponding to the approximate horizontal plateau near ultimate conditions in Fig. 12-17 which shows typical test data under RTC stress path for the Ottawa sand (44).

It is interesting to note from Tables 12-1 and 12-2 that the values of the six functions at ultimate are nearly invariant. However, the degree of invariance is different for different functions. Function f_6 shows essentially a constant value at ultimate around 0.55 for both materials. These results indicate the possibility of choosing the best function(s) based on test data for a given material through a process of optimization.

The generality of the foregoing approach was also indicated by studying the incremental forms of constitutive models from hypoelastic behavior. For instance, the idea that yielding or failure can be identified when the constitutive matrix C_{ijkl} in

$$d\sigma_{ij} = C_{ijkl}\, d\epsilon_{kl} \qquad (12\text{-}48a)$$

or

$$\{d\sigma\} = [C]\{d\epsilon\} \qquad (12\text{-}48b)$$

becomes singular, was used [see Thomas (46), Tokuoka (47), Coon and Evans (48), Davis and Mullenger (49)]. Singularity of $[C]$ implies that its determinant tends to zero

$$|C| \to 0$$

TABLE 12-1 Values of f at Ultimate for Ottawa Sand [36]

Stress Path	$\sigma_0; \gamma_0^a$	$\dfrac{J_1}{J_2^{1/2}}$	$\dfrac{J_1^3}{J_3}$	$\dfrac{J_1^2 J_2^{1/2}}{J_3}$	$\dfrac{(J_2^{1/2})^3}{J_3}$	$\dfrac{(J_2^{1/2})^3}{J_1^2 J_3^{1/3}}$	$\dfrac{(J_2^{1/2})^2}{J_1 J_3^{1/3}}$
CTC	5(34.5); 1.73	1.79	6.46	3.61	1.12	0.32	0.58
	10(69); 1.73	1.84	7.05	3.84	1.14	0.31	0.57
RTC	20(138); 1.72	1.68	5.17	3.08	1.11	0.36	0.61
PL	10(69); 1.73	1.87	7.52	4.02	1.15	0.30	0.56
TC	5(34.5); 1.74	1.68	5.17	3.08	1.09	0.36	0.61
	10(69); 1.70	1.74	5.84	3.35	1.10	0.34	0.59
	20(138); 1.72	1.84	7.05	3.84	1.14	0.31	0.57
TE	5(34.5); 1.73	2.07	13.30	6.43	1.51	0.27	0.56
	10(69); 1.73	2.07	13.30	6.43	1.51	0.27	0.55
SS	5(34.5); 1.74	1.93	9.56	4.95	1.33	0.29	0.57
	20(138); 1.73	2.23	11.04	5.49	1.36	0.27	0.55
Lowest value of f: a		1.68	5.17	3.08	1.09	0.27	0.55
Largest value of f: b		2.23	13.30	6.43	1.51	0.36	0.61
Mean value of f: c		1.89	8.32	4.38	1.23	0.31	0.57
Percentage $(c-a)/c \times 100$	11	38	30	11	13	4
Difference $(c-b)/c \times 100$	18	60	47	23	16	11
f for initial hydrostatic pressure, σ_0		2.45	27.00	11.00	1.84	0.20	0.50

$^a\sigma_0$, initial hydrostatic pressure, psi (kN/m^2); γ_0, initial density, g/cm^3.

TABLE 12-2 Values of f at Ultimate for Artifical Soil [36]

Stress Path	$\sigma_0; \gamma_0^a$	$\dfrac{J_1}{J_2^{1/2}}$	$\dfrac{J_1^3}{J_3}$	$\dfrac{J_1^2 J_2^{1/2}}{J_3}$	$\dfrac{(J_2^{1/2})^3}{J_3}$	$\dfrac{(J_2^{1/2})^3}{J_1^2 J_3^{1/3}}$	$\dfrac{(J_2^{1/2})^2}{J_1 J_3^{1/3}}$
RTC	10(69); 2.00	1.8	6.55	3.64	1.13	0.32	0.58
RTE	10(69); 1.87	2.00	12.00	6.00	1.50	0.29	0.57
TC	10(69); 2.04	1.84	7.23	3.91	1.15	0.31	0.57
	28(193); 2.00	2.22	15.70	7.07	1.43	0.23	0.51
TE	10(69); 1.89	2.13	14.74	6.91	1.52	0.25	0.54
SS	10(69); 2.05	1.97	10.30	5.22	1.34	0.28	0.56
Lowest value of f: a		1.80	6.55	3.64	1.13	0.25	0.51
Largest value of f: b		2.22	15.70	7.07	1.52	0.32	0.58
Mean value of f: c		1.99	11.09	5.46	1.35	0.28	0.56
Percentage $(c-a)/c \times 100$	10	41	33	16	11	9
Difference $(c-b/c) \times 100$	11	42	29	13	14	4
f for initial hydrostatic pressure, σ_0		2.45	27.00	11.00	1.84	0.20	0.50

$^a\sigma_0$, initial hydrostatic pressure, psi (kN/m^2); γ_0, initial density, g/cm^3.

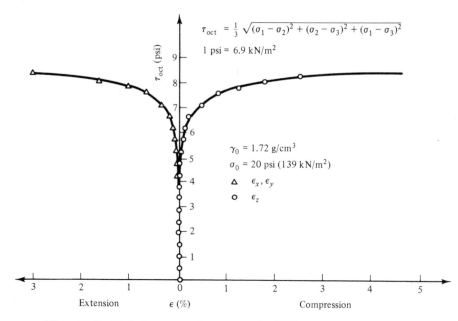

Figure 12-17 Typical stress–strain response for RTC stress path: Ottawa sand (Ref. 36).

or

$$|C_n||C_s| \rightarrow 0 \tag{12-48c}$$

or

$$|C_n| \rightarrow 0$$
$$|C_s| \rightarrow 0$$

It was shown that use of Eq. (12-48c) with respect to the zero-grade hypoelastic law (Chapter 7) results in the Drucker–Prager yield criterion [Eq. (10-6)], which is a special case of Eq. (12-45). Similarly, the first-grade hypoelastic law can lead to the criterion in Eq. (12-46a).

Analysis and Evaluation of f_6, Eq. (12-47f)

One of the foregoing functions, [Eq. (12-47f)] was studied in detail by expressing it in terms of J_1, J_{2D} and J_3 as

$$\bar{f}_6 = J_{2D} + \alpha J_1 - \beta J_1 J_3^{1/3} - \gamma J_1 - k^2 = 0 \tag{12-49a}$$

where α, γ and k are material parameters; the latter is the measure of cohesion. β is treated as the hardening or growth function. The function in Eq. (12-49a) plots convex and continuous in the stress space. Figures 12-18(a) and (b) show

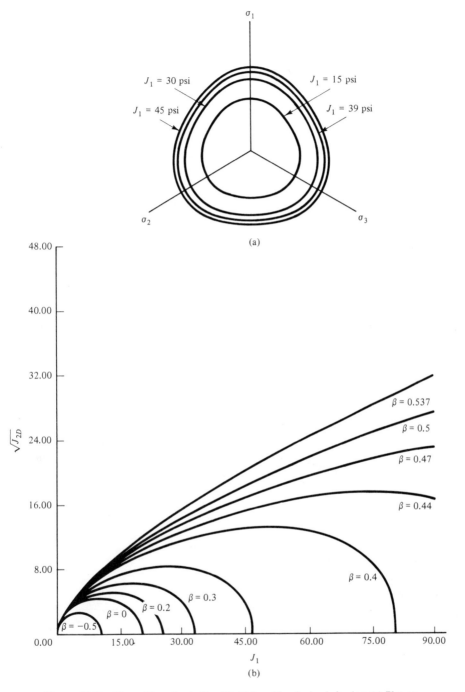

Figure 12-18 Plots of function in Eq. (12-49a) and hardening behavior: (a) Plot on octahedral planes; (b) Plots on $\sqrt{J_{2D}} - J_1$ plane for given values of J_{3D}; (c) Ultimate yield or failure envelope; (d) Plot of β vs λ for CTC test, $\sigma_0 = 10$ psi; (e) Plots of β vs λ, λ_v and λ_D for CTC test, $\sigma_0 = 20$ psi (Refs. 36, 45).

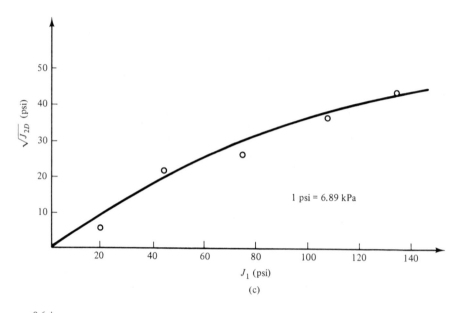

(c)

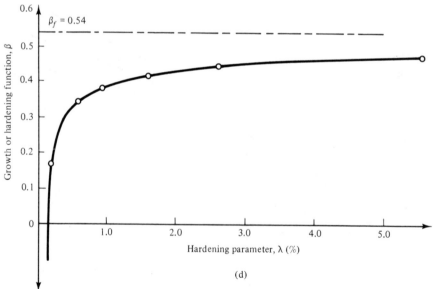

(d)

Figure 12-18 (*continued*)

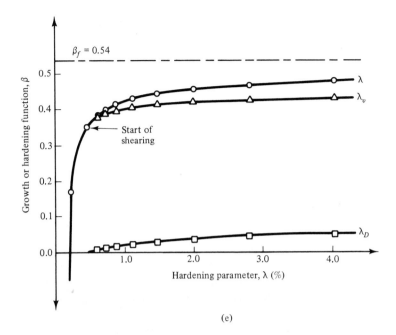

Figure 12-18 (*continued*)

the plots in the octahedral and $\sqrt{J_{2D}} - J_1$ planes, respectively, for the silty sand described in Chapters 10 and 11.

It is interesting and useful to note that the model [Eq. (12-49a)] plots as a series of single continuous functions which include both the yielding and ultimate yield (or failure) behavior. This is a significant simplification compared to the models such as cap and critical state (Chapter 11) that involve two separate functions for yielding and failure. For $\beta = 0$, the function assumes elliptical shape and for $J_{3D} = 0$, it represents circular shape. The value of β at failure, β_f, in Eq. (12-49a) for the silty sand is found to be about 0.54.

The parameters α, γ and k can be calculated from ultimate yield envelope [Fig. 12-18(c)], obtained from the laboratory test data. For the silty sand they were found to be: $\alpha = 0.18$, $\gamma = 3.70$ and $k = 0$. Note that $3\alpha = \beta_f$.

Based on the test results for the silty sand, β is assumed in the form

$$\beta = \beta_f\left(1 - \frac{\beta_a}{\lambda^\eta}\right) \tag{12-49b}$$

or

$$\beta = \beta_f\left[1 - \frac{1}{\bar{\beta}_a\lambda_v^{\eta_1}(1 + \beta_b\lambda_D^{\eta_2})}\right] \tag{12-49c}$$

where β_f, β_a, η, $\bar{\beta}_a$, β_b, η_1 and η_2 are material parameters, λ is trajectory of plastic strains [Eq. (12-21b)], and λ_v and λ_D are volumetric and deviatoric parts of λ. Plots of β vs λ for CTC test with $\sigma_0 = 10$ psi [Fig. 10-8(a)], and β vs λ_v and λ_D for CTC test with $\sigma_0 = 20$ psi [Fig. 10-8(b)], are shown in Figs. 12-18(d) and 12-18(e), respectively.

Both of the forms, Eq. (12-49b) and (12-49c) or other suitable forms can be employed. If the first were used, the values of the parameters can be found by plotting $\ln(\beta_a)$ vs $\ln(1 - \beta/\beta_f)$ as: $\beta_a = 0.0255$ and $\eta = 0.545$, and $\beta_f = 3\alpha = 0.54$.

Advantages

Many previously proposed models (Chapter 11 and this chapter) were derived on the basis of test data for specific material(s). As a result, their applicability is often limited to the specific material(s), and for application to other materials, they need modifications. On the other hand, the concept presented above (36, 45) can provide a general basis for developing appropriate model(s) for a given material.

It is possible to identify functions such as in Eq. (12-49a) that involve only a single function for both yielding and ultimate or failure behavior. Such single-function models are much simpler than the previously proposed two-surface models such as critical state, cap and that proposed by Lade (described subsequently). For instance, the model, Eqs. (12-49a) and (12-49b) require only five parameters to describe plastic behavior of isotropic materials, this number is smaller than that for the two-surface models that allow for hardening in frictional materials. Also, the parameters can be found from relatively simple plots obtained from laboratory test data. Further work by Desai and co-workers (36, 45) on implementation of this concept and its extension for nonassociative behavior, induced anisotropy, inclusion of pore water pressure, and cyclic behavior is in progress.

Nonassociative Characteristics

Since many geologic materials exhibit nonassociative properties, the yield function f is different from the plastic potential function Q. It was proposed (36, 37) that the foregoing functions from Eq. (12-45) can be used to develop plastic potential functions Q as

$$Q = f_1(J_i) + \kappa f_2(J_i) \tag{12-50}$$

where κ is a material parameter. A rather new idea of the use of correction functions to allow for nonassociative characteristic was proposed by Desai and Siriwardane (37). In general, the plastic potential function Q can be expressed as composed of yield function and a correction function, A:

$$Q(J_i, I_i^p, K_j) = f(J_i, I_i^p, K_j) + A(J_i, I_i^p, K_j) \tag{12-51}$$

A special case of Eq. (12-51),

$$Q = A_1(J_1, I_i^p)f + A_2(J_{2D}, I_i^p)f \tag{12-52a}$$

was used in the context of the theory of plasticity as

$$d\epsilon_{ij}^p = \lambda A_1 \frac{\partial f}{\partial S_{ij}} + \lambda A_2 \frac{\partial f}{\partial \sigma_{nn}} \delta_{ij} \tag{12-52b}$$

in order to examine laboratory behavior of a saturated cohesive soil (37).

Many geological materials exhibit nonassociative characteristics only through the hydrostatic behavior, whereas the deviatoric response exhibits associative properties (50). Figure 12-19 shows that the behavior for two cohesionless and one clayey material is associative in deviatoric response (40, 51, 52). For such materials Eq. (12-51) can reduce to

$$Q(J_i) = f_1(J_i) + A(J_1) \tag{12-53}$$

that is, the correction function is dependent only on the first stress invariant, J_1. An example supporting Eq. (12-53) can be cited from Frydman et al. (53), who found the following relations from experiments on a sand:

$$f_1 = \frac{J_{2D}}{J_1} \tag{12-54a}$$

$$Q = \frac{J_{2D}}{J_1} + \beta \ln J_1 \tag{12-54b}$$

where β is a material parameter.

The idea of correction function can have significant potential in application for many (geological) materials, and further investigation of this idea is desirable.

Isotropic Work Hardening Elastic–Plastic Model by Lade

Lade (39–43) has proposed an elastic–plastic work hardening model based on expanding yield surfaces and a failure or ultimate surface expressed in terms of the first and the third invariants of the stress tensor. This model is found from tests with cohesionless soils and normally consolidated (remolded) clays. The following description is based on the work reported by Lade in Refs. 39 to 43. Since this model has been developed with support of comprehensive laboratory test data and documentation, we have described it in greater detail.

In the initial stage, the following function was proposed for the failure surface based on experimental observation for Monterey O sand:

$$J_1^3 - \kappa_1 J_3 = 0 \tag{12-55}$$

$$s = \frac{\sigma_1 + 2\sigma_3}{\sqrt{3}} \qquad t = \frac{2}{3}(\sigma_1 - \sigma_3) \qquad v^p = \frac{\epsilon_1^p + 2\epsilon_3^p}{3} \qquad \gamma^p = \frac{2}{3}(\epsilon_1^p - \epsilon_3^p)$$

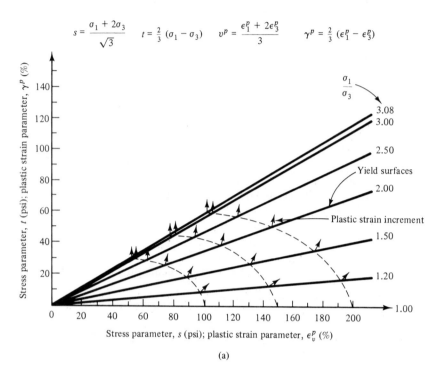

(a)

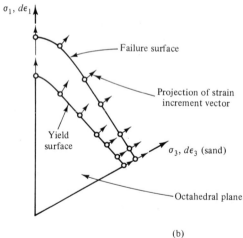

(b)

Figure 12-19   Test results showing deviatoric normality: (a) results from Pooroshasb et al. (Ref. 51); (b) results from Lade (Refs. 39–42); (c) results from Young and McKyes (Ref. 52).

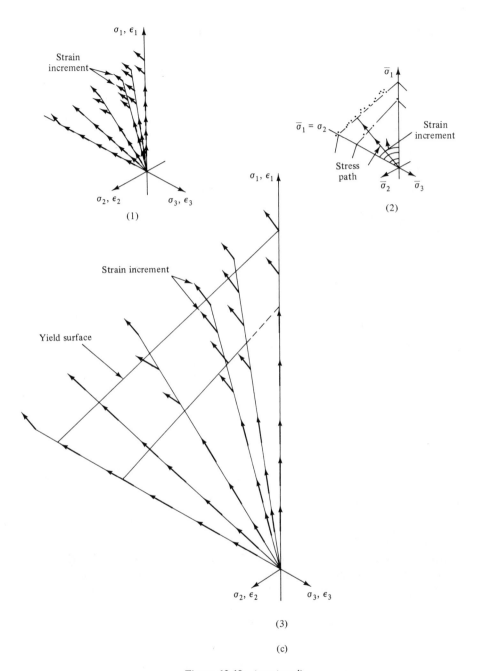

Figure 12-19 (*continued*)

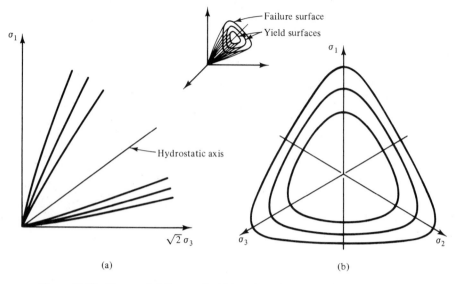

Figure 12-20 Traces of failure and yield surfaces in (a) triaxial plane; (b) octahedral plane (Refs. 40–42).

where $J_1 = \sigma_{ii} = \sigma_{xx} + \sigma_{yy} + \sigma_{zz} = \sigma_1 + \sigma_2 + \sigma_3$, $J_3 = \sigma_x\sigma_y\sigma_z + 2\tau_{xy}\tau_{yz}\tau_{xz} - (\sigma_x\tau_{xz}^2 + \sigma_y\tau_{zx}^2 + \sigma_z\tau_{xy}^2) = \sigma_1\sigma_2\sigma_3$ and $\kappa_1 =$ constant dependent on the density of the sand. The model based on this expression is referred to as κ-theory.

The expanding yield surfaces [Fig. 12-20(a)] are defined by function f given by

$$f = \frac{J_1^3}{J_3} \tag{12-56}$$

whose values vary with loading and reach κ_1 at failure. In the κ-theory the plastic potential function is expressed as

$$Q = J_1^3 - \kappa_2 J_3 \tag{12-57}$$

where κ_2 is a constant for a given f. The plastic potential function Q has the same shape as f in the octahedral plane (Fig. 12-20); however, f and Q can be different. Hence the theory allows for nonassociative characteristics.

The projection of f and Q on the σ_1 versus $\sqrt{2}\,\sigma_3$ space is a straight line. Subsequently, the theory was modified to include curved surfaces and was referred to as η-theory, which includes the κ-theory as a special case. According to the η-theory, the yield surfaces are expressed as

$$f_p = \left(\frac{J_1^3}{J_3} - 27\right)\left(\frac{J_1}{p_a}\right)^m \tag{12-58a}$$

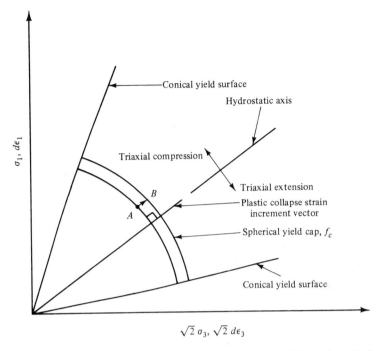

Figure 12-21 Location of yield cap in relation to conical yield surface (Refs. 40–42).

and for the failure or ultimate surface

$$f_p = \eta_1 \tag{12-58b}$$

Here p_a is the atmospheric pressure and m is the material parameter. The plastic potential function in the modified theory is also expressed in a similar fashion as

$$Q_p = J_1^3 - \left[27 + \eta_2 \left(\frac{p_a}{J_1}\right)^m\right] J_3 \tag{12-59}$$

where η_2 is a constant for given f_p and σ_3.

In addition to the foregoing modifications, the η-theory also included a spherical yield cap f_c, as shown in Fig. 12-21:

$$f_c = I_1^2 + 2I_2 \tag{12-60}$$

where

$$I_2 = \tau_{xy}^2 + \tau_{yz}^2 + \tau_{xz}^2 - (\sigma_x\sigma_y + \sigma_y\sigma_z + \sigma_x\sigma_z) = -(\sigma_1\sigma_2 + \sigma_2\sigma_3 + \sigma_1\sigma_3)$$

Strain increments. With the foregoing definitions of f_p, f_c, and Q_p, the total strain increment $d\epsilon_{ij}$ was assumed to be composed of three parts:

$$d\epsilon_{ij} = d\epsilon_{ij}^e + d\epsilon_{ij}^c + d\epsilon_{ij}^p \tag{12-61}$$

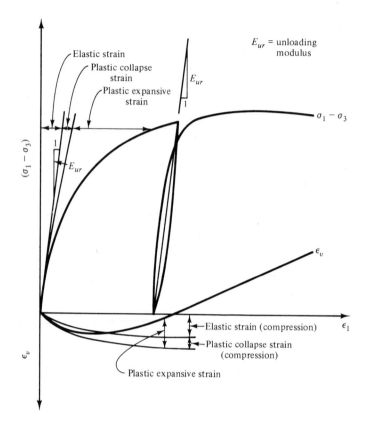

Figure 12-22 Elastic, plastic collapse, and plastic expansive strain components (Refs. 40–42).

where $d\epsilon_{ij}^e$, $d\epsilon_{ij}^c$, and $d\epsilon_{ij}^p$ are the elastic, plastic collapse, and plastic expansive components, respectively. Figure 12-22 shows a pictorial representation of these three components. The two plastic strain increments are evaluated by using Eq. (12-6) with f_c and Q_p, respectively, whereas the elastic part is found from the unloading–reloading modulus, E_{ur}.

Elastic strains. These are evaluated by using the (incremental) Hooke's law based on the elastic modulus E_{ur} and Poisson's ratio ν; a value of 0.2 is often used for the latter. The expression used for E_{ur} is given by

$$E_{ur} = K_{ur} p_a \left(\frac{\sigma_3}{p_a} \right)^n \tag{12-62}$$

where K_{ur} is the nondimensional modulus number and n is the exponent, which is a parameter. Thus three parameters, K_{ur}, n, and ν, are required to define the elastic behavior.

Plastic collapse strains.   Consider point A (Fig. 12-21) on the current yield surface f_c [Eq. (12-60)]. Under a stress increment, the material will yield plastically to point B and harden to the corresponding yield surface if

$$df_c > 0 \qquad (12\text{-}63\text{a})$$

If the stress change is such that

$$df_c \leqslant 0 \qquad (12\text{-}63\text{b})$$

the material experiences elastic deformations. If we assume associated flow rule (and normality), the plastic potential function Q_c that would define flow during yielding is equal to f_c [Eq. (12-60)]. Then, invoking the normality rule, the increments of plastic collapse strains are given by

$$d\epsilon_{ij}^c = \Delta\lambda_c \frac{\partial Q_c}{\partial \sigma_{ij}} = \Delta\lambda_c \frac{\partial f_c}{\partial \sigma_{ij}} \qquad (12\text{-}64)$$

where $\Delta\lambda_c$ is the constant of proportionality. By taking appropriate partial derivatives of f_c with respect to σ_{ij}, the following expression is obtained:

$$\{d\epsilon^c\} = 2\,\Delta\lambda_c\{\bar{\sigma}\} \qquad (12\text{-}65)$$

where $\{d\epsilon^c\}^{\mathrm{T}} = [d\epsilon_{xx}^c \; d\epsilon_{yy}^c \; d\epsilon_{zz}^c \; d\gamma_{xy}^c \; d\gamma_{yz}^c \; d\gamma_{xz}^c]$ and $\{\bar{\sigma}\}^{\mathrm{T}} = [\sigma_{xx} \;\; \sigma_{yy} \;\; \sigma_{zz} \;\; 2\tau_{xy}$ $2\tau_{yz} \;\; 2\tau_{xz}]$.

The value of $\Delta\lambda_c$ is expressed in terms of the increment of plastic work dW_c and the potential function Q_c as

$$\Delta\lambda_c = \frac{dW_c}{2Q_c} \qquad (12\text{-}66\text{a})$$

where

$$dW_c = Cpp_a \left(\frac{p_a^2}{f_c} \right)^{1-p} d\left(\frac{f_c}{p_a^2} \right) \qquad (12\text{-}66\text{b})$$

Here C and p are material parameters. The values of C and p are found by plotting W_c/p_a versus f_c/p_a^2 as shown in Fig. 12-23.

Plastic expansive strains.   The following expression permits computation of increments of plastic expansive strains $d\epsilon_{ij}^p$:

$$d\epsilon_{ij}^p = \Delta\lambda_p \frac{\partial Q_p}{\partial \sigma_{ij}} \qquad (12\text{-}67)$$

where $\Delta\lambda_p$ is a constant of proportionality and Q_p is given in Eq. (12-59). The expression for $\Delta\lambda_p$ is obtained as

$$\Delta\lambda_p = \frac{dW_p}{3Q_p + m\eta_2(p_a/J_1)^m J_3} \qquad (12\text{-}68)$$

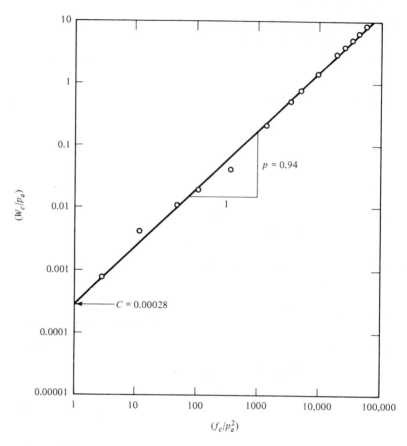

Figure 12-23 Plastic collapse work, W_c versus f_c for loose Sacramento River sand (Refs. 40–42).

where dW_p is the increment of plastic work due to an increase in the stress level df_p given by

$$\Delta\lambda_p = \frac{df_p}{f_p} \frac{1}{(1/qw_p) - b} \tag{12-69}$$

where q and b are material constants. The value of f_p can be approximately expressed as

$$f_p = ae^{-bW_p}\left(\frac{W_p}{p_a}\right)^{1/q}, \qquad q > 0 \tag{12-70}$$

Here a is a material constant. In the foregoing q is defined as

$$q = \frac{\log\left[W_{p(\text{peak})}/W_{p(60)}\right] - \left[1 - W_{p(60)}/W_{p(\text{peak})}\right]0.434}{\log\left[\eta_1/f_{p(60)}\right]} \tag{12-71a}$$

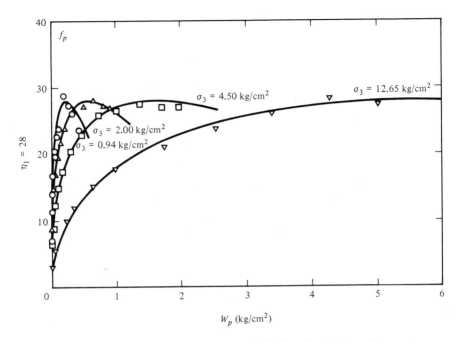

Figure 12-24 Variation of total plastic work with f_p and σ_3 for loose Sacramento River sand (Refs. 40–42).

where $[W_{p(60)}, f_{p(60)}]$ and $(W_{p(\text{peak})}, \eta_1)$ are two points in the plot of f_p versus W_p (Fig. 12-24), and e is the base for natural logarithm. Here the peak point and the point corresponding to 60% of η_1 are chosen. A simplified expression for q can also be used:

$$q = \alpha + \beta \frac{\sigma_3}{p_a} \qquad (12\text{-}71b)$$

which is shown in Fig. 12-25.

A summary of the parameters needed for finding $d\epsilon_{ij}^p$ is given below:

η_1 and m for defining f_p
η_2 for defining Q_p
a and b

Lade has derived the following expressions and has proposed procedures for their evaluation:

$\boldsymbol{\eta_1}$ **and** $\boldsymbol{m}$**:** The parameters are obtained from a plot of $(J_1^3/J_3 - 27)$ versus p_a/J_1 (Fig. 12-26). Here η_1 is the intercept of the ordinate when $p_a/J_1 = 1$, and m is the slope of the curve in Fig. 12-26.

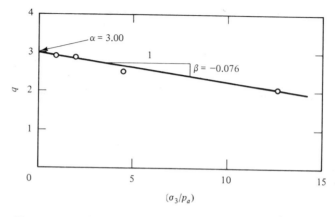

Figure 12-25 Variation of q with confining pressure σ_3 (Refs. 40–42).

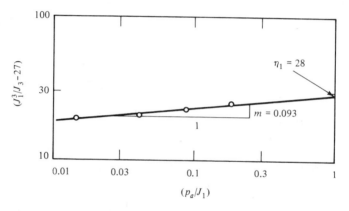

Figure 12-26 Determination of the values of η_1 and m involved in failure criterion for loose Sacramento River sand (Refs. 40–42).

η_2: The expression for η_2 is given by

$$\eta_2 = Sf_p + R\sqrt{\frac{\sigma_3}{p_a}} + t \qquad (12\text{-}72)$$

The value of S is found from the plots of η_2 versus f_p [Fig. 12-27(a)] for different σ_3 and the values of R and t are obtained from the plot of the intercept on η_2 in Fig. 12-27(b) versus $\sqrt{\sigma_3/p_a}$.

a and b: The expression for a and b are

$$a = \eta_1 \left(\frac{ep_a}{W_{p(\text{peak})}}\right)^{1/q} \qquad (12\text{-}73\text{a})$$

$$b = \frac{1}{qW_{p(\text{peak})}} \qquad (12\text{-}73\text{b})$$

where $W_{p(\text{peak})} = Pp_a(\sigma_3/p_a)^l$ (Fig. 12-28).

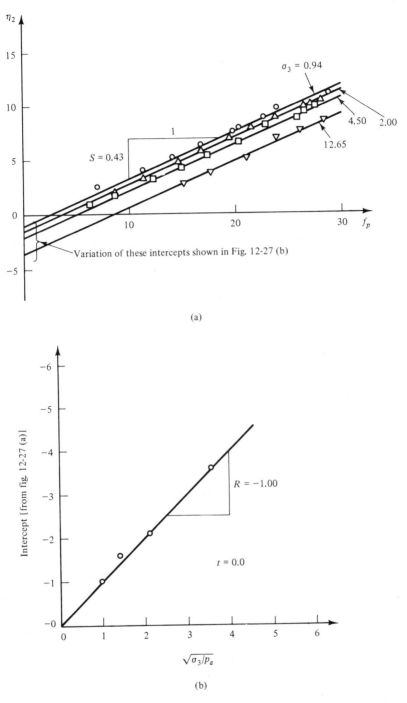

Figure 12-27 (a) Variation of η_2 with f_p and σ_3 for loose Sacramento River sand: (b) variation of intercepts from (a) with the value of σ_3 for loose Sacramento River sand (Refs. 40–42).

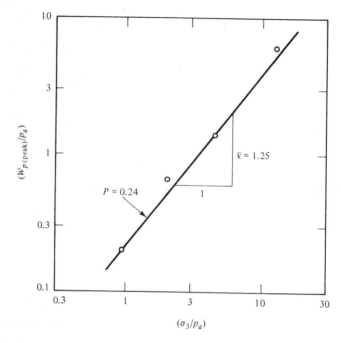

Figure 12-28 Variation of $W_{p(\text{peak})}$ with confining pressure σ_3 (Refs. 40–42).

Thus the constants needed for finding $d\epsilon_{ij}^p$ are η_1, m, R, s, t, α, β, P, and l.

Lade evaluated the foregoing parameters for a number of soils. These parameters for Sacramento River sand, Crushed Napa basalt, and Painted Rock material are reproduced in Table 12-3. These parameters are determined for CTC tests for the foregoing materials tested at two different (initial) relative densities.

The stress–strain response was predicted by using the incremental Hooke's law for the elastic part and Eqs. (12-65) and (12-67) for the plastic collapse and expansive parts, respectively. The predicted responses were then compared with observed behavior. Figure 12-29 shows such typical comparisons for dense Sacramento River sand.

Comments. The model has been cast in a form that permits easy evaluation of the required parameters from CTC tests only. These parameters possess physical meanings, which facilitates understanding of the model. It appears to possess significant potential for application.

The model is applicable for isotropic work hardening behavior, and is at this time not applicable for anisotropic hardening; hence it cannot account for inherent or induced material anisotropy. The parameters of the model are

TABLE 12-3 Summary of Soil Parameters for Sacramento River Sand, Crushed Napa Basalt, and Painted Rock Material (40–42)

Parameter	Sacramento River Sand $D_r = 100\%$	Sacramento River Sand $D_r = 38\%$	Crushed Napa Basalt $D_r = 100\%$	Crushed Napa Basalt $D_r = 70\%$	Painted Rock Material $D_r = 100\%$	Painted Rock Material $D_r = 70\%$	Strain Component
Void ratio, e	0.61	0.87	0.53	0.66	0.40	0.48	
Modulus number, K_{ur}	1680	960	1520	900	1580	730	Elastic
Exponent, n	0.57	0.57	0.34	0.38	0.49	0.66	
Poisson's ratio, ν	0.20	0.20	0.20	0.20	0.20	0.20	
Collapse modulus, C	0.00023	0.00028	0.00075	0.00120	0.00100	0.00140	Plastic
Collapse exponent, p	0.86	0.94	0.74	0.775	0.63	0.644	collapse
Yield constant, η_1	80	28	280	130	101	67	Plastic
Yield exponent, m	0.23	0.093	0.423	0.30	0.21	0.16	expansive
Plastic potential constant, R	−2.95	−1.00	−5.90	−3.03	−2.34	−2.21	
Plastic potential constant, S	0.44	0.43	0.41	0.40	0.44	0.44	
Plastic potential constant, t	8.45	0.00	0.00	0.00	2.80	3.10	
Work hardening constant, α	3.00	3.00	2.22	2.35	3.45	3.28	
Work hardening constant, β	0.060	−0.076	−0.023	−0.046	−0.033	−0.029	
Work hardening constant, P	0.12	0.24	0.50	0.35	0.12	0.080	
Work hardening exponent, l	1.16	1.25	1.09	1.23	1.38	1.61	

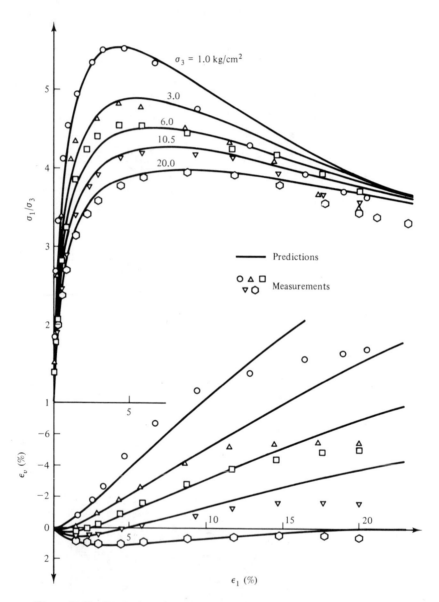

Figure 12-29 Comparison between measured and predicted stress–strain and volume change behavior for dense Sacramento River sand in drained triaxial compression tests (Refs. 40–42).

based essentially on CTC tests. It is possible that if the parameters were referred to other stress paths, their values may be different. Furthermore, the number of material parameters required is considered to be large.

A Model for Inherent Anisotropy

Ghaboussi and Momen (54, 55) proposed the following representation for yield function f:

$$f = f_1(J_{2D}) - f_2(J_1, \theta, \epsilon_{ij}^p) = 0 \qquad (12\text{-}74a)$$

where

$$f_1(J_{2D}) = \sqrt{3J_{2D}} \qquad (12\text{-}74b)$$

$$f_2(J_1, \theta, \epsilon_{ij}^p) = kpR(\theta) \qquad (12\text{-}74c)$$

$$k = f_3(\epsilon_{ij}^p) \qquad (12\text{-}74d)$$

$$p = \frac{J_1}{3} \qquad (12\text{-}74e)$$

$$\theta = \frac{1}{3}\sin^{-1}\left[-\frac{27J_{3D}}{2\left(\sqrt{3J_{2D}}\right)^3}\right] \qquad (12\text{-}74f)$$

Here k is the hardening function and θ is the Lode angle. The simplified form [Eqs. (12-74b) and (12-74c)] implies that the shape of the yield surface remains the same during plastic straining. Various forms of $R(\theta)$ can define different yield surfaces; a form proposed by William and Warnke (56) was used by Ghaboussi and coworkers.

The failure surface f_f is also expressed in a similar form:

$$f_f = \sqrt{3J_{2D}} - MpR(\theta) = 0 \qquad (12\text{-}75)$$

where M is a material constant dependent on the angle of friction ϕ.

The inherent anisotropy is incorporated by writing the yield surface f as

$$f = f_1(S_{ij}, a_{ij}) - kpR(\theta) = 0 \qquad (12\text{-}76a)$$

with

$$f_1(S_{ij}, a_{ij}) = \bar{q} \qquad (12\text{-}76b)$$

$$\bar{q} = \{S\}^T[N]\{S\} \qquad (12\text{-}76c)$$

where $\{S\}$ is the vector of the components of the deviatoric stress tensor and $[N]$ is the matrix of anisotropy. It may be noted that Eq. (12-76) can be considered to be a special case of the model [Eq. (12-41c)] proposed by Baltov and Sawczuk (7). Ghaboussi and Momen used Eq. (12-74) for simulating a special case of inherent anisotropy in vertically deposited specimens of sand with horizontal bedding planes.

HYPOELASTICITY AND PLASTICITY

A constitutive model represents the behavior of a given material. Hence, a number of such models for the material, if they are rooted in and satisfy the natural laws, should lead toward its "unique" behavior. An example of this idea was given previously in this chapter, where by using the singularity condition in the constitutive matrix of hypoelastic models [Eq. (12-48)], it was possible to derive various yield criteria used in plasticity. The following description also takes this observation into account and deals with the possible common grounds between hypoelasticity and plasticity.

One of the early investigators in the study of hypoelasticity and plasticity was Thomas (46). He studied a special form of the hypoelastic equations (Chapter 7) and the Prandtl–Reuss equations (Chapter 9) for perfectly plastic flow. He proposed a concept for plastic yield without a yield surface based on a continuous transition from elastic to perfectly plastic behavior during loading. Thomas presented incremental form of a constitutive model applicable to infinitesimal elasticity starting from zero state of stress and to finite deformations with the Prandtl–Reuss equations. In this formulation the von Mises criterion does not act as a yield surface but as an ultimate asymptotic condition which is approached with increasing deviatoric deformations.

Green (57, 58) presented an approach in which the general hypoelastic relation was expressed in a tensorial polynomial form. He defined the loading and unloading processes by using the sign of the deviatoric stress power Φ_D, given by the trace of the product of the deviatoric stress tensor S_{ij} and the increment of the deviatoric strain tensor dE_{ij}:

$$\Phi_D = S_{ij} dE_{ij} \quad \text{or} \quad \text{tr}(S\,dE) \qquad (12\text{-}77)$$

This quantity represents the increment in deviatoric strain energy or rate of internal mechanical work. The loading and unloading were then defined as (Fig. 12-30):

$$\text{Loading:} \quad S_{ij} dE_{ij} > 0 \qquad (12\text{-}78a)$$

That is, the rate of internal mechanical work done per unit time and per unit volume of the present configuration is positive; here the material is assumed to be incompressible. For unloading and neutral loading, the proposed criteria were

$$\text{Unloading:} \quad S_{ij} dE_{ij} < 0 \qquad (12\text{-}78b)$$

$$\text{Neutral state:} \quad S_{ij} dE_{ij} = 0 \qquad (12\text{-}78c)$$

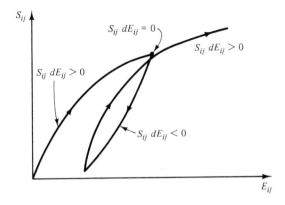

Figure 12-30 Schematic of loading, unloading, and neutral states.

It is required that the constitutive equations for loading and unloading should coincide when the stress power is zero, which implies a neutral state and a smooth transition from loading to unloading, and vice versa. This requirement may not be necessary because experimental results for many materials indicate discontinuity of derivatives at points of transition. Green also attempted to reconcile this proposition with the classical idea of a yield surface; here a specific form of the constitutive coefficients was required in order to satisfy the yield condition and to maintain the plastic states until unloading occurred. In other words, if the material reaches the yield surface, it must flow plastically until unloading occurs. For nonyielding states, the material undergoes elastic deformations. Green's work presented a true generalization of the Prandtl–Reuss theory in an invariant form for finite deformations.

The foregoing concepts were related essentially to metals. Also, the assumption by Green (57, 58) that the constitutive equations should coincide for loading and unloading at neutral states may not be realistic for many materials; Romano (59) did not follow this requirement and for granular materials, he expressed the hypoelastic equations (Chapter 7) by assuming constitutive parameters as linear functions of the rate of deformation tensor. In our notation for stress and strain, Romano's equations can be expressed as follows:

$$d\sigma_{ij} = [h_1 \, dE_{kk} + h_2 \sigma_{mn} \, dE_{mn}]\delta_{ij} + [h_4 \, dE_{kk} + h_5 \sigma_{mn} \, dE_{mn}]\sigma_{ij} + h_{10} E_{ij}$$

$$(12\text{-}79)$$

Here we have maintained Romano's notation for the constitutive parameters $h_1, h_2, \ldots, h_{10}$, whereas in Chapter 7 we have used different notation for these parameters. These quantities are dependent on the invariants of stress and state parameters such as density. Equation (12-79) can be decomposed into mean or volumetric, and deviatoric components as

$$dp = \left[h_1 + \tfrac{1}{3}h_{10} - h_4 p - h_2 p + h_5 p^2 \right] dE_{kk} - \left[h_5 p - h_2 \right] S_{ij} dE_{ij}$$

(12-80a)

$$dS_{ij} = h_{10} dE_{ij} + \left[(h_4 - h_5 p) d\epsilon_{kk} + h_5 S_{mn} dE_{mn} \right] S_{ij}$$

(12-80b)

If it is assumed that the deviatoric stress is independent of the volumetric deformation, implying that S_{ij} is not a function of dE_{kk}, then Eq. (12-80b) leads to

$$h_4 - h_5 p = 0$$

(12-81)

or

$$h_4 = h_5 p$$

Hence Eqs. (12-80) reduce to

$$dp = \left[h_1 + \tfrac{1}{3}h_{10} - h_2 p \right] d\epsilon_{kk} - \left[h_4 - h_2 \right] S_{ij} dE_{ij}$$

(12-82a)

$$dS_{ij} = h_{10} dE_{ij} + \frac{h_4}{p} (S_{mn} dE_{mn}) S_{ij}$$

(12-82b)

Davis and Mullenger (60, 61) used a concept similar to that of Romano (59) in combining the hypoelasticity formulations and the concept of critical state with plasticity (Chapter 11); this formulation is called rate-type fluid (RTF) model. The descriptions presented here follow the developments by Davis and Mullenger (60, 61).

Rate-Type Fluid (RTF) Model

For small deformations, the second term on the right-hand side of Eq. (12-80b) can be ignored, and the equation reduces to

$$dS_{ij} = h_{10} dE_{ij}$$

(12-83a)

Note that this equation is analogous to Eq. (7-25f) for linearly elastic materials with

$$h_{10} = 2G$$

(12-83b)

where G is the shear modulus assumed to be a monotonic (increasing) function of the pressure p; Davis and Mullenger (60) assumed the following simple linear relation

$$G = g_0 + g_1 p$$

(12-84)

Furthermore, for pure volumetric deformations, that is, $E_{ij} = 0$, Eq. (12-82a) reduces to

$$dp = \left(h_1 + \tfrac{1}{3}h_{10} - h_2 p \right) d\epsilon_{kk}$$

(12-85a)

or

$$dp = \left(h_1 + \tfrac{1}{3}h_{10} - h_2 p \right) \frac{dv}{v} \tag{12-85b}$$

where $d\epsilon_{kk} = dv/v$, $v = 1/\rho$ is the specific volume and $\rho =$ density.

As discussed in Chapter 10, the volume logarithm of pressure (v–$\log p$) response is assumed to be linear and parallel to the critical volume–$\log p$ relation; this is considered to be valid if the current value of pressure equals or exceeds all past pressures experienced by the material. Based on this consideration, the function involving coefficients h in Eq. (12-85) can be written as

$$h_1 + \tfrac{1}{3}h_{10} - h_2 p = \kappa p v \tag{12-86}$$

Then Eq. (12-85) becomes

$$dp = \kappa p \, dv \tag{12-87}$$

The expression for κ (Chapter 11) is given by

$$\kappa = \kappa_L \left(\frac{p_{vc}}{p} \right)^{\pi} \tag{12-88}$$

where p_{vc} is the virgin consolidation pressure

$$p_{vc} = p_0 e^{\kappa_L (v_0 - v)} \tag{12-89}$$

Here v_0 is a reference specific volume at reference pressure p_0 (Fig. 12-31) and π is a dimensionless material constant. The critical state line (Fig. 12-31) can be expressed as

$$v_c = v_{c0} - \frac{1}{\kappa_L} \log \left(\frac{p}{p_0} \right) \tag{12-90}$$

where v_c denotes critical specific volume as a function of pressure, and v_{c0} is the reference critical specific volume.

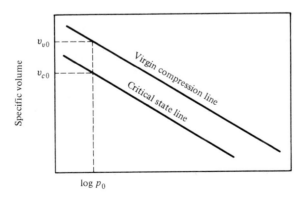

Figure 12-31 Virgin compression and critical state lines.

With the foregoing developments, Eqs. (12-80) now become

$$dp = \kappa p \, dv - (h_4 - h_2) S_{ij} \, dE_{ij} \tag{12-91a}$$

and

$$dS_{ij} = 2G \, dE_{ij} + \frac{h_4}{p} (S_{mn} \, dE_{mn}) S_{ij} \tag{12-91b}$$

Since κ and G have already been defined, it now remains to adopt functional relations for the parameters h_2 and h_4 based on the critical state concept.

At the critical conditions, the state of stress is independent of the rate of deviatoric deformation; hence the second term in Eq. (12-91a) vanishes, resulting in $h_2 = h_4$. Furthermore, at critical conditions, the specific volume reaches the critical value v_c, and

$$S_{ij} S_{ij} = M^2 \tag{12-92}$$

where M is a function of pressure and defines critical states similar to the idea of yield surface in classical plasticity. Then by assuming that M is independent of the deviatoric stress power $S_{ij} \, dE_{ij}$, an expression for h_4 at critical state was found to be (60, 61)

$$h_4 = \frac{-2Gp}{M^2} \tag{12-93}$$

For states before the critical state, the second term in Eq. (12-91a) can contribute toward volumetric deformation. For constant pressure $dp = 0$; hence Eq. (12-91a) gives

$$\kappa p \, dv = (h_4 - h_2) S_{ij} \, dE_{ij} \tag{12-94}$$

Loading

Since during *loading*, the deviatoric stress power $S_{ij} \, dE_{ij} > 0$ and since κ and p are positive, the sign of dv will be the same as that for $(h_4 - h_2)$. As a dense soil dilates, dv and $(h_4 - h_2)$ are negative when the current specific volume $v > v_c$. In the case of a loose soil, dv and $(h_4 - h_2)$ are positive and $v < v_c$.

Based on the foregoing observations, the expression for h_2 and h_4 can be written as

$$h_2 = \frac{2Gp}{M^2} \tag{12-95a}$$

$$h_4 = \left(\frac{v}{v_c}\right)^{\gamma} \frac{2Gp}{M^2} \tag{12-95b}$$

where γ is a material parameter and can be a function of p. Then the final form of Eqs. (12-91) for loading were obtained as

$$dp = \kappa p\, dv - \frac{2Gp}{M^2}\left[1 - \left(\frac{v}{v_c}\right)^{\gamma}\right]S_{ij}\, dE_{ij} \tag{12-96a}$$

$$dS_{ij} = 2G\left[E_{ij} - \left(\frac{v}{v_c}\right)^{\gamma}\frac{S_{mn}\, dE_{mn}}{M^2}S_{ij}\right] \tag{12-96b}$$

Equations (12-96) are derived by assuming that the total stress power $\Phi_T = \sigma_{ij}\, d\epsilon_{ij}$ and the deviatoric stress power Φ_D are positive. In order to generalize these equations for unloading, it is necessary to determine which parameters control the mean and deviatoric stress effects. By assuming that κ depends only on the sign of Φ_T, volumetric strains are irreversible, and (infinitesimal) shear deformations are reversible, Davis and Mullenger used the following relations:

$$\kappa = \begin{cases} \kappa_L\left(\dfrac{p_{vc}}{p}\right)^{\pi} & \text{for } \Phi_T > 0 \\[2ex] \kappa_U & \text{for } \Phi_T < 0 \end{cases} \tag{12-97a}$$

$$h_4 = b\left(\frac{v}{v_c}\right)^{\gamma}\frac{2Gp}{M^2} \tag{12-97b}$$

$$h_2 = b\frac{2Gp}{M^2} \tag{12-97c}$$

where κ_U = constant, $\kappa_U > \kappa_L$, $b = -1$ when $\Phi_D > 0$, $b \geqslant 0$ when $\Phi_D \leqslant 0$, and at the critical state h_2 and h_4 coincide. Then the general constitutive equations for the RTF model were obtained as

$$dp = \kappa p\, dv - \frac{2Gb}{M^2}\left[1 - \left(\frac{v}{v_c}\right)^{\gamma}\right]S_{ij}\, dE_{ij} \tag{12-98a}$$

$$dS_{ij} = 2G\left[E_{ij} + b\left(\frac{v}{v_c}\right)^{\gamma}\frac{S_{mn}\, dE_{mn}}{M^2}S_{ij}\right] \tag{12-98b}$$

Davis and Mullenger specialized the RTF model for various test configurations, such as volumetric deformation, isochoric or constant volume deformation, biaxial stress and deformation, uniaxial strain, conventional, and triaxial compression tests. They also evaluated material parameters, such as κ, v_{v0}, v_{c0}, π, M, γ, and b, for various geologic materials.

For example, let us consider isochoric deformation for which Eqs. (12-96) lead to the following relation between the increments of second invariant of the deviatoric stress tensor dJ_{2D} and mean pressure dp:

$$dJ_{2D} = \frac{M^2 - 2(v_0/v_c)^{\gamma}J_{2D}}{1 - (v_0/v_c)^{\gamma}}\frac{dp}{p} \tag{12-99}$$

Here M, v_c, and γ are functions of p as

$$M = M_1 - M_2 e^{-\beta p} \tag{12-100a}$$

$$\gamma = \gamma_0 p \tag{12-100b}$$

where M_1, M_2, and γ_0 are material constants. The relation between v_c and p is given in Eq. (12-90).

Integration of Eq. (12-99) yields stress paths associated with the constant volume deformation. Figure 12-32 shows various stress paths for different values of γ_0. For loose materials ($v_0 > v_{c0}$) the stress path approaches the (ultimate) yield surface from below with continuous decrease in the pressure. If the material is initially dense ($v_0 < v_{c0}$), the stress paths cross the yield surface and then approach it from above with continuous increase in pressure. The behavior in Fig. 12-32 illustrates that the critical state concept (Chapter 11) is contained in the RTF model.

Davis and Mullenger (49) have derived a failure criterion for the RTF model by using the idea of singularity of C_{ijkl} [Eq. (12-48)] proposed by Tokuoka (47). This concept, in a generalized form for granular materials is presented by Mullenger and Davis (62).

Mroz (63) has compared the nested or multisurface plasticity models with the above rate-type model based on hypoelasticity. He has pointed out the drawbacks of the hypoelastic models: (a) since the hypoelastic model does not allow translation of the yield surface, it cannot account for (induced) anisotropy; (b) the continuity condition for neutral loading is violated, hence the

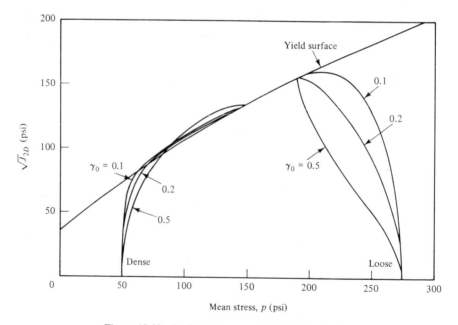

Figure 12-32 Isochoric stress paths (Refs. 49, 60, 61).

model may not accurately predict unloading; and (c) the loading–unloading condition based on the sign of the stress power or work rate may not be applicable for some materials.

We feel that the subject of constitutive laws is in a continuing state of development, and we will see future research which will modify and improve existing models. For instance, the hypoelastic model can be modified by introducing internal constraints or barriers in order to reduce or eliminate the foregoing drawbacks.

BRIEF REVIEW OF OTHER RECENT DEVELOPMENTS

In Chapters 6 to 12 we have covered details of a number of those constitutive laws that have been proposed and developed in the past and many of those that have been implemented in solution procedures. The subject of constitutive modeling has seen significant recent activities, and it is not our intention to cover their details in this introductory text. However, for a sense of completeness, we have provided brief details of some of these recent developments; this includes models based on plasticity, viscoplasticity, and endochronic theory as relevant to geological materials such as soils, rock, and granular media. Similar reviews are also included for cyclic loading and for rock behavior.

General

A number of investigators have modified and/or improved models based on the theory of plasticity and have implemented them for solution of boundary value problems. Here factors special to geological materials, such as stress history, dilatancy, nonassociativeness, softening, inherent and induced anisotropy, drained and undrained behavior, and large strains, have been considered: Nayak and Zienkiewicz (64), Rudnicki and Rice (65), Gudehus et al. (66, 67), Nova and Wood (68), Banerjee and Stipho (69), Maier (70), Desai (71), Chen (72), Desai and Saxena (73), *Proceedings, Workshop* (74), Ko and Sture (75), Gazetas (76), and Mizuno and Chen (77).

Various investigators have proposed models for porous, partially and fully saturated, geological materials: Nur and Byerlee (78), Shipman et al. (79), Johnson et al. (80), Carroll (81), and Smith and Pattillo (82). Zienkiewicz et al. (83–85) have proposed a general framework for drained and undrained behavior of porous (saturated) materials and have applied it to static and cyclic loading conditions.

Viscoplastic

Zienkiewicz, Cormeau, et al. (86–89) proposed a viscoplastic model similar to that developed by Perzyna (90). In this model an explicit time-integration scheme is used to evaluate plastic deformations in time, and the initial stress approach is used. The explicit method may cause numerical instability; to reduce this, Cormeau (87) proposed limits on the size of time step which can provide stable solutions for a limited class of yield functions in plasticity. Katona (91) has used Perzyna's theory to develop a viscoplastic cap model which results in a set of nonlinear differential equations in stress and strain vectors. A one-parameter Crank–Nicolson procedure is used with provision for both explicit and implicit time integration. Other viscoplastic models for geological materials have been presented by Akai et al. (92).

Granular Materials

Based on a statistical averaging procedure on the behavior of individual grains at microscale, Nemat-Nasser et al. (93–95) have proposed models for granular materials for static and cyclic loading to account for factors such as densification, liquefaction, and dilatancy. Details of other models for granular materials are available in Cowin and Satake (96).

Endochronic Model

This theory is derived from the laws of thermodynamics, that is, conservation of energy and dissipation (Clausius–Duhem) inequality (97–100). It is based on the idea of internal variable developments (100). The theory was proposed by Valanis (97–99) by choosing the internal variable to be an intrinsic time scale. It was shown by Valanis that factors such as hysteresis and cross-hardening in metals can be predicted accurately by using the model. Subsequently, Bazant et al. (101–103) and others (104–106) have applied the model for soils and concrete. The endochronic theory is still in the development stage and very few cases of implementation are available. It has been shown by Sandler (106) that in the present form, the endochronic model can cause difficulties in (numerical) implementation, particularly in accounting for unloading, which may be remedied by introducing internal barriers. In that case, the theory may exhibit some features as the yield conditions in the plasticity theory.

Rocks

Many of the models described in the text are suitable for rocks. A number of investigators have developed models specifically for rocks; some of these are: Bieniawski (107), Miller and Cheatham (108), Green et al. (109), Pariseau (110), Shipman et al. (79), Johnson et al. (80), Brady (111), Brady et al. (112), Hueckel and Maier (113), Zienkiewicz and Pande (114), Sture and Ko (115), Kachanov (116), Smith and Cheatham (117), Tan and Kang (118), and Richter et al. (119).

Cyclic Loading

A number of models discussed in this chapter and the preceding chapters are relevant for both static and cyclic loading (e.g., Refs. 10 to 20 in Chapter 11). Some of the additional works that consider cyclic loading are: Seed and Idriss (120), Silver and Seed (121), Castro (122), Ceullar et al. (123), Pender (124), *Proceedings, Specialty Conference* (125), Hardin (126), Idriss et al. (127), Egan and Sangrey (128), Pyke (129), Finn (130), Baladi and Rohani (131), Anderson et al. (132), Drucker and Palgen (133), Zienkiewicz et al. (134), Prakash (135), Darve and Labanieh (136), and Pande and Zienkiewicz (137).

Papers published in a number of recent conferences can be relevant to the subject of constitutive modeling. For geologic materials, books such as Desai and Christian (138), proceedings (74, 125, 135), cited above, and the forthcoming proceedings (139, 140, 141) include many useful works. Implementation of some plasticity models for finite element analysis is given in Ref. 142. For general engineering materials, including metals and composites, geologic materials, discontinuous media (interfaces and joints), concrete, granular materials and aggregates, and implementation of various models based on elasticity, hypoelasticity, viscoelasticity, viscoplasticity, rate type, and endochrony, reference can be made to Desai and Gallagher (143, 144).

REFERENCES

1. Hill, R., *The Mathematical Theory of Plasticity*, Oxford University Press, London, 1950.

2. Hodge, P. G., "The Theory of Piecewise Linear Isotropic Plasticity," *IUTAM Colloq.*, Madrid, Spain, 1955.

3. Ishlinski, A. Yu., "General Theory of Plasticity with Linear Strain Hardening," *Ukr. Mat. Zh.*, Vol. 6, 1954, p. 314.

4. Prager, W., "A New Method of Analyzing Stresses and Strains in Work-Hardening Plastic Solids," *J. Appl. Mech.*, Vol. 78, 1956, p. 493.

5. Shield, R. T., and Ziegler, H., "On Prager's Hardening Rule," *Z. Angew. Math. Phys.*, Vol. 9, 1958, pp. 260–276.

6. Ziegler, H., "A Modification of Prager's Hardening Rule," *Quart. Appl. Math.*, Vol. 17, 1959, p. 55.

7. Baltov, A., and Sawczuk, A., "A Rule for Anisotropic Hardening," *Acta Mech.*, Vol. 1, No. 2, 1965, pp. 81–92.

8. Phillips, A., and Sierakowski, R. L., "On the Concept of the Yield Surface," *Acta Mech.*, Vol. 1, 1965, p. 29.

9. Justusson, J. W., and Phillips, A., "Stability and Convexity in Plasticity," *Acta Mech.*, Vol. 2, 1966, p. 251.

10. Eisenberg, M. A., and Phillips, A., "A Theory of Plasticity with Non-coincident Yield and Loading Surfaces," *Acta Mech.*, Vol. 11, 1971, pp. 247–260.

11. Green, A. E., and Naghdi, P. M., "A General Theory of an Elastic–Plastic Continuum," *Arch. Ration. Mech. Anal.*, Vol. 18, No. 4, 1965, pp. 252–281.

12. Kadashevitch, Iu. I., and Novozhilov, V. V., "The Theory of Plasticity Which Takes into Account Residual Microstresses," *Appl. Math. Mech.*, Vol. 22, 1959, p. 104.

13. Eisenberg, M. A., and Phillips, A., "On Nonlinear Kinematic Hardening," *Acta Mech.*, Vol. 5, 1968, p. 1.

14. Koiter, W. T., "Stress–Strain Relations, Uniqueness, and Variational Theorems for Elastic–Plastic Materials with a Singular Yield Surface," *Quart. Appl. Math.*, Vol. 11, 1953, p. 350.

15. Batdford, S. B., and Budiansky, B., "A Mathematical Theory of Plasticity Based on the Concept of Slip," *Tech. Note 1871*, NACA, 1949.

16. Mroz, Z., "On the Description of Anisotropic Work Hardening," *J. Mech. Phys. Solids*, Vol. 15, 1967, pp. 163–175.

17. Iwan, W. D., "On a Class of Models for the Yielding Behavior of Continuous and Composite Systems," *J. Appl. Mech., Trans. ASME*, Vol. 34, No. E3, Sept. 1967,. pp. 612–617.

18. Prevost, J. H., "Mathematical Modeling of Monotonic and Cyclic Undrained Clay Behavior," *Int. J. Numer. Anal. Methods Geomech.*, Vol. 1, No. 2, 1977, pp. 195–216.

19. Prevost, J. H., "Anisotropic Undrained Stress–Strain Behavior of Clays," *J. Geotech. Eng. Div., ASCE*, Vol. 104, No. GT8, Aug. 1978, pp. 1075–1090.

20. Prevost, J. H., "Plasticity Theory for Soil Stress–Strain Behavior," *J. Eng. Mech. Div., ASCE*, Vol. 104, No. EM5, Oct. 1978, pp. 1177–1194.

21. Prevost, J. H., "Mathematical Modeling of Soil Stress–Strain–Strength Behavior," *Proc. 3rd Int. Conf. Numer. Methods Geomech.*, Aachen, W. Germany, W. Wittke (Ed.), Vol. 1, Balkema Press, Rotterdam, 1979, pp. 347–361.

22. Prevost, J. H., "Nonlinear Anisotropic Stress–Strain–Strength Behavior of Soils," Report, Dept. of Civil Eng., Princeton Univ., Princeton, N.J., 1980.

23. Dafalias, Y. F., and Popov, E. P., "A Model for Nonlinearly Hardening Materials for Complex Loading," *Acta Mech.*, Vol. 21, No. 3, 1975, pp. 173–192.

24. Dafalias, Y. F., and Popov, E. P., "Cyclic Loading for Materials with a Vanishing Elastic Region," *Nucl. Eng. Des.*, Vol. 41, No. 2, 1977, pp. 293–302.

25. Dafalias, Y. F., "A Bounding Surface Plasticity Model," *Proc. 7th Can. Congr. Appl. Mech.*, Sherbrooke, Canada, 1979.

26. Dafalias, Y. F., and Popov, E. P., "Plastic Internal Variables Formalism of Cyclic Plasticity," *J. Appl. Mech.*, Vol. 98, No. 4, Dec. 1976, pp. 645–651.

27. Krieg, R. D., "A Practical Two-Surface Plasticity Theory," *J. Appl. Mech.*, Vol. 42, 1975, pp. 641–646.

28. Sture, S., "Model by Prevost," in *Evaluation of Constitutive Parameters for Geological Materials*, C. S. Desai (Ed.), *Workshop Session, Symp. on Implementa-*

tion of Computer Procedures and Stress–Strain Laws in Geotech. Eng., Chicago, Aug. 1981.

29. Mroz, Z., Norris, V. A., and Zienkiewicz, O. C., "An Anisotropic Hardening Model for Soils and Its Application to Cyclic Loading," *Int. J. Numer. Anal. Methods Geomech.*, Vol. 2, 1978, pp. 203–221.

30. Mroz, Z., Norris, V. A., and Zienkiewicz, O. C., "Application of an Anisotropic Hardening Model in the Analysis of Elasto-plastic Deformation of Soils," *Geotechnique*, Vol. 29, No. 1, 1979, pp. 1–34.

31. Dafalias, Y. F., "A Model for Soil Behavior under Monotonic and Cyclic Loading Conditions," *Proc. 5th Int. Conf. Struct. Mech. Reactor Technol.*, Berlin, W. Germany, Aug. 1979.

32. Dafalias, Y. F., and Herrmann, L. R., "Bounding Surface Formulation of Soil Plasticity," in *Soils under Cyclic and Transient Load*, G. N. Pande and O. C. Zienkiewicz (Eds.), John Wiley & Sons Ltd., Chichester, England, 1982.

33. Prevost, J. H., "Two Surface versus Multi-surface Plasticity Theories," *Int. J. Numer. Anal. Methods Geomech.*, 1982, in press.

34. Rivlin, R. S. and Ericksen, J. I., "Stress Deformation Relations for Isotropic Materials," *J. Ratl. Mech. Analy.*, Vol. 4, 1955, pp. 323–425.

35. Baker, R. and Desai, C. S., "Induced Anisotropy During Plastic Straining," *Int. J. Numer. Anal. Methods Geomech.*, in press.

36. Desai, C. S., "A General Basis for Yield, Failure and Potential Functions in Plasticity," *Int. J. Numer. Anal. Methods Geomech.*, Vol. 4, 1980, pp. 361–375.

37. Desai, C. S., and Siriwardane, H. J., "A Concept of Correction Functions to Account for Non-associative Characteristics of Geologic Media," *Int. J. Numer. Anal. Methods Geomech.*, Vol. 4, 1980, pp. 377–387.

38. Truesdell, C. and Noll, W., "The Nonlinear Field Theories of Mechanics," *Encyclopedia of Physics*, J. Flugge (Ed.), Vol. 3/1, Springer-Verlag, Berlin, 1965.

39. Lade, P. V. and Duncan, J. M., "Cubical Triaxial Tests on Cohesionless Soil," *J. Soil Mech. Found. Div., ASCE*, Vol. 99, No. SM10, Oct. 1973, pp. 793–812.

40. Lade, P. V. and Duncan, J. M., "Elastoplastic Stress–Strain Theory for Cohesionless Soil," *J. Geotech. Eng. Div., ASCE*, Vol. 101, No. GT10, Oct. 1975, 1037–1053.

41. Lade, P. V., "Elasto-Plastic Stress–Strain Theory for Cohesionless Soil with Curved Yield Surfaces," *Rep. UCLA-ENG 7594*, Univ. of Calif., Los Angeles, Nov. 1975.

42. Lade, P. V., "Elastic-Plastic Stress–Strain Theory for Cohesionless Soil with Curved Yield Surfaces," *Int. J. Solids Struct.*, Vol. 13, 1977, pp. 1019–1035.

43. Matsuoka, H. and Nakai, T., "Stress-Deformation and Strength Characteristics of Soil Under Three Different Principal Stresses," *Proc. Japan Soc. Civil Engrs.*, No. 232, 1974, pp. 59–70.

44. Mould, J., "Constitutive Characterization of Granular Materials at Low Effective Stress Levels," *M.S. Thesis*, VPI & SU, Blacksburg, VA, 1979.

45. Desai, C. S. and Faruque, M. O., "Further Development of Generalized Basis for Modelling of Geological Materials," *Report*, Dept. of Civil Eng. and Eng. Mech., Univ. of Arizona, Tucson, AZ, 1982.

46. Thomas, T. Y., "Combined Elastic and Prandtl–Reuss Stress–Strain Relations," *Proc. Nat. Academy Sci.*, USA, Vol. 41, 1955, pp. 720–726.

47. Tokuoka, T., "Yield Conditions and Flow Rules Derived from Hypoelasticity," *Arch. Rat. Mech. Anal.*, Vol. 42, 1971, pp. 239–252.

48. Coon, M. D. and Evans, R. J., "Incremental Constitutive Laws and Their Associated Failure Criteria with Application to Plain Concrete," *Int. J. Solids Struct.*, Vol. 9, 1972, pp. 1169–1183.

49. Davis, R. O. and Mullenger, G., "Derived Failure Criteria for Granular Media," *Int. J. Numer. Anal. Methods Geomech.*, Vol. 3, No. 3, 1979, pp. 279–283.

50. Baker, R. and Desai, C. S., "Consequences of Deviatoric Normality in Plasticity with Isotropic Strain Hardening," *Int. J. Numer. Anal. Methods Geomech.*, Vol. 6, No. 3, 1982, pp. 383–390.

51. Pooroshasb, H. B., Holubec, I. and Sherbourne, A. N., "Yielding and Flow of Sand in Triaxial Compression, Parts 2 and 3," *Can. Geotech. J.*, Vol. 4, 1967, pp. 376–397.

52. Young, R. N. and McKyes, E., "Yield and Failure of Clay Under Triaxial Stress," *J. Soil. Mech. Found. Div., ASCE*, Vol. 97, 1971, pp. 159–176.

53. Frydman, S., Zeitlin, J. G., and Alpan, I., "The Yielding Behavior of Particulate Media," *Can. Geotech. J.*, Vol. 10, 1973, pp. 361–362.

54. Ghaboussi, J., and Momen, H., "Plasticity Model for Cyclic Behavior of Sands," *Proc. 3rd Int. Conf. Numer. Methods Geomech.*, Aachen, W. Germany, W. Wittke (Ed.), Vol. 1, Balkema Press, Rotterdam, 1979, pp. 423–434.

55. Ghaboussi, J., and Momen, H., "Plasticity Model for Inherently Anisotropic Behavior of Sands," *Int. J. Numer. Anal. Methods Geomech.*, in press.

56. Willam, K. J., and Warnke, E. P., "Constitutive Model for the Triaxial Behavior of Concrete," *Seminar on Concrete Structures Subjected to Triaxial Stress*, ISMES, Bergamo, Italy, 1974.

57. Green, A. E., "Hypoelasticity and Plasticity," *Proc. Roy. Soc.*, Vol. A234, 1956, pp. 46–59.

58. Green, A. E., "Hypoelasticity and Plasticity, II," *J. Ration. Mech. Anal.*, Vol. 5, No. 5, 1956, pp. 725–734.

59. Romano, M., "A Continuum Theory for Granular Media with a Critical State," *Arch. Mech.*, Vol. 26, 1974, pp. 1011–1028.

60. Davis, R. O., and Mullenger, G., "Rate-Type Fluid Model for Granular Media with a Critical State," *Rep. AFWL-TR-77-143*, Air Force Weapons Lab., Kirtland AFB, N.Mex., Feb. 1978.

61. Davis, R. O., and Mullenger, G., "A Rate-Type Constitutive Model for Soil with a Critical State," *Int. J. Numer. Anal. Methods Geomech.*, Vol. 12, 1978, pp. 255–282.

62. Mullenger, G., and Davis, R. O., "A Unified Yield Criterion for Cohesionless Granular Materials," *Int. J. Numer. Anal. Methods Geomech.*, Vol. 5, No. 3, 1981, pp. 285–294.

63. Mroz, Z., "On Hypoelasticity and Plasticity Approaches to Constitutive Modelling of Inelastic Behaviour of Soils," *Int. J. Numer. Anal. Methods Geomech.*, Vol. 4, No. 1, 1980, pp. 45–55.

64. Nayak, G. C., and Zienkiewicz, O. C., "Elasto-plastic Stress Analysis: A Generalization of Various Constitutive Relations Including Strain Softening," *Int. J. Numer. Methods Eng.*, Vol. 5, 1972, pp. 113–135.

65. Rudnicki, J. W., and Rice, J. R., "Conditions for the Localization of Deformation in Pressure-Sensitive Dilatant Materials," *J. Mech. Phy. Solids*, Vol. 23, 1975, pp. 371–394.

66. Gudehus, G., Goldscheider, M., and Winter, H., "Mechanical Properties of Sand and Clay and Numerical Integration Methods: Some Sources of Errors, Bounds and Accuracy," in *Finite Elements in Geomechanics*, G. Gudehus (Ed.), John Wiley & Sons, Ltd., Chichester, England, 1977.

67. Gudehus, G., "A Comparison of Some Constitutive Laws for Soils under Radially Symmetric Loading and Unloading," *Proc. 3rd Int. Conf. Numer. Methods Geomech.*, Aachen, W. Germany, W. Wittke (Ed.), Vol. 4, Balkema Press, Rotterdam, 1979, pp. 1309–1323.

68. Nova, R., and Wood, D. M., "A Constitutive Model for Sand in Triaxial Compression," *Int. J. Numer. Anal. Methods Geomech.*, Vol. 3, No. 3, 1979, pp. 255–278.

69. Banerjee, P. K., and Stipho, A. S., "Associated and Non-associated Constitutive Relations for Undrained Behaviour of Isotropic Clays," *Int. J. Numer. Anal. Methods Geomech.*, Vol. 2, No. 1, 1978, pp. 35–36.

70. Maier, G., "Nonassociated and Coupled Flow Rules of Elastoplasticity for Geotechnical Media," *Proc. 9th Int. Conf. Soil Mech. Found. Eng.*, Tokyo, 1977.

71. Desai, C. S., "A Consistent Finite Element Technique for Work-Softening Behavior," *Proc. Int. Conf. Comput. Methods Nonlinear Mech.*, J. T. Oden (Ed.), TICOM, Univ. of Texas, Austin, 1974.

72. Chen, W. F., "Plasticity in Soil Mechanics and Landslides," *J. Eng. Mech. Div.*, *ASCE*, Vol. 106, No. EM3, June 1980, pp. 443–464.

73. Desai, C. S., and Saxena, S. K. (Eds.), *Implementation of Computer Procedures and Stress–Strain Laws in Geotech. Eng.*, Proc. of Symp. held in Chicago, Illinois, Aug. 1981.

74. *Proc., Workshop on Limit Equilibrium, Plasticity and Generalized Stress–Strain in Geotech. Eng.*, McGill Univ., Montreal, May 1980, ASCE, New York, 1981.

75. Ko, H. Y., and Sture, S., "Data Reduction and Application for Analytical Modelling in Laboratory Shear Strength of Soils," *Spec. Tech. Publ. 74*, ASTM, Philadelphia, 1981.

76. Gazetas, G., "Deformational Soil Anisotropy: Experimental Evaluation and Mathematical Modeling," *Publ. 8102*, Soil Mech. Res. Lab., Case Institute of Technology, Cleveland, Ohio, Apr. 1981.

77. Mizuno, E., and Chen, W. F., "Plasticity Analysis of Slope with Different Flow Rules," *Rep. CE-STR-82-3*, School of Civil Eng., Purdue Univ., West Lafayette, Ind., 1982.

78. Nur, A., and Byerlee, J. D., "An Exact Effective Stress Law for Elastic Deformation of Rock with Fluids," *J. Geophy. Res.*, Vol. 76, No. 26, 1971, pp. 6414–6419.

79. Shipman, F. H., Johnson, J. N., and Green, J. J., "Mechanical Properties of Two Highly Porous Geologic Materials," *AMMRC CTR 74-25*, Terra Tek, Salt Lake City, Utah, 1973.

80. Johnson, J. N., Scott, R. F., Brace, W. F., Jones, A. H., and Green, S. J., "Plasticity Theory for Partially and Fully Saturated Porous Geological Materials," *Rep. TR 74-53*, Terra Tek, Salt Lake City, Utah, 1974.

81. Carroll, M. M., "Mechanical Response of Fluid-Saturated Porous Materials," *Proc. 15th Int. Congr. Theor. Appl. Mech.*, Univ. of Toronto, Toronto, Aug. 1980, pp. 251–262.

82. Smith, M. B., and Pattillo, P. D., "A Material Model for Inelastic Rock Deformation with Spatial Variation of Pore Pressure," *Int. J. Numer. Anal. Methods Geomech.*, in press.

83. Zienkiewicz, O. C., "Basic Formulation of Static and Dynamic Behaviour of Soil and Other Porous Media," *Proc. NATO Workshop Numer. Methods Geomech.*, Vimeiro, Portugal, J. B. Martins (Ed.), D. Reidel Publishing Co., Dordrecht, The Netherlands, 1982.

84. Zienkiewicz, O. C., "Constitutive Laws and Numerical Analysis for Soil Foundations under Static, Transient or Cyclic Loads," *J. Appl. Ocean Res.*, Vol. 2, No. 1, 1980, pp. 23–31.

85. Zienkiewicz, O. C., Chang, C. T., and Bettess, P., "Drained, Undrained, Consolidating and Dynamic Behaviour Assumptions in Soils: Limits of Validity," *Rep. C/R/349/79*, Institute For Numerical Methods in Engineering, Univ. of Wales, Swansea, 1979.

86. Zienkiewicz, O. C., and Cormeau, I. C., "Viscoplasticity, Plasticity, and Creep in Elastic Solids: A Unified Numerical Approach," *Int. J. Numer. Methods Eng.*, Vol. 8, 1976, pp. 821–845.

87. Cormeau, I., "Numerical Stability in Quasi-static Elasto/viscoplasticity," *Int. J. Numer. Methods Eng.*, Vol. 9, 1975, pp. 109–127.

88. Zienkiewicz, O. C., Humpheson, C., and Lewis, R. W., "Associated and Nonassociated Viscoplasticity and Plasticity in Soil Mechanics," *Geotechnique*, Vol. 25, No. 4, 1975, pp. 671–689.

89. Zienkiewicz, O. C., and Humpheson, C., "Viscoplasticity: A Generalized Model for Description of Soil Behavior," in *Numerical Methods in Geotechnical Engineering*, C. S. Desai and J. T. Christian (Eds.), McGraw-Hill Book Company, New York, 1977, Chapter 3.

90. Perzyna, P., "Fundamental Problems in Viscoplasticity," *Adv. Appl. Mech.*, Vol. 9, 1966, pp. 244–368.

91. Katona, M. G., "A Viscoplastic Model for Soils and Rocks," *Proc. Int. Conf. Constitutive Laws Eng. Materials: Theory and Application*, Univ. of Arizona, Tucson, Jan. 1983.

92. Akai, K., Adachi, T., and Nishi, K., "Mechanical Properties of Soft Rocks," *Proc. 9th Int. Conf. Soil Mech. Found. Eng.*, Vol. 1, Tokyo, Japan, 1977, pp. 7–10.

93. Nemat-Nasser, S., and Shokooh, A., "A Unified Approach to Densification and Liquefaction of Cohesionless Sand in Cyclic Shearing," *Can. Geotech. J.*, Vol. 16, No. 4, 1979, pp. 659–678.

94. Nemat-Nasser, S., "On Dynamic and Static Behavior of Granular Materials," *Proc. Int. Symp. Soils under Cyclic and Transient Loading*, Univ. of Wales, Swansea, Jan. 1980.

95. Mehrabadi, M. M., and Nemat-Nasser, S., "On Statistical Description of Stress and Fabric in Granular Materials," *Int. J. Numer. Anal. Methods Geomech.*, Vol. 6, No. 1, 1982, pp. 95–108.

96. Cowin, S. C., and Satake, M. (Eds.), *Proc. U.S.–Japan Seminar on Continuum—Mechanical and Statistical Approaches in the Mechanics of Granular Materials*, Gakujutsu Bunken Fukjukai, Tokyo, 1978.

97. Valanis, K. C., "A Theory of Viscoplasticity without a Yield Surface: Parts I and II," *Arch. Mech.*, Warsaw, Vol. 23, 1971, pp. 517–534 and 535–551.

98. Valanis, K. C., "On the Foundations of the Endochronic Theory of Viscoplasticity," *Arch. Mech.*, Warsaw, Vol. 27, 1975, pp. 857–868.

99. Valanis, K. C., "Some Recent Developments in the Endochronic Theory of Plasticity—The Concept of Internal Barriers," in *Constitutive Equations in Viscoplasticity: Phenomenological and Physical Aspects*, AMD, Vol. 21, ASME, New York, 1976, pp. 15–32.

100. Shapery, R. A., "On a Thermodynamic Constitutive Theory and Its Applications to Various Nonlinear Materials," *Proc. IUTAM Symp.*, B. A. Boley (Ed.), Springer-Verlag, New York, 1968, p. 259.

101. Bazant, Z. P., "Endochronic Inelasticity and Incremental Plasticity," *Int. J. Solids Struct.*, Vol. 14, 1978, pp. 691–714.

102. Bazant, Z. P., and Krizek, R. J., "Endochronic Constitutive Law for Liquefaction of Sand," *J. Eng. Mech. Div., ASCE*, Vol. 102, 1976, pp. 225–232.

103. Bazant, Z. P., and Bhat, P., "Prediction of Hysteresis of Reinforced Concrete Beams," *J. Struct. Eng. Div., ASCE*, Vol. 103, 1977, pp. 153–167.

104. Finn, W. D. L., and Bhatia, S., "Endochronic Theory of Sand Liquefaction," *Proc. 7th World Conf. Earthquake Eng.*, Vol. 3, Istanbul, 1980, pp. 149–156.

105. Wu, H. C., "Loading and Unloading of Drained Sand—An Endochronic Theory," *Proc. Int. Conf. Constitutive Laws Eng. Materials*, Univ. of Arizona, Tucson, Jan. 1983.

106. Sandler, I., "On the Uniqueness and Stability of Endochronic Theories of Material Behavior," *J. Appl. Mech.*, Vol. 45, 1978, pp. 263–278.

107. Bieniawski, Z. T., "Deformational Behavior of Fractured Rock under Multiaxial Compression," in *Structure, Solid Mechanics, and Engineering Design*, Part 1, M. Te'eni (Ed.), Wiley-Interscience, New York, 1971.

108. Miller, T. W., and Cheatham, J. B., "A New Yield Condition and Hardening Rule for Rocks," *Int. J. Rock Mech. Miner. Sci.*, Vol. 9, 1972, pp. 453–474.

109. Green, S. J., Leasia, J. D., Perkins, R. D., and Jones, A. H., "Triaxial Stress Behavior of Solehofen Limestone and Westerly Granite at High Strain Rates," *J. Geophy. Res.*, Vol. 77, No. 20, 1972, pp. 3711–3724.

110. Pariseau, W. G., "Plasticity Theory for Anisotropic Rocks and Soils," *Proc. 10th U.S. Symp. Rock Mech.* (Basic and Applied Rock Mechanics), SME/AIME, New York, 1972, pp. 267–293.

111. Brady, B. T., "A Mechanical Equation of State for Brittle Rock, Parts I and II," *Int. J. Rock Mech. Miner. Sci.*, Vol. 7, 1970, pp. 385–421; Vol. 10, No. 4, 1973, pp. 291–309.

112. Brady, B. T., Duvall, W. I., and Horino, F. G., "An Experimental Determination of the True Triaxial Stress–Strain Behavior of Brittle Rock," *Rock Mech.*, Vol. 5, 1973, pp. 107–120.

113. Hueckel, T., and Maier, G., "Incremental Boundary Value Problems in the Presence of Coupling of Elastic and Plastic Deformations: A Rock Mechanics Oriented Theory," *Int. J. Solids Struct.*, Vol. 13, 1977, pp. 1–15.

114. Zienkiewicz, O. C., and Pande, G. N., "Time-Dependent Multilaminate Model of Rocks—A Numerical Study of Deformation and Failure of Rock Masses," *Int. J. Numer. Anal. Methods Geomech.*, Vol. 1, No. 3, 1977, pp. 219–247.

115. Sture, S., and Ko, H. Y., "Strain Softening of Brittle Geologic Materials," *Int. J. Numer. Anal. Methods Geomech.*, Vol. 2, 1978, pp. 237–253.

116. Kachanov, M. L., "Microcrack Model for Rock Inelasticity," *Rep. 20*, Div. of Eng., Brown Univ., Providence, R.I., Oct. 1980.

117. Smith, M. B., and Cheatham, J. B., "An Anisotropic Compacting Yield Condition Applied to Porous Limestone," *Int. J. Rock Mech. Miner. Sci.*, Vol. 17, 1980, pp. 159–165.

118. Tan, T. K., and Kang, W. F., "Locked in Stresses, Creep and Dilatancy of Rocks and Constitutive Equations," *Rock Mech.*, Vol. 13, 1980, pp. 5–22.

119. Richter, T., Wulf, A., and Borchert, K. M., "Use of the Endochronic Theory on Rock Salt," *Rock Mech.*, Vol. 13, No. 3, 1981, pp. 131–143.

120. Seed, H. B., and Idriss, I. M., "Soil Moduli and Damping Factors for Dynamic Response Analysis," *Rep. EERC-70-10*, Earthquake Eng. Res. Center, Univ. of California, Berkeley, 1970.

121. Silver, M. L., and Seed, H. B., "Deformation Characteristics of Sands under Cyclic Loading," *J. Soil Mech. Found. Div.*, *ASCE*, Vol. 97, No. SM8, 1971, pp. 1081–1098.

122. Castro, G., "Liquefaction and Cyclic Mobility of Saturated Sands," *J. Geotech. Eng. Div.*, *ASCE*, Vol. 101, No. GT6, 1975, pp. 551–569.

123. Ceullar, V., Bazant, Z. P., Krizek, R. J., and Silver, M. L., "Densification and Hysteresis of Sand under Cyclic Shear," *J. Geotech. Eng. Div.*, *ASCE*, Vol. 103, No. GT5, May 1977, pp. 399–416.

124. Pender, M. J., "Modelling of Soil Behavior under Cyclic Loading," *Proc. 9th Int. Conf. Soil Mech. Found. Eng.*, Vol. 2, Tokyo, 1977, pp. 325–331.

125. *Proc. Specialty Conf. Earthquake Eng. Soil Dyn.*, *Geotech. Eng. Div.*, Pasadena, Calif., ASCE, 1978.

126. Hardin, B. O., "The Nature of Stress–Strain Behavior of Soils," *Proc. Specialty Conf. Earthquake Eng. Soil Dyn.*, Vol. 1, Pasadena, Calif., ASCE, 1978.

127. Idriss, I. M., Dobry, R., and Singh, R. D., "Nonlinear Behavior of Soft Clay during Cyclic Loading," *J. Geotech. Eng. Div.*, *ASCE*, Vol. 104, 1978, pp. 1427–1447.

128. Egan, J. A., and Sangrey, D. A., "Critical State Model for Cyclic Load Pore Pressure," *Proc. Specialty Conf. Earthquake Eng. Soil Dyn.*, Vol. 1, Pasadena, Calif., ASCE, 1978, pp. 410–424.

129. Pyke, R. M., "Nonlinear Soil Models for Irregular Cyclic Loading," *J. Geotech. Eng. Div., ASCE*, Vol. 105, 1979, pp. 715–726.

130. Finn, W. D. L., "Critical Review of Dynamic Effective Stress Analysis," *Proc. 2nd U.S. Natl. Conf. Earthquake Eng.*, Palo Alto, Calif., Aug. 1979.

131. Baladi, G. Y., and Rohani, B., "Elastic–Plastic Model for Saturated Sand," *J. Geotech. Eng. Div., ASCE*, Vol. 105, No. GT4, 1979, pp. 465–480.

132. Anderson, K. H., Pool, J. H., Brown, S. F., and Rosenbrand, Wim F., "Cyclic and Static Laboratory Tests for Dramman Clay," *J. Geotech. Eng. Div., ASCE*, Vol. 106, No. GT5, 1980, pp. 499–529.

133. Drucker, D. C., and Palgen, L., "On Stress–Strain Relations Suitable for Cyclic and Other Loading," *T&AM Rep. 443*, Dept. of Theor. Appl. Mech., Univ. of Illinois, Urbana-Champaign, Ill., 1980.

134. Zienkiewicz, O. C., Chang, C. T., Hinton, E., and Leung, K. H., "Effective Stress Dynamic Modelling for Soil Structures Including Drainage and Liquefaction," *Proc. Int. Symp. Soils under Cyclic and Transient Loading*, Univ. of Wales, Swansea, Jan. 1980.

135. Prakash, S. (Ed.), *Proc. Int. Conf. Recent Adv. Geotech. Earthquake Eng. Soil Dyn.*, St. Louis, Mo., Apr. 1981.

136. Darve, F., and Labanieh, S., "Incremental Constitutive Law for Sands and Clays: Simulation of Monotonic and Cyclic Tests," *Int. J. Numer. Anal. Methods Geomech.*, Vol. 6, No. 2, 1982, pp. 243–275.

137. Pande, G. N., and Zienkiewicz, O. C. (Ed.), *Soil Mechanics—Transients and Cyclic Loads: Constitutive Relations and Numerical Treatment*, John Wiley & Sons Ltd., Chichester, England, 1982.

138. Desai, C. S., and Christian, J. T. (Eds.), *Numerical Methods in Geotechnical Engineering*, McGraw-Hill Book Company, New York, 1977.

139. Gudehus, G., et al. (Eds.), *Proc. Int. Workshop Constitutive Behavior of Soils*, Grenoble, France, 1982.

140. Bazant, Z. (Ed.), *Mechanics of Geomaterials* (Proc. William Prager Symp.), IUTAM, Northwestern Univ., Evanston, Ill., 1983.

141. *Proc. U.S.–Japan Seminar on New Models and Constitutive Relations in the Mechanics of Granular Materials*, Cornell Univ., Ithaca, N.Y., 1982.

142. Owen, D. R. J., and Hinton, E., *Finite Elements in Plasticity, Theory and Practice*, Pineridge Press Ltd., Swansea, U.K., 1980.

143. Desai, C. S., and Gallagher, R. H. (Eds.), *Proc. Int. Conf. on Constitutive Laws Eng. Materials: Theory and Applications*, Univ. of Arizona, Tucson, Jan. 1983.

144. Desai, C. S., and Gallagher, R. H. (Eds.), *Mechanics of Engineering Materials*, John Wiley & Sons Ltd., Chichester, England, 1983, in press.

APPENDIX 1

REVIEW OF INDICIAL AND TENSOR NOTATION

One of the main advantages of using indicial and tensor notation is the economy of writing complicated and long mathematical expressions in a compact manner. Here we have given brief descriptions of various topics useful for this text; additional details are available in various References (1, 2).

INDICIAL NOTATION

One subscript or first-order notation. If three variables, say coordinates of a point, are denoted by X, Y, Z or X_1, X_2, X_3, they can be expressed compactly as

$$X_i, \quad i = 1, 2, 3$$

or

$$X^i$$

(A1-1)

Two subscripts or second-order notation

$$X_{ij} \quad i, j = 1, 2, 3$$

or

$$X_i^j$$

(A1-2)

represents nine terms. It implies

$$\text{set} \quad \begin{matrix} i = 1, & j = 1,2,3 \\ i = 2, & j = 1,2,3 \\ i = 3, & j = 1,2,3 \end{matrix} \Rightarrow \begin{pmatrix} X_{11} & X_{12} & X_{13} \\ X_{21} & X_{22} & X_{23} \\ X_{31} & X_{32} & X_{33} \end{pmatrix}$$

Three subscripts or third-order notation

$$X_{ijk}, \quad i, j, k = 1,2,3 \tag{A1-3}$$

This third-order tensor contains 27 terms.

In general, the number of terms N in an indicial notation of nth order can be expressed as

$$N = (3)^n \tag{A1-4}$$

Free index. A free index appears once in *each* term in the notation. For example,

$$\left. \begin{matrix} \text{Index} \quad i \quad \text{in } A_i \\ \text{in } X_i = Y_i + Z_i \\ i \text{ and } j \text{ in } \sigma_{ij} \quad = \alpha_1 \delta_{ij} + \alpha_2 \epsilon_{ij} \end{matrix} \right\} \tag{A1-5}$$

Here $i, j = 1,2,3$.

Dummy index. A dummy index appears twice in a term. For instance,

$$A_{ii} \tag{A1-6a}$$

A dummy index implies that the term is summed over the range of the index—for example:

$$A_{ii} = A_{11} + A_{22} + A_{33}, \quad i = 1,2,3 \tag{A1-6b}$$

$$\sigma_{pp} = \sigma_{11} + \sigma_{22} + \sigma_{33}, \quad p = 1,2,3 \tag{A1-6c}$$

$$a_m X_m = a_r X_r = a_p X_p \quad (m = 1,2,3; r = 1,2,3; p = 1,2,3)$$

$$= a_1 X_1 + a_2 X_2 + a_3 X_3 \tag{A1-6d}$$

In the expression

$$X_i = Y_j Z_{ij} \tag{A1-7a}$$

i is free and j is a dummy index.

In Eq. (A1-7a) j is also called a connecting index. Similarly, j in

$$Y_j = D_{ij} C_i \tag{A1-7b}$$

is a connecting index. If Eqs. (A1-7a) and (A1-7b) were combined, we would obtain

$$X_i = D_{ij} C_i Z_{ij} \tag{A1-7c}$$

This equation has no meaning because it is not consistent with the rules of indicial notation. The index i appears three times in the right-hand-side term. To obtain a correct expression, it is necessary to *overhaul* the dummy index i in Eq. (A1-7b). Because the use of a particular letter does not matter, we write

$$Y_j = D_{kj}C_k \tag{A1-7d}$$

Now its substitution in Eq. (A1-7a) gives

$$X_i = D_{kj}C_k Z_{ij} \tag{A1-7e}$$

which is a valid expression. If we expand the dummy indices (k and j) in Eq. (A1-7e) over their range of $1, 2, 3$, we obtain

$$
\begin{aligned}
X_i &= D_{k1}C_k Z_{i1} + D_{k2}C_k Z_{i2} + D_{k3}C_k Z_{i3} \\
&= D_{11}C_1 Z_{i1} + D_{12}C_1 Z_{i2} + D_{13}C_1 Z_{i3} \\
&\quad + D_{21}C_2 Z_{i1} + D_{22}C_2 Z_{i2} + D_{23}C_2 Z_{i3} \\
&\quad + D_{31}C_3 Z_{i1} + D_{32}C_3 Z_{i2} + D_{33}C_3 Z_{i3}
\end{aligned}
\tag{A1-8a}
$$

This equation has three components, X_1, X_2, X_3: for example,

$$
\begin{aligned}
X_2 &= D_{11}C_1 Z_{21} + D_{12}C_1 Z_{22} + D_{13}C_1 Z_{23} \\
&\quad + D_{21}C_2 Z_{21} + D_{22}C_2 Z_{22} + D_{23}C_2 Z_{23} \\
&\quad + D_{31}C_3 Z_{21} + D_{32}C_3 Z_{22} + D_{33}C_3 Z_{23}
\end{aligned}
\tag{A1-8b}
$$

Partial differentiation. Consider a function $f(X_1, X_2, X_3)$ in the Cartesian coordinate system, x_1, x_2, x_3; here X_1, X_2, X_3 denote general coordinates in the reference system, Fig. (A1-1a). The partial derivative of f is given by

$$\frac{\partial f}{\partial X_i} = \frac{\partial f}{\partial X_1}\left(\frac{\partial X_1}{\partial X_i}\right) + \frac{\partial f}{\partial X_2}\left(\frac{\partial X_2}{\partial X_i}\right) + \frac{\partial f}{\partial X_3}\left(\frac{\partial X_3}{\partial X_i}\right) \tag{A1-9a}$$

or in compact notation,

$$\frac{\partial f}{\partial X_i} = \frac{\partial f}{\partial X_k}\left(\frac{\partial X_k}{\partial X_i}\right), \qquad k = 1, 2, 3 \tag{A1-9b}$$

A partial derivative is often denoted by a comma as

$$\frac{\partial f}{\partial X_i} \equiv f,_{X_i} \tag{A1-10a}$$

$$\frac{\partial u_i}{\partial X_j} \equiv u_{i,j} \tag{A1-10b}$$

$$\frac{\partial u_i}{\partial X_k}\left(\frac{\partial u_i}{\partial X_j}\right) \equiv u_{i,k} u_{i,j} \tag{A1-10c}$$

Here i is a dummy and k and j are free indices.

Expansion of $u_{i,k}u_{i,j}$ in Eq. (A1-10c) gives

$$
\begin{aligned}
u_{i,k}u_{i,j} &\equiv u_{1,k}u_{1,j} + u_{2,k}u_{2,j} + u_{3,k}u_{3,j} \\
&\equiv u_{1,1}u_{1,j} + u_{2,1}u_{2,j} + u_{3,1}u_{3,j} \qquad \text{for } k = 1; \\
&\quad\; u_{1,2}u_{1,j} + u_{2,2}u_{2,j} + u_{3,2}u_{3,j} \qquad \text{for } k = 2; \\
&\quad\; u_{1,3}u_{1,j} + u_{2,3}u_{2,j} + u_{3,3}u_{3,j} \qquad \text{for } k = 3
\end{aligned}
$$

Now set $j = 1$, $j = 2$, $j = 3$; for example, $j = 1$ leads to

$$
\begin{aligned}
&u_{1,1}u_{1,1} + u_{2,1}u_{2,1} + u_{3,1}u_{3,1} \qquad \text{for } k = 1, j = 1; \\
&u_{1,2}u_{1,1} + u_{2,2}u_{2,1} + u_{3,2}u_{3,1} \qquad \text{for } k = 2, j = 1; \qquad \text{(A1-10d)} \\
&u_{1,3}u_{1,1} + u_{2,3}u_{2,1} + u_{3,3}u_{3,1} \qquad \text{for } k = 3, j = 1.
\end{aligned}
$$

and so on.

Kronecker delta.   Consider unit vectors i_1, i_2, i_3 [Fig. A1-1(a)]. Then taking the dot product as

$$
\begin{aligned}
i_1 i_1 &= i_2 i_2 = i_3 i_3 = 1 \\
i_1 i_2 &= i_1 i_3 = i_2 i_3 = 0
\end{aligned} \qquad \text{(A1-11a)}
$$

or in general

$$
i_m i_n = \begin{cases} 1 & \text{for } m = n \\ 0 & \text{for } m \neq n \end{cases} \qquad \text{(A1-11b)}
$$

This is often expressed by using a single notation as

$$
\delta_{ij} = \begin{cases} 1 & \text{for } i = j \\ 0 & \text{for } i \neq j \end{cases} \qquad \text{(A1-11c)}
$$

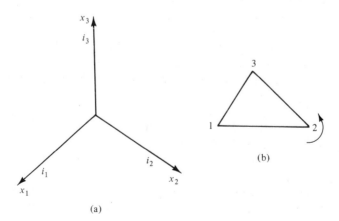

Figure A1-1   (a) Kronecker delta; (b) permutation symbol.

Here δ_{ij} is called the Kronecker delta; it is the matrix counterpart of $[\mathbf{I}]$:

$$[\mathbf{I}] = \begin{bmatrix} 1 & 0 & 0 \\ 0 & 1 & 0 \\ 0 & 0 & 1 \end{bmatrix} \qquad \text{(A1-11d)}$$

Special Result

$$\begin{aligned} \delta_{ii} &= \delta_{11} + \delta_{22} + \delta_{33} = & 3 \\ \delta_{ij}\delta_{ji} &= & 3 \end{aligned} \qquad \text{(A1-12)}$$

Permutation Symbol.   This is given by ϵ_{ijk}, which contains 27 terms. It implies [Fig. A1-1(b)] that

$$\epsilon_{ijk} = \begin{cases} 1 & \text{if } ijk \text{ is even permutation of 123} \\ -1 & \text{if } ijk \text{ is odd permutation of 123} \\ 0 & \text{if } ijk \text{ is not a permutation of 123} \end{cases} \qquad \text{(A1-13)}$$

For example,

$$\epsilon_{123} = \epsilon_{231} = \epsilon_{312} = 1$$

$$\epsilon_{213} = \epsilon_{132} = \epsilon_{321} = -1$$

$$\epsilon_{112} = \epsilon_{221} = \epsilon_{331} = \epsilon_{121} = \epsilon_{212} = \epsilon_{313} = 0$$

TENSOR NOTATION

Scalar or zero-order tensor.   A zero-order tensor is a quantity that carries no free index. For instance,

$$\begin{aligned} A & \qquad \text{is a scalar} \\ A_{nn} & \qquad \text{is a scalar} \\ A_{ij}B_{ij} & \qquad \text{is a scalar} \\ X_{ij}X_{jk}X_{ki} & \qquad \text{is a scalar} \end{aligned} \qquad \text{(A1-14)}$$

First-order tensor (vector).   A first-order tensor carries one free index, for example,

A_i, referred to coordinate system or coordinate axes x_i [Fig. A1-2(a)]

$$\text{(A1-15)}$$

A first-order tensor is governed by the following transformation rule:

$$A'_i = l_{ij}A_j \qquad \text{(A1-16a)}$$

where A'_i is referred to the x'_i coordinate system [Fig. A1-2(b)], and l_{ij} is the

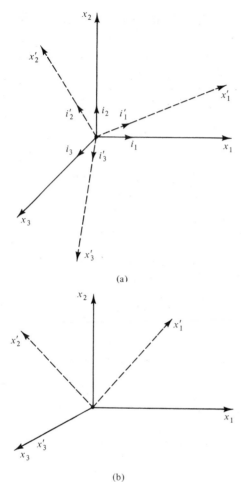

(a)

(b)

Figure A1-2 Coordinate transformations.

transformation tensor (matrix) given as

$$l_{ij} \equiv \begin{pmatrix} l_{11} & l_{12} & l_{13} \\ l_{21} & l_{22} & l_{23} \\ l_{31} & l_{32} & l_{33} \end{pmatrix} \equiv \cos(x_i', x_j) \qquad \text{(A1-16b)}$$

The second-order tensor, l_{ij}, denotes the direction cosines of angles between x_i' and x_j as follows:

	x_1	x_2	x_3
x_1'	l_{11}	l_{12}	l_{13}
x_2'	l_{21}	l_{22}	l_{23}
x_3'	l_{31}	l_{32}	l_{33}

where

$$l_{11} = \cos(x_1', x_1)$$
$$l_{12} = \cos(x_1', x_2) \tag{A1-17}$$

and so on.

Transformation Matrix. Consider vector **A** with components (X_1, X_2, X_3) in Fig. A1-2(a). It can be expressed as

$$\mathbf{A} = X_1 \mathbf{i}_1 + X_2 \mathbf{i}_2 + X_3 \mathbf{i}_3 = X_j \mathbf{i}_j \tag{A1-18a}$$

If the coordinate system x_i was rotated to x_i' with the origin fixed, then

$$\mathbf{A} = X_m' i_m' \tag{A1-18b}$$

From Eqs. (A1-18a) and (A1-18b), we have

$$X_j i_j = X_m' i_m' \tag{A1-18c}$$

Now take a dot product with respect to i_n; then

$$X_j i_j i_n = X_m' i_m' i_n \tag{A1-19a}$$

Because $i_j i_n = \delta_{jn}$, Eq. (A1-19a) becomes

$$X_n = X_m' i_m' i_n \tag{A1-19b}$$

By definition we have

$$i_m' i_n = \cos(x_m', x_n) = l_{mn} \tag{A1-20}$$

Then Eq. (A1-19b) gives

$$X_n = l_{mn} X_m' \tag{A1-21a}$$

Similarly, by taking the dot product with respect to i_n', we have

$$X_n' = l_{nm} X_m \tag{A1-21b}$$

where l_{mn} is given in the foregoing table.

Properties of Transformation Matrix. Differentiation of Eq. (A1-21a) with respect to X_i gives

$$X_{n,i} = l_{mn} X_{m,i}' \tag{A1-22a}$$

Because, $X_{1,1} = 1$, $X_{1,2} = 0$, and so on, we have

$$X_{n,i} = \delta_{ni} \tag{A1-22b}$$

Then

$$\delta_{ni} = l_{mn} X_{m,i}' \tag{A1-22c}$$

Similarly, from Eq. (A1-21b), we have

$$X_{n,i}' = l_{nm} X_{m,i} = l_{nm} \delta_{mi} \tag{A1-22d}$$

If Eq. (A1-22d) is inserted in Eq. (A1-22c), we obtain

$$\delta_{ni} = l_{mn}l_{mk}\delta_{ki} \tag{A1-22e}$$

By using the definition of δ_{ki}, Eq. (A1-22e) becomes

$$\delta_{ni} = l_{mn}l_{mi} \tag{A1-23a}$$

By differentiation of Eq. (A1-21b), we have

$$l_{nj}l_{ij} = \delta_{ni} \tag{A1-23b}$$

Therefore,

$$l_{mn}l_{mi} = l_{nj}l_{ij} = \delta_{ni} \tag{A1-23c}$$

Now expansion of Eq. (A1-23a) gives

$$l_{11}^2 + l_{21}^2 + l_{31}^2 = \delta_{11} = 1$$
$$l_{12}^2 + l_{22}^2 + l_{32}^2 = \delta_{22} = 1 \tag{A1-24a}$$
$$l_{13}^2 + l_{23}^2 + l_{33}^2 = \delta_{33} = 1$$

$$l_{11}l_{12} + l_{21}l_{22} + l_{31}l_{32} = \delta_{12} = 0$$
$$l_{11}l_{13} + l_{21}l_{23} + l_{31}l_{33} = \delta_{13} = 0 \tag{A1-24b}$$
$$l_{12}l_{13} + l_{22}l_{23} + l_{32}l_{33} = \delta_{23} = 0$$

Similarly, expansion of Eq. (A1-23b) gives

$$l_{11}^2 + l_{12}^2 + l_{13}^2 = 1$$
$$l_{21}^2 + l_{22}^2 + l_{23}^2 = 1 \tag{A1-25a}$$
$$l_{31}^2 + l_{32}^2 + l_{33}^2 = 1$$

$$l_{11}l_{21} + l_{12}l_{22} + l_{13}l_{23} = 0$$
$$l_{11}l_{31} + l_{12}l_{32} + l_{13}l_{33} = 0 \tag{A1-25b}$$
$$l_{21}l_{31} + l_{22}l_{32} + l_{23}l_{33} = 0$$

Equations (A1-24a) and (A1-25a) are called *normalization conditions*. They state that the sum of the squares of the elements of any row or column of the transformation matrix is unity.

Equations (A1-24b) and (A1-25b) denote the *orthogonality conditions*, which state that the sum of the products of elements in any two different rows or columns vanishes.

Example A1-1

Consider a rotation angle of 45° between x_1 and x_1', and x_2 and x_2' and zero between x_3 and x_3'; then [Fig. A1-2(b)]

$$l_{mn} = \cos(x_m', x_n) = \begin{pmatrix} \dfrac{\sqrt{2}}{2} & \dfrac{\sqrt{2}}{2} & 0 \\[2mm] -\dfrac{\sqrt{2}}{2} & \dfrac{\sqrt{2}}{2} & 0 \\[2mm] 0 & 0 & 1 \end{pmatrix} \tag{A1-26}$$

Example of a First-Order Tensor. Consider the vector

$$X_i = \begin{Bmatrix} X_1 \\ X_2 \\ X_3 \end{Bmatrix} = \begin{Bmatrix} 2 \\ 5 \\ 8 \end{Bmatrix} \tag{A1-27a}$$

The magnitude of $X_i = \sqrt{X_i X_i} = \sqrt{(2)^2 + (5)^2 + (8)^2} = \sqrt{93}.$ For a rotated primed x_i' system with the angle of rotation equal to $45°$,

$$X_1' = l_{11}X_1 + l_{12}X_2 + l_{13}X_3$$

$$= \frac{\sqrt{2}}{2}(2) + \frac{\sqrt{2}}{2}(5) + 0(8) = \frac{\sqrt{2}}{2}(7)$$

$$X_2' = l_{21}X_1 + l_{22}X_2 + l_{23}X_3$$

$$= -\frac{\sqrt{2}}{2}(2) + \frac{\sqrt{2}}{2}(5) + 0(8) = \frac{\sqrt{2}}{2}(3) \tag{A1-27b}$$

$$X_3' = l_{31}X_1 + l_{32}X_2 + l_{33}X_3$$

$$= 0(2) + 0(5) + 1(8) = 8$$

Second-order tensor. A second-order tensor carries *two* free indices. It is governed by the following transformation law:

$$T_{ij}' = l_{im}l_{jn}T_{mn} = l_{ik}l_{js}T_{ks} \tag{A1-28a}$$

Here T_{ij} is referred to the x_i' system. Also,

$$T_{ij} = l_{mi}l_{nj}T_{mn}' \tag{A1-28b}$$

Expansion of T_{ij} leads to

$$T_{ij}' = l_{i1}l_{js}T_{1s} + l_{i2}l_{js}T_{2s} + l_{i3}l_{js}T_{3s}$$

$$= l_{i1}l_{j1}T_{11} + l_{i1}l_{j2}T_{12} + l_{i1}l_{j3}T_{13}$$

$$+ l_{i2}l_{j1}T_{21} + l_{i2}l_{j2}T_{22} + l_{i2}l_{j3}T_{23}$$

$$+ l_{i3}l_{j1}T_{31} + l_{i3}l_{j2}T_{32} + l_{i3}l_{j3}T_{33} \tag{A1-28c}$$

Example A1-2

For a rotation angle of $45°$, we have

$$l_{ij} = \cos(x_i', x_j)$$

or

$$l_{ij} \equiv \begin{vmatrix} \dfrac{\sqrt{2}}{2} & \dfrac{\sqrt{2}}{2} & 0 \\ -\dfrac{\sqrt{2}}{2} & \dfrac{\sqrt{2}}{2} & 0 \\ 0 & 0 & 1 \end{vmatrix} \tag{A1-29a}$$

Verification as in Eq. (A1-23) gives

$$l_{ik}l_{jk} = \delta_{ij}$$

$$l_{11}l_{31} + l_{12}l_{32} + l_{13}l_{33} = \delta_{13} = 0 + 0 + 0 = 0 \qquad \text{(A1-29b)}$$

$$l_{11}l_{11} + l_{12}l_{12} + l_{13}l_{13} = \delta_{11} = \tfrac{1}{2} + \tfrac{1}{2} + 0 = 1$$

Now consider a symmetric tensor

$$T_{mn} \equiv \begin{pmatrix} 5 & 2 & 1 \\ 2 & 6 & 0 \\ 1 & 0 & 10 \end{pmatrix} \qquad \text{(A1-30)}$$

Note that usually we deal with symmetrical tensors. Now we compute

$$T'_{ij} = l_{im}l_{jn}T_{mn} \qquad \text{(A1-28a)}$$

Properties of Second-Order Tensor. By setting $i = j$ in Eq. (A1-28a), we have

$$T'_{ii} = l_{im}l_{in}T_{mn} = \delta_{mn}T_{mn}$$

$$= T_{mm} \qquad \text{(A1-31a)}$$

On expansion, we have

$$T'_{ii} = \delta_{1n}T_{in} + \delta_{2n}T_{2n} + \delta_{3n}T_{3n}$$

$$= \delta_{11}T_{11} + 0 + 0 + 0 + \delta_{22}T_{22} + 0 + 0 + 0 + \delta_{33}T_{33}$$

$$= T_{11} + T_{22} + T_{33} = T_{kk} \qquad \text{(A1-31b)}$$

That is,

$$T'_{ii} = T_{kk} = T_{ii}$$

$$= \text{tr}(T_{ij}) \qquad \text{(A1-31c)}$$

This is the *first* property of a second-order tensor. It means that the first *trace* of a second-order tensor is *invariant* to transformation of spatial coordinates. Here "tr" denotes the trace of T_{ij}.

Multiplication. Consider the multiplication of A_{ij} as

$$A_{ip}A_{jp} \qquad \text{(A1-32a)}$$

which results in a second-order tensor, $A_{ip}A_{jp}$. Now

$$\text{tr}(A_{ij}) = A_{ii} = A_{pp}$$

$$\text{tr}(A_{ip}A_{jp}) = A_{mp}A_{mp} \qquad \text{(A1-32b)}$$

In general, higher-order multiplications lead to

$$\text{tr}(A_{ij})^2 = A_{ik}A_{ki}$$

$$\text{tr}(A_{ij})^3 = A_{ik}A_{km}A_{mi} \qquad \text{(A1-32c)}$$

Consider

$$T'_{ij}T'_{ij} = \left(l_{im}l_{jn}T_{mn}\right)\left(l_{is}l_{jt}T_{st}\right)$$
$$= \delta_{ms}\delta_{nt}T_{mn}T_{st}$$
$$= T_{st}T_{st}$$

This implies that the second trace of a second-order tensor is invariant to coordinate transformation. Also

$$T'_{im}T'_{mn}T'_{in} = \left(l_{ik}l_{ms}T_{ks}\right)\left(l_{mp}l_{nq}T_{pq}\right)\left(l_{ir}l_{nt}T_{rt}\right)$$
$$= l_{ik}l_{ms}T_{ks}l_{mp}l_{nq}T_{pq}l_{ir}l_{nt}T_{rt}$$
$$= \delta_{kr}T_{ks}\delta_{sp}T_{pq}\delta_{qt}T_{rt}$$
$$= T_{rs}T_{sq}T_{rq}$$
$$= T_{im}T_{mn}T_{in}$$

This implies that the third trace of a second-order tensor is invariant to coordinate transformation.

PROBLEMS

A1-1. Show that $l_{mn}l_{mj} = \delta_{nj}$.

A1-2. Consider

$$A_{ij} = \begin{pmatrix} 5 & 2 & 1 \\ 2 & 6 & 0 \\ 1 & 0 & 10 \end{pmatrix}$$

Find $A'_{ij} = l_{ik}l_{jp}A_{kp}$. Use coordinate transformation as in Example A1-1.

A1-3. Find
(a) First tr(A_{ii})
    Second tr$(A_{ij}A_{ij})$
    Third tr$(A_{im}A_{mn}A_{in})$
(b) Find $A'_{ii}, A'_{ij}A'_{ij}$, and $A'_{im}A'_{mn}A'_{in}$.
Verify that

$$\text{tr}(A_{ii}) = \text{tr}(A'_{ii})$$
$$\text{tr}(A_{ij}A_{ij}) = \text{tr}(A'_{ij}A'_{ij})$$
$$\text{tr}(A_{im}A_{mn}A_{in}) = \text{tr}(A'_{im}A'_{mn}A'_{in})$$

REFERENCES

1. Sokolnikoff, I. S., *Tensor Analysis*, Second edition, John Wiley & Sons, Inc., New York, 1964.
2. Jeffreys, H., *Cartesian Tensors*, Cambridge University Press, Cambridge, England, 1965.

APPENDIX 2

CAYLEY–HAMILTON THEOREM AND INVARIANTS OF TENSORS

CAYLEY–HAMILTON THEOREM

This theorem* states that "any square matrix will satisfy its own characteristic equation."

Let us consider a 3×3 matrix [a]. Its characteristic equation is expressed as

$$\lambda^3 - I_{1a}\lambda^2 + I_{2a}\lambda - I_{3a} = 0 \tag{A2-1a}$$

where I_{1a}, I_{2a}, and I_{3a} are the coefficients of the characteristic equation, given by

$$I_{1a} = \text{tr}[\mathbf{a}] \tag{A2-1b}$$

$$I_{2a} = \begin{vmatrix} a_{11} & a_{12} \\ a_{21} & a_{22} \end{vmatrix} + \begin{vmatrix} a_{22} & a_{23} \\ a_{32} & a_{33} \end{vmatrix} + \begin{vmatrix} a_{11} & a_{13} \\ a_{31} & a_{33} \end{vmatrix} \tag{A2-1c}$$

$$I_{3a} = |a| \tag{A2-1d}$$

According to the theorem, Eq. (A2-1) will be satisfied when [a] is substituted for λ.

---

*The proof of this theorem can be rather lengthy (1). Here we will illustrate its proof by using specific examples only.

Example A2-1

Consider a 3 × 3 matrix

$$[\mathbf{a}] = \begin{bmatrix} 10 & 0 & 0 \\ 0 & -5 & 0 \\ 0 & 0 & -5 \end{bmatrix} \tag{A2-2}$$

Use of Eqs. (A2-1b) to (A2-1d) leads to

$$I_{1a} = 0$$

$$I_{2a} = -50 + 25 - 50 = -75$$

$$I_{3a} = 250$$

Substitution of these values in Eq. (A2-1a) gives

$$\lambda^3 - 75\lambda^2 - 250 = 0 \tag{A2-3a}$$

To verify the theorem, substitute [a] for λ; therefore,

$$[\mathbf{a}]^3 - 75[\mathbf{a}]^2 - 250[\mathbf{I}] = [0] \tag{A2-3b}$$

Check:

$$[\mathbf{a}]^3 = \begin{bmatrix} 10 & 0 & 0 \\ 0 & -5 & 0 \\ 0 & 0 & -5 \end{bmatrix} \cdot \begin{bmatrix} 10 & 0 & 0 \\ 0 & -5 & 0 \\ 0 & 0 & -5 \end{bmatrix} \cdot \begin{bmatrix} 10 & 0 & 0 \\ 0 & -5 & 0 \\ 0 & 0 & -5 \end{bmatrix}$$

$$= \begin{bmatrix} 100 & 0 & 0 \\ 0 & 25 & 0 \\ 0 & 0 & 25 \end{bmatrix} \cdot \begin{bmatrix} 10 & 0 & 0 \\ 0 & -5 & 0 \\ 0 & 0 & -5 \end{bmatrix}$$

$$[\mathbf{a}]^3 = \begin{bmatrix} 1000 & 0 & 0 \\ 0 & -125 & 0 \\ 0 & 0 & -125 \end{bmatrix}$$

$$75[\mathbf{a}]^2 = \begin{bmatrix} 750 & 0 & 0 \\ 0 & -375 & 0 \\ 0 & 0 & -375 \end{bmatrix}$$

Hence the left side of Eq. (A2-3b) can be written as

$$\begin{bmatrix} 1000 & 0 & 0 \\ 0 & -125 & 0 \\ 0 & 0 & -125 \end{bmatrix} + \begin{bmatrix} -750 & 0 & 0 \\ 0 & 375 & 0 \\ 0 & 0 & 375 \end{bmatrix} + \begin{bmatrix} -250 & 0 & 0 \\ 0 & -250 & 0 \\ 0 & 0 & -250 \end{bmatrix}$$

$$= \begin{bmatrix} 0 & 0 & 0 \\ 0 & 0 & 0 \\ 0 & 0 & 0 \end{bmatrix}$$

It can be seen that [a] satisfies its own characteristic equation.

Example A2-2

Let

$$[\mathbf{a}] = \begin{bmatrix} 10 & 2 & 0 \\ 3 & 4 & 1 \\ 5 & 2 & 3 \end{bmatrix} \tag{A2-4}$$

Then

$$I_{1a} = 17$$
$$I_{2a} = 34 + 10 + 30 = 74$$
$$I_{3a} = 10(10) - 2(4) = 100 - 8 = 92$$

The characteristic equation can be written as

$$\lambda^3 - 17\lambda^2 + 74\lambda - 92 = 0 \qquad \text{(A2-5a)}$$

Now

$$[\mathbf{a}]^2 = \begin{bmatrix} 10 & 2 & 0 \\ 3 & 4 & 1 \\ 5 & 2 & 3 \end{bmatrix} \begin{bmatrix} 10 & 2 & 0 \\ 3 & 4 & 1 \\ 5 & 2 & 3 \end{bmatrix}$$

$$= \begin{bmatrix} 106 & 28 & 2 \\ 47 & 24 & 7 \\ 71 & 24 & 11 \end{bmatrix} \qquad \text{(A2-5b)}$$

Therefore,

$$[\mathbf{a}]^3 = \begin{bmatrix} 106 & 28 & 2 \\ 47 & 24 & 7 \\ 71 & 24 & 11 \end{bmatrix} \begin{bmatrix} 10 & 2 & 0 \\ 3 & 4 & 1 \\ 5 & 2 & 3 \end{bmatrix}$$

$$= \begin{bmatrix} 1154 & 328 & 34 \\ 577 & 204 & 45 \\ 837 & 260 & 57 \end{bmatrix} \qquad \text{(A2-5c)}$$

To verify the theorem, let us substitute $[a]$ in place of λ in Eq. (A2-5a). That is,

$$[\mathbf{a}]^3 - 17[\mathbf{a}]^2 + 74[\mathbf{a}] - 92[\mathbf{I}] = [0] \qquad \text{(A2-5d)}$$

Substituting the values of [a], we can obtain

$$\begin{bmatrix} 1154 & 328 & 34 \\ 577 & 204 & 45 \\ 837 & 260 & 57 \end{bmatrix} + \begin{bmatrix} -1802 & -476 & -34 \\ -799 & -408 & -119 \\ -1207 & -408 & -187 \end{bmatrix} + \begin{bmatrix} 740 & 148 & 0 \\ 222 & 296 & 74 \\ 370 & 148 & 222 \end{bmatrix}$$

$$+ \begin{bmatrix} -92 & 0 & 0 \\ 0 & -92 & 0 \\ 0 & 0 & -92 \end{bmatrix} = \begin{bmatrix} 0 & 0 & 0 \\ 0 & 0 & 0 \\ 0 & 0 & 0 \end{bmatrix}$$

The examples above verify the Cayley–Hamilton theorem.

INVARIANTS OF A TENSOR

Consider a vector $\langle \mathbf{A} \rangle$ with respect to the coordinate system denoted by x_k (Fig. A2-1). Vector $\langle \mathbf{A} \rangle$ will have three components:

$$\langle \mathbf{A} \rangle = \begin{Bmatrix} A_1 \\ A_2 \\ A_3 \end{Bmatrix} \qquad \text{(A2-6)}$$

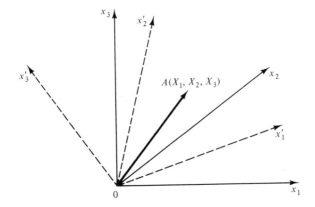

Figure A2-1 Rotation of coordinate systems.

The magnitude of the vector is the length of OA, given by

$$OA^2 = A_1^2 + A_2^2 + A_3^2 \tag{A2-7}$$

If we rotate the coordinate axes to new system denoted by x_k', then the components of the vector $\{A\}$ will be different, and given by

$$\{A'\} = \begin{Bmatrix} A_1' \\ A_2' \\ A_3' \end{Bmatrix} \tag{A2-8}$$

However, the magnitude of the vector will still be the length OA, and it does not depend on the coordinate axes employed. Since the length does not vary or change with the coordinate system, it is known as an *invariant*.

A vector is a first-order tensor, and a first-order tensor has one invariant.

A second-order tensor such as A_{ij} which is symmetric posesses three such invariant quantities. These are given by

First invariant:

$$\bar{I}_{1A} = \operatorname{tr}(A) = A_{ii} \tag{A2-9a}$$

Second invariant:

$$\bar{I}_{2A} = \tfrac{1}{2}\operatorname{tr}(A)^2 = \tfrac{1}{2}A_{ik}A_{ki} \tag{A2-9b}$$

Third invariant:

$$\bar{I}_{3A} = \tfrac{1}{3}\operatorname{tr}(A)^3 = \tfrac{1}{3}A_{ik}A_{kj}A_{ji} \tag{A2-9c}$$

To illustrate these invariants, we now consider the example of the second-order and symmetric tensor A_{ij}, given by

$$(\mathbf{A}) = \begin{pmatrix} A_{11} & A_{12} & A_{13} \\ A_{12} & A_{22} & A_{23} \\ A_{13} & A_{23} & A_{33} \end{pmatrix} \tag{A2-10}$$

First we express the coefficients of the characteristic equation [Eqs. (A2-1b) to (A2-1d)

$$I_{1A} = A_{11} + A_{22} + A_{33} \tag{A2-11a}$$

$$I_{2A} = \left(A_{11}A_{22} - A_{12}^2 \right) + \left(A_{22}A_{33} - A_{23}^2 \right) + \left(A_{11}A_{33} - A_{13}^2 \right) \tag{A2-11b}$$

$$I_{3A} = |\mathbf{A}| \tag{A2-11c}$$

Then the invariants [Eqs. (A2-9)] can be expressed in terms of these coefficients as

$$\bar{I}_{1A} = I_{1A} \tag{A2-12a}$$

$$\begin{aligned}
\bar{I}_{2A} &= \tfrac{1}{2} A_{ik} A_{ki} \\
&= \tfrac{1}{2} \left(A_{1k} A_{k1} + A_{2k} A_{k2} + A_{3k} A_{k3} \right) \\
&= \tfrac{1}{2} \left(A_{11}A_{11} + A_{12}A_{21} + A_{13}A_{31} \right) + \tfrac{1}{2} \left(A_{21}A_{12} + A_{22}A_{22} + A_{23}A_{32} \right) \\
&\quad + \tfrac{1}{2} \left(A_{31}A_{13} + A_{32}A_{23} + A_{33}A_{33} \right)
\end{aligned} \tag{A2-12b}$$

Since $A_{ij} = A_{ji}$,

$$\bar{I}_{2A} = \tfrac{1}{2} \left(A_{11}^2 + A_{22}^2 + A_{33}^2 + 2A_{12}^2 + 2A_{23}^2 + 2A_{13}^2 \right) \tag{A2-12c}$$

or

$$\begin{aligned}
I_A^2 - 2I_{2A} &= A_{11}^2 + A_{22}^2 + A_{33}^2 + 2A_{11}A_{22} + 2A_{22}A_{33} + 2A_{11}A_{33} \\
&\quad - 2A_{22}A_{33} - 2A_{11}A_{33} - 2A_{11}A_{22} \\
&\quad + 2\left(A_{23}^2 + A_{13}^2 + A_{12}^2 \right) \\
&= A_{11}^2 + A_{22}^2 + A_{33}^2 + 2A_{12}^2 + 2A_{23}^2 + 2A_{13}^2
\end{aligned} \tag{A2-12d}$$

Therefore,

$$\bar{I}_{2A} = \tfrac{1}{2} \left(I_{1A}^2 - 2I_{2A} \right) \tag{A2-13}$$

Now, by using Cayley–Hamilton theorem, we have

$$[\mathbf{A}]^3 - I_{1A}[\mathbf{A}]^2 + I_{2A}[\mathbf{A}] - I_{3A}[\mathbf{I}] = [\mathbf{0}] \tag{A2-14a}$$

Therefore,

$$[\mathbf{A}]^3 = I_{1A}[\mathbf{A}]^2 - I_{2A}[\mathbf{A}] + I_{3A}[\mathbf{I}] \tag{A2-14b}$$

In indicial form, Eq. (A2-14b) becomes

$$A_{ik}A_{km}A_{mj} = I_{1A}A_{ik}A_{kj} - I_{2A}A_{ij} + I_{3A}\delta_{ij} \tag{A2-14c}$$

By contracting i and j, that is, by taking the trace of the tensor above, we have

$$A_{ik}A_{km}A_{mi} = I_{1A}A_{ik}A_{ki} - I_{2A}A_{ii} + 3I_{3A} \tag{A2-14d}$$

Hence

$$3\bar{I}_{3A} = 2I_{1A}\bar{I}_{2A} - I_{2A}I_{1A} + 3I_{3A}$$

or

$$3\bar{I}_{3A} = I_{1A}\left(I_{1A}^2 - 2I_{2A}\right) - I_{1A}I_{2A} + 3I_{3A} \qquad \text{(A2-14e)}$$

or

$$\bar{I}_{3A} = \tfrac{1}{3}\left(I_{1A}^3 - 3I_{1A}I_{2A} + 3I_{3A}\right)$$

Example A2-3

Let us consider an example

$$A_{ij} = \begin{pmatrix} 10 & 5 & 0 \\ 5 & 2 & 3 \\ 0 & 3 & 4 \end{pmatrix} \qquad \text{(A2-15)}$$

Coefficients of the characteristic equation are

$$I_{1A} = 10 + 2 + 4 = 16 \qquad \text{(A2-16a)}$$

$$I_{2A} = (20 - 25) + (8 - 9) + (40) = 34 \qquad \text{(A2-16b)}$$

$$I_{3A} = |A_{ij}| = 10(-1) - 5(20) = -110 \qquad \text{(A2-16c)}$$

Invariants of the tensor are

$$\bar{I}_{1A} = \text{tr}(A) = 16 \qquad \text{(A2-17a)}$$

$$\bar{I}_{2A} = \tfrac{1}{2}\,\text{tr}(A)^2 = \tfrac{1}{2}\left(A_{11}^2 + A_{22}^2 + A_{33}^2 + 2A_{12}^2 + 2A_{13}^2 + 2A_{23}^2\right)$$

$$= \tfrac{1}{2}(100 + 4 + 16 + 50 + 0 + 18)$$

$$= \tfrac{188}{2} = 94 \qquad \text{(A2-17b)}$$

Check:

$$\bar{I}_{2A} = \tfrac{1}{2}\left(I_{1A}^2 - 2I_{2A}\right)$$

$$= \tfrac{1}{2}(16^2 - 2 \times 34) = 94$$

Similarly,

$$\bar{I}_{3A} = \tfrac{1}{3}\left[16^3 - 3 \cdot 16 \cdot 34 + 3(-110)\right]$$

$$= \tfrac{2134}{3} \qquad \text{(A2-17c)}$$

Example A2-4

Consider

$$A_{ij} = \begin{pmatrix} 2 & 1 & 0 \\ 1 & 3 & 2 \\ 0 & 2 & 4 \end{pmatrix} \qquad \text{(A2-18)}$$

Coefficients of the characteristic equation are

$$I_{1A} = 2 + 3 + 4 = 9 \qquad \text{(A2-19a)}$$

$$I_{2A} = 5 + 8 + 8 = 21 \qquad \text{(A2-19b)}$$

$$I_{3A} = 2(8) - 1(4) = 12 \qquad \text{(A2-19c)}$$

The invariants of the tensor are

$$\bar{I}_{1A} = 9 \tag{A2-20a}$$

$$\bar{I}_{2A} = \tfrac{1}{2}(9^2 - 2\cdot 21) = \tfrac{39}{2} \tag{A2-20b}$$

$$\bar{I}_{3A} = \tfrac{1}{3}[9^3 - 3\cdot 9\cdot 21 + 3\cdot 12] = \tfrac{198}{3} = 66 \tag{A2-20c}$$

The invariant property implies that the values of the invariants for a given tensor are unchanged for any transformed coordinate system. To illustrate this, we need to consider first the subject of transformation of tensors from one coordinate system to another.

TRANSFORMATION OF TENSORS

The first-order tensor A_i refers to coordinate system or coordinate axes x_i (Fig. A2-2). The first-order tensor is governed by the following transformation law:

$$A'_i = l_{ij} A_j \tag{A2-21}$$

where A'_i is referred to the x'_i coordinate system and l_{ij} is the transformation tensor (matrix) given as

$$l_{ij} = \begin{pmatrix} l_{11} & l_{12} & l_{13} \\ l_{21} & l_{22} & l_{23} \\ l_{31} & l_{32} & l_{33} \end{pmatrix} = \cos(x'_i, x_j) \tag{A2-22a}$$

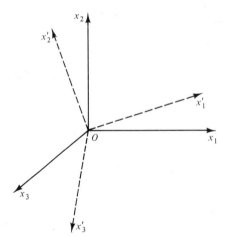

Figure A2-2 Transformation of coordinates (tensors).

l_{ij} are the direction cosines of angles between x_i' and x_i:

	x_1	x_2	x_3
x_1'	l_{11}	l_{12}	l_{13}
x_2'	l_{21}	l_{22}	l_{23}
x_3'	l_{31}	l_{32}	l_{33}

(A2-2b)

where

$$l_{11} = \cos(x_1', x_1)$$
$$l_{12} = \cos(x_1', x_2) \qquad \text{and so on}$$

Consider vector **A** with components (X_1, X_2, X_3) in Fig. A2-1. It can be expressed as

$$A = X_1 \mathbf{i}_1 + X_2 \mathbf{i}_2 + X_3 \mathbf{i}_3 = X_j \mathbf{i}_j \qquad (A2\text{-}23)$$

With the fixed origin, if the coordinate system x_i was rotated to x_i', then

$$A = X_m' \mathbf{i}_m' \qquad (A2\text{-}24)$$

From Eqs. (A2-23) and (A2-24) we have

$$X_j \mathbf{i}_j = X_m' \mathbf{i}_m' \qquad (A2\text{-}25)$$

Now taking the dot product with respect to $\mathbf{i}_n$, we obtain

$$X_j \mathbf{i}_j \cdot \mathbf{i}_n = X_m' \mathbf{i}_m' \cdot \mathbf{i}_n \qquad (A2\text{-}26)$$

Because $\mathbf{i}_j \cdot \mathbf{i}_n = \delta_{jn}$, Eq. (A2-26) becomes

$$X_n = X_m' \mathbf{i}_m' \cdot \mathbf{i}_n \qquad (A2\text{-}27)$$

By definition, we have

$$\mathbf{i}_m' \cdot \mathbf{i}_n = \cos(x_m', x_n) \qquad (A2\text{-}28)$$
$$= l_{mn}$$

Then Eq. (A2-27) gives

$$X_n = l_{mn} X_m' \qquad (A2\text{-}29)$$

Similarly by taking dot product with respect to $\mathbf{i}_n'$, we have

$$X_n' = l_{nm} X_m \qquad (A2\text{-}30)$$

l_{mn} is given by Eq. (A2-22b).

Properties of the Transformation Matrix

Differentiation of Eq. (A2-29) with respect to X_i gives

$$X_{n,i} = l_{mn} X_{m,i}' \qquad (A2\text{-}31)$$

Because $X_{1,1} = 1$, $X_{1,2} = 0$, and so on.

$$X_{n,i} = \delta_{ni} \qquad (A2\text{-}32)$$

Then Eq. (A2-31) becomes

$$\delta_{ni} = l_{mn} X'_{m,i} \tag{A2-33}$$

Similarly, from Eq. (A2-30), we have

$$X'_{n,i} = l_{nm} X_{m,i} = l_{nm} \delta_{mi} \tag{A2-34}$$

If we substitute Eq. (A2-34) into Eq. (A2-31), we obtain

$$\delta_{ni} = l_{mn} l_{mk} \delta_{ki} \tag{A2-35}$$

Using the definition of δ_{ki}, Eq. (A2-35) becomes

$$\delta_{ni} = l_{mn} l_{mi} \tag{A2-36}$$

Similarly, by differentiation of Eq. (A2-30), we can obtain

$$l_{nj} l_{ij} = \delta_{ni} \tag{A2-37}$$

Therefore,

$$l_{mn} l_{mj} = l_{nj} l_{ij} = \delta_{ni} \tag{A2-38}$$

Now expansion of Eq. (A2-36) gives

$$l_{11}^2 + l_{21}^2 + l_{31}^2 = 1$$
$$l_{12}^2 + l_{22}^2 + l_{32}^2 = 1 \tag{A2-39a}$$
$$l_{13}^2 + l_{23}^2 + l_{33}^2 = 1$$

and

$$l_{11} l_{12} + l_{21} l_{22} + l_{31} l_{32} = \delta_{12} = 0$$
$$l_{11} l_{13} + l_{21} l_{23} + l_{31} l_{33} = \delta_{13} = 0 \tag{A2-39b}$$
$$l_{12} l_{13} + l_{22} l_{23} + l_{32} l_{33} = 0$$

Similarly, expansion of Eq. (A2-37) gives

$$l_{11}^2 + l_{12}^2 + l_{13}^2 = 1$$
$$l_{21}^2 + l_{22}^2 + l_{23}^2 = 1 \tag{A2-40a}$$
$$l_{31}^2 + l_{32}^2 + l_{33}^2 = 1$$

and

$$l_{11} l_{21} + l_{12} l_{22} + l_{13} l_{23} = 0$$
$$l_{11} l_{31} + l_{12} l_{32} + l_{13} l_{33} = 0 \tag{A2-40b}$$
$$l_{21} l_{31} + l_{22} l_{32} + l_{23} l_{33} = 0$$

Example A2-5

Consider the tensor A_{ij} in Example A2-4, and the transformation of the system x to x' obtained by rotation of $45°$ about x_3 (Fig. A2-3). The transformation tensor a_{ij} for the

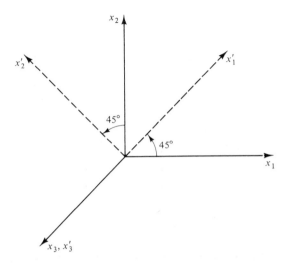

Figure A2-3 Coordinate transformation for Example A2-5.

rotation above is as follows:

	x_1	x_2	x_3
x_1'	$\dfrac{1}{\sqrt{2}}$	$\dfrac{1}{\sqrt{2}}$	0
x_2'	$-\dfrac{1}{\sqrt{2}}$	$\dfrac{1}{\sqrt{2}}$	0
x_3'	0	0	1

That is,

$$l_{ij} = \begin{pmatrix} \dfrac{1}{\sqrt{2}} & \dfrac{1}{\sqrt{2}} & 0 \\[2mm] -\dfrac{1}{\sqrt{2}} & \dfrac{1}{\sqrt{2}} & 0 \\[2mm] 0 & 0 & 1 \end{pmatrix} \qquad \text{(A2-41)}$$

The tensor A_{ij} in the new coordinate system x_i' can be found as follows:

$$A_{ij}' = l_{im}l_{jn}A_{mn}$$

$$= l_{im}l_{j1}A_{m1} + l_{im}l_{j2}A_{m2} + l_{im}l_{j3}A_{m3} \qquad \text{(A2-42a)}$$

$$A_{ij}' = l_{i1}l_{j1}A_{11} + l_{i2}l_{j1}A_{21} + l_{i3}l_{j1}A_{31}$$

$$+ l_{i1}l_{j2}A_{12} + l_{i2}l_{j2}A_{22} + l_{i3}l_{j2}A_{32}$$

$$+ l_{i1}l_{j3}A_{13} + l_{i2}l_{j3}A_{23} + l_{i3}l_{j3}A_{33} \qquad \text{(A2-42b)}$$

Various components of A_{ij}' can be found by assigning values to $i, j = 1, 2, 3$.

The expression in Eq. (A2-42b) can be written in matrix notation as

$$[A'] = [l] \quad [A] \quad [l]^T \qquad (A2\text{-}43a)$$
$$(3\times 3) \quad (3\times 3)(3\times 3)(3\times 3)$$

$$= \begin{bmatrix} \dfrac{1}{\sqrt{2}} & \dfrac{1}{\sqrt{2}} & 0 \\[2mm] -\dfrac{1}{\sqrt{2}} & \dfrac{1}{\sqrt{2}} & 0 \\[2mm] 0 & 0 & 1 \end{bmatrix} \begin{bmatrix} 2 & 1 & 0 \\ 1 & 3 & 2 \\ 0 & 2 & 4 \end{bmatrix} \begin{bmatrix} \dfrac{1}{\sqrt{2}} & -\dfrac{1}{\sqrt{2}} & 0 \\[2mm] \dfrac{1}{\sqrt{2}} & \dfrac{1}{\sqrt{2}} & 0 \\[2mm] 0 & 0 & 1 \end{bmatrix}$$

$$= \begin{bmatrix} \dfrac{3}{\sqrt{2}} & \dfrac{4}{\sqrt{2}} & \dfrac{2}{\sqrt{2}} \\[2mm] -\dfrac{1}{\sqrt{2}} & \dfrac{2}{\sqrt{2}} & \dfrac{2}{\sqrt{2}} \\[2mm] 0 & 2 & 4 \end{bmatrix} \begin{bmatrix} \dfrac{1}{\sqrt{2}} & -\dfrac{1}{\sqrt{2}} & 0 \\[2mm] \dfrac{1}{\sqrt{2}} & \dfrac{1}{\sqrt{2}} & 0 \\[2mm] 0 & 0 & 1 \end{bmatrix}$$

$$[A'] = \begin{bmatrix} \dfrac{7}{2} & \dfrac{1}{2} & \dfrac{2}{\sqrt{2}} \\[2mm] \dfrac{1}{2} & \dfrac{3}{2} & \dfrac{2}{\sqrt{2}} \\[2mm] \dfrac{2}{\sqrt{2}} & \dfrac{2}{\sqrt{2}} & 4 \end{bmatrix} \qquad (A2\text{-}43b)$$

Now, the coefficients of the characteristic equations for [A'] are

$$I_{1A'} = \tfrac{7}{2} + \tfrac{3}{2} + 4 = 9 \qquad (A2\text{-}44a)$$

$$I_{2A'} = 5 + 4 + 12 = 21 \qquad (A2\text{-}44b)$$

$$I_{3A'} = \tfrac{7}{2}(4) - 0 + \sqrt{2}\left(-2/\sqrt{2}\right) = 12 \qquad (A2\text{-}44c)$$

and the invariants for [A'] are

$$\bar{I}_{1A'} = 9 \qquad (A2\text{-}45a)$$

$$\bar{I}_{2A'} = \tfrac{1}{2}\left[9^2 - 2(21)\right] = \tfrac{39}{2} \qquad (A2\text{-}45b)$$

$$\bar{I}_{3A'} = \tfrac{1}{3}\left[9^3 - 3(9)(21) + 3(12)\right] = 66 \qquad (A2\text{-}45c)$$

It can be seen that the above coefficients of the characteristic equation as well as the three quantities denoted by $\bar{I}_{1A'}$, $\bar{I}_{2A'}$, and $\bar{I}_{3A'}$ and those in Example A2-4 do not change or are invariant with the coordinate transformation. That is why the above quantities are known as the *invariants*.

REFERENCE

1. Franklin, J. N., *Matrix Theory*, Prentice-Hall, Inc., Englewood Cliffs, N.J. 1965.

APPENDIX 3

LEAST-SQUARES-FIT PROGRAM

A description of the least-squares-fit procedure used in this text is given in Chapter 6. This appendix presents a listing of the computer code/program, user's guide, and a sample of input data for a stress–strain curve simulated by using a cubic polynomial:

$$y_i = a_0 + a_1 x_i + a_2 x_i^2 + a_3 x_i^3 \qquad \text{(A3-1)}$$

where $i = 1, 2, \ldots, n$, n = number of points chosen on the curve and $a_0, a_1, \ldots, a_3$ are the parameters to be determined. The expanded form can be written as

$$\begin{Bmatrix} y_1 \\ y_2 \\ \vdots \\ y_n \end{Bmatrix} = \begin{bmatrix} 1 & x_1 & x_1^2 & x_1^3 \\ 1 & x_2 & x_2^2 & x_2^3 \\ \vdots & & & \vdots \\ 1 & x_n & x_y^2 & x_n^3 \end{bmatrix} \begin{Bmatrix} a_0 \\ a_1 \\ \vdots \\ a_n \end{Bmatrix} \qquad \text{(A3-2a)}$$

or

$$\{y\} = [x]\{A\} \qquad \text{(A3-2b)}$$

To assign appropriate weights to the data points, a diagonal weighting matrix [W] can be introduced by premultiplying Eq. (A3-2b) as

$$[x]^T [W]\{y\} = [x]^T [W][x]\{A\}$$

or

$$\{X\} = [Y]\{A\} \qquad \text{(A3-3)}$$

or

$$\{A\} = [Y]^{-1}\{X\}$$

Here $\langle A \rangle$ is the vector of $a_0, a_1, \ldots, a_n$ that is obtained by solving Eq. (A3-3). Substitution of computed values of $a_0, a_1, \ldots, a_n$ in a polynomial such as in Eq. (A3-1) yields the required least-squares-fit approximation to the observed stress–strain data.

USER'S GUIDE

Description	Symbol
Card Set 1: Format (3I5)	
Number of points (n) on a curve	NPT
(e.g., $n = 9$ in Fig. A3-1)	
Number of parameters to be found	NPM
[e.g., 4 in Eq. (A3-1)]	
Option for weighting matrix:	IFLAG
IFLAG = 1, supply weight(s)	
other than unity;	
$\ne$ 1, computer assigns unity as all weights	
Card Set 2: Format (8 F10.5)	
First row of matrix [x] and first	A(1, J), B(I)
element of vector $\langle y \rangle$	J = 1, 2, \ldots, m,
(continue on more than one card,	where m = number of
if necessary)	a's in the polynomial
Second row of [x] and second element of $\langle y \rangle$	A(2, J), B(2)
	J = 1, 2, \ldots, m
nth row of [x] and nth element of $\langle y \rangle$	A(n, J), B(n)
	J = 1, 2, \ldots, m
Card Set 3: Format (8F10.5) If IFLAG $\ne$ 1,	
these data not needed	
Weights for data points:	W(I)
I = 1, 2, \ldots, NPT	

Input data for the curve in Fig. A3-1 for [x] and $\langle y \rangle$ corresponding to Eq. (A3-1) are as follows:

		[x]		
1	x	x^2	x^3	$\langle y \rangle$
1	0.20	0.04	0.008	1.00
1	0.40	0.16	0.064	2.00
1	0.60	0.36	0.216	2.50
1	1.15	1.32	1.52	3.00
1	1.50	2.25	3.38	3.25
1	1.95	3.80	7.42	3.50
1	2.50	6.25	15.63	3.75
1	4.00	16.00	64.0	4.00
1	7.00	49.00	343.0	4.15

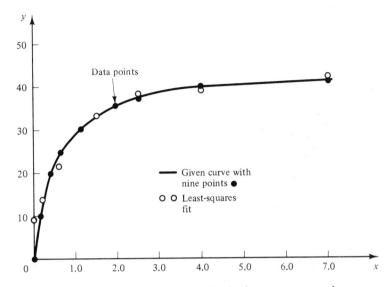

Figure A3-1 Example of curve fitted using least-squares procedure.

For the above, the values of a's computed are

$$a_0 = 0.970 \qquad a_1 = 2.298$$
$$a_2 = -0.563 \qquad a_4 = 0.043$$

LISTING OF CODE

```
C
C**********************************************************************
C
C ***   DETERMINATION OF CONSTITUTIVE PARAMETERS
C ***   BY LEAST SQUARE METHOD
C
C**********************************************************************
C
        DIMENSION A(50,25),B(50),W(50),AWA(25,25),AWB(25)
C
        READ(5,1)NPT,NPM,IFLAG
        WRITE(6,2) NPT,NPM,IFLAG
  1     FORMAT(3I5)
  2     FORMAT(/////5X,'NPT=',I8,' NPM=',I8,' IFLAG=',I8//)
C
C ***   NPT = NO. OF POINTS USED
C ***   NPM = NO. OF UNDETERMINED PARAMETERS
C ***   IFLAG = AN INDEX WHICH DETERMINES WHEN TO READ THE WEIGHTS
C                 W(I)
C               IFLAG = 1 ; READ THE WEIGHTS
C               IFLAG = ANY OTHER NUMBER ; DO NOT READ THE WEIGHTS
C               DO 88 I=1,NPT
        READ(5,89)(A(I,J),J=1,NPM),B(I)
 88     CONTINUE
 89     FORMAT(8F10.5)
        WRITE(6,200)
```

```
200     FORMAT(//5X,'COEFFICIENT MATRIX A(I,J) AND VECTOR B(I)'/)
        DO 6 I=1,NPT
        WRITE(6,250)(A(I,J),J=1,NPM),B(I)
250     FORMAT(1X,8E13.5)
  6     CONTINUE
        IF(IFLAG.EQ.1)GO TO 15
        DO 10 I=1,NPT
 10     W(I)=1.0
        GO TO 20
 15     READ(5,89)(W(I),I=1,NPT)
 20     CALL GENERI(A,W,B,NPT,NPM,AWA,AWB,50,25)
        CALL GAUSSE(AWA,AWB,25,NPM)
        WRITE(6,300)
300     FORMAT(//7X,'VALUES OF PARAMETERS'/)
        WRITE(6,400)(AWB(I),I=1,NPM)
400     FORMAT(10X,E12.5/)
        STOP
        END
C
C*********************************************************************
C
        SUBROUTINE GENERI(A,W,B,NPT,NPM,AA,BB,NA,NB)
C
        DIMENSION A(NA,NB),W(NA),B(NA),AA(NB,NB),BB(NB),AT(25,25)
C
        DO 20 I=1,NPM
        DO 20 K=1,NPT
 20     AT(I,K)=A(K,I)*W(K)
        DO 40 I=1,NPM
        SUM=0.0
        DO 30 J=1,NPM
        SUM1=0.0
        DO 25 K=1,NPT
 25     SUM1=SUM1+AT(I,K)*A(K,J)
 30     AA(I,J)=SUM1
        DO 35 K=1,NPT
 35     SUM=SUM+AT(I,K)*B(K)
 40     BB(I)=SUM
        RETURN
        END
C
C*********************************************************************
C
        SUBROUTINE GAUSSE(A,B,NN,N)
C
        DIMENSION A(NN,NN),B(1)
C
        NM1=N-1
        DO 30 I=1,NM1
        XLARGE=1.E-15
        JCOUNT=0
        DO 15 J=I,N
        IF(ABS(A(J,I)).GT.XLARGE) GO TO 10
        GO TO 15
 10     XLARGE=ABS(A(J,I))
        JCOUNT=J
 15     CONTINUE
        IF(JCOUNT.EQ.0) GO TO 60
        DO 20 K=I,N
        TEMP=A(I,K)
        A(I,K)=A(JCOUNT,K)
        A(JCOUNT,K)=TEMP
```

```
20      CONTINUE
        TEMP=B(I)
        B(I)=B(JCOUNT)
        B(JCOUNT)=TEMP
        IP1=I+1
        DO 30 K=IP1,N
        Q=-A(K,I)/A(I,I)
        B(K)=Q*B(I)+B(K)
        DO 30 J=IP1,N
        A(K,J)=Q*A(I,J)+A(K,J)
30      CONTINUE
        IF(A(N,N).EQ.0.) GO TO 60
        B(N)=B(N)/A(N,N)
        NP1=N+1
        DO 50 K=1,NM1
        Q=0.0
        NMK=N-K
        DO 40 J=1,K
        Q=Q+A(NMK,(NP1-J))*B(NP1-J)
40      CONTINUE
        B(NMK)=(B(NMK)-Q)/A(NMK,NMK)
50      CONTINUE
        IERROR=1
        RETURN
60      IERROR=2
        WRITE(6,300)
300     FORMAT(/////'  ***** PROBLEM ! ***** MATRIX IS SINGULAR *****'/)
        RETURN
        END
```

APPENDIX 4

COMPUTER CODE FOR
PARAMETERS OF CAP MODEL

This appendix gives details of the computer code/program for finding parameters for the cap model described in Chapter 11. Similar computer codes can be developed for other models, and the work for evaluation of parameters can be made significantly easy. For example, such a procedure is being developed for the new model, Eq. (12-49a) and Fig. 12-18, which is much simpler than the procedure presented here for the two-surface Cap model.

A listing of the program is included in the following pages, together with user's guide for preparing input data for given sets of stress–strain curves (Fig. 11-29). The use of the program is illustrated by finding the parameters for the silty sand described in Chapter 10; typical input/out data are included for this problem. Details of this procedure are given by Zaman, Desai, and Faruque (see Ref. 21 of Chapter 11).

Evaluation of Cap Parameters for Silty Sand

The stress–strain curves for the silty sand are given in Chapter 10. Table A4-1 shows the parameters as printed out by the computer. Figure A4-1 shows symbolic plots of yield surfaces as loci of equal volumetric plastic strain that can be drawn by the computer by fitting the ellipses f_2 [Eq. (11-44b)] through the intersections of the loci shown by dashed curves; the solid curves show the fitted ellipses. The fixed surface, f_1, is drawn by the computer by a least-squares-fit procedure (Appendix 3) on the "ultimate" values of $\sqrt{J_{2D}}$ for various stress–strain curves. Figure A4-2 shows the computed plots of the vectors of incremental plastic strains; curves drawn normal to these provide a plastic potential function Q.

TABLE A4-1   **Parameters for Silty Sand**

Elastic parameters:
 $E = 14,966$ psi ($103,115$ kPa)
 $\nu = 0.36$
Parameters for fixed envelope:
 $\alpha = 37.22$ psi (256.5 kPa)
 $\beta = -0.00255$ (psi)$^{-1}$ [-0.00037 (kPa)$^{-1}$]
 $\gamma = 37.22$ psi (256.5 kPa)
 $\theta = 0.426$
Shape factor and hardening parameters:
 $R = 1.9301$
 $W = 7.37$
 $D = 0.0028$ (psi)$^{-1}$ [0.0004 (kPa)$^{-1}$]
 $Z = 0$ assumed

USER'S GUIDE

All input is in free format.

	Description	Symbol
I.	Total number of tests to be used (IN MAIN) other than HC test	NTEST
II.	Stress components at ultimate (or failure) for each test (IN MAIN) (Fig. A4-3)	
	σ_1 component of failure stress	STRF1(IA)
	σ_2 component of failure stress	STRF2(IA)
	σ_3 component of failure stress	STRF3(IA)
	Index or flag for calculating plastic strain	INDEX(IA)[a]
Note:	Input as many cards as the number of tests used ($=$ NTEST).	
Note:	Sets III, IV, and V are read in HYDST.	
III.	Title or heading for hydrostatic compression (HC) test	NAME(I)
IV.	Number of points chosen on the HC curve	NPH
	Unloading slope for ϵ_1	UMH1
	Unloading slope for ϵ_2	UMH2
	Unloading slope for ϵ_3	UMH3
	(See Fig. A4-4 for an explanation of unloading moduli.)	
V.	Mean stress $\left(\dfrac{\sigma_1 + \sigma_2 + \sigma_3}{3}\right)$ for a selected point on the HC curve	SMIN(I)
	ϵ_1 component corresponding to SMIN(I)	EXHC
	ϵ_2 component corresponding to SMIN(I)	EYHC
	ϵ_3 component corresponding to SMIN(I)	EZHC
Note:	Input as many cards as number of selected points on HC curve ($=$ NPH).	

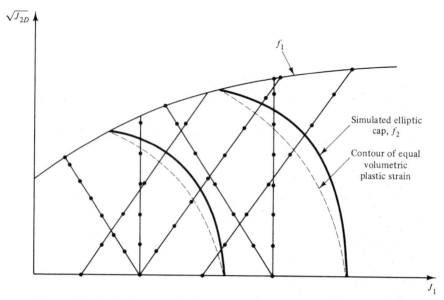

Figure A4-1 Typical stress path plots with fixed envelope, f_1, and yielding caps, f_2.

	Description	Symbol
Note:	Sets VI, VII, and VIII read in MAIN. These data pertain to other stress–strain curves.	
VI.	Title or heading for a given set of curves for a stress–strain curve	NAME(I)
VII.	Number of selected points for a given test data curves	NPT(IA)
	Unloading slope for ϵ_1 (E_{1r})	UMOD1(IA)
	Unloading slope for ϵ_2 (E_{2r})	UMOD2(IA)
	Unloading slope for ϵ_3 (E_{3r})	UMOD3(IA)
VIII.	$(\sigma_1, \epsilon_1), (\sigma_2, \epsilon_2), (\sigma_3, \epsilon_3)$ are stress–strain components at selected point for the given test [Fig. A4-4(a)]	
	Value of σ_1 component at selected point IB	SX(IB)
	Value of σ_2 component at selected point IB	SY(IB)
	Value of σ_3 component at selected point IB	SZ(IB)
	Corresponding value of ϵ_1 component	EX(IB)
	Corresponding value of ϵ_2 component	EY(IB)
	Corresponding value of ϵ_3 component	EZ(IB)
Note:	Input as many cards as the number of selected points NPT(IA).	
Note:	Repeat VI, VII, and VIII for all the tests (= NTEST).	
IX.	Data for plotting fixed surface, stress paths, incremental plastic strain vectors, and yield caps (IN FAILP)	

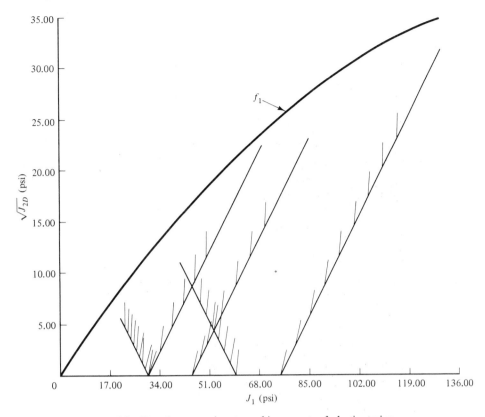

Figure A4-2 Plot of computed vectors of increments of plastic strains.

Description	Symbol
Option for plotting: = 1: plot ≠ 1: no plot	KFLAG
If KFLAG ≠ 1, the following data are not needed *(see Fig. A4-5):*	
Number of divisions or segments for drawing a curve such as fixed surface and cap (usually 50 to 100)	ND
Constant length of increment of plastic strain vector (say 0.5 in.)	XL
Length of x-axis (say, 8 in.)	XLEN
Length of y-axis (say, 8 in.)	YLEN
Height of letters for adding legend on figures (say, 0.2 in.)	HGT

Subroutine FAILP has a provision for input of weights
   W(I) for failure points for each of NTEST. This can
   be activated by inputting IFLAG = 1 instead of IFLAG = 0,
   which is set right now. Weights are used in the least-	

Description	Symbol
squares program to obtain the failure envelope. For details of the least-squares procedure, see Chapter 6 and Appendix 3.	
X. Number of caps to be used for computing shape factor R for the yielding caps (IN CAPMOD)	NCAP

[a]*Explanation for INDEX(IA).* This quantity specifies the stress components with respect to which the elastic strain increments are computed by using appropriate unloading slopes [Fig. A4-4(b)], E_{1r}, E_{2r}, E_{3r}; here subscripts $1,2,3$ relate to strain components. For instance, for the CTC test, the user should plot σ_1 versus $\epsilon_1, \epsilon_2, \epsilon_3$ and specify three unloading slopes; then INDEX(IA) = 1 because σ_1 is the variable stress. Then the elastic strain increments are found as

$$d\epsilon_1^e = \frac{d\sigma_1}{E_{1r}}$$

$$d\epsilon_2^e = \frac{d\sigma_1}{E_{2r}}$$

$$d\epsilon_3^e = \frac{d\sigma_1}{E_{3r}}$$

In the case of the CTE test, σ_1 does not change and σ_2 and σ_3 are increased equally. Here the user can plot σ_2 (or σ_3) versus ϵ_1, ϵ_2, and ϵ_3 and specify INDEX(IA) = 2 (or 3). Then

$$d\epsilon_1^e = \frac{d\sigma_2}{E_{1r}}$$

$$d\epsilon_2^e = \frac{d\sigma_2}{E_{2r}}$$

$$d\epsilon_3^e = \frac{d\sigma_2}{E_{3r}}$$

Thus the values of INDEX(IA) refer to the component of stress with respect to which strain components are plotted.

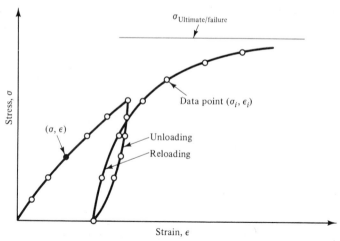

Figure A4-3   Typical stress–strain plot showing unloading and reloading.

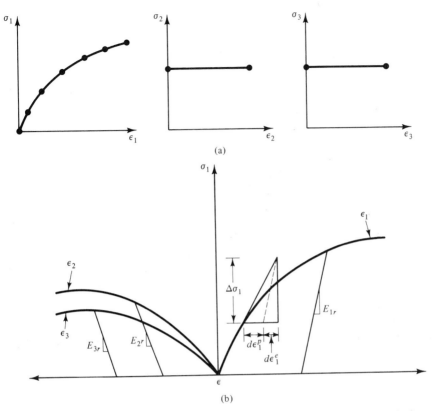

(a)

(b)

Figure A4-4 Example of input for CTC test: (a) typical input set VIII for CTC test; (b) designation of INDEX (IA) and computation of $d\epsilon^e$ and $d\epsilon^{p-}$

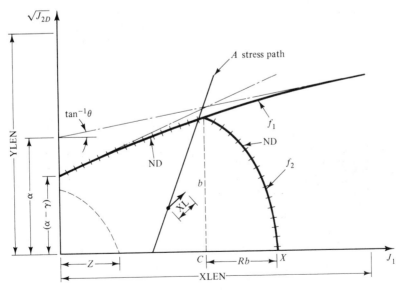

Figure A4-5 Data for plotting.

LISTING OF CODE

```
C   ***   THIS PROGRAM DETERMINES THE PARAMETERS FOR CAP MODEL IN
C   ***   WHICH THE FAILURE SURFACE IS ASSUMED IN THE FORM
C   ***   SQRT(J2D) = ALPHA + THETA*J1 - GAMA*EXP(-BETA*J1)
C   ***   OPTION FOR PLOTTING THE FAILURE SURFACE, STRESS PATHS, PLASTIC
C   ***   STRAIN INCREMENT VECTORS AND THE YIED CAPS ARE ALSO INCLUDED.
C
C
C       ======================================================================
C
C
C                               DEVELOPED BY:
C                               C. S. DESAI
C                               M. M. ZAMAN
C                               M. O. FARUQUE
C
C       ======================================================================
C
C
C
        DIMENSION STRF1(20),STRF2(20),STRF3(20),INDEX(20),XJ1F(20),
       +SQJ2DF(20),NPT(20),UMOD1(20),UMOD2(20),UMOD3(20),SX(20),SY(20),
       +SZ(20),EX(20),EY(20),EZ(20),PJ1(20,20),PJ2D(20,20)
C
        DIMENSION DI1P(20,20),DI2DP(20,20),TI1P(20,20),EVPH(20)
        DIMENSION HVPH(20),SMIN(20),BB(20),PJ1M(20),NAME(20)
C
        OPEN(UNIT=5,FILE='INPUT')
        OPEN(UNIT=6,FILE='OUTPUT')
        READ(5,*)NTEST
        WRITE(6,1001)NTEST
 1001 FORMAT(///T5,'NO. OF TESTS USED  = ',I3)
        NTEST = NTEST-1
        XJMX=0.0
        DO 10 IA=1,NTEST
        READ(5,*)STRF1(IA),STRF2(IA),STRF3(IA),INDEX(IA)
        XJ1F(IA)=STRF1(IA)+STRF2(IA)+STRF3(IA)
        IF(XJMX.LT.XJ1F(IA))XJMX=XJ1F(IA)
   10 SQJ2DF(IA)=SQRT(0.5*(STRF1(IA)**2+STRF2(IA)**2+STRF3(IA)**2-3.*
       +(XJ1F(IA)/3.)**2))
        WRITE(6,1004)
 1004 FORMAT(//
       +T5,'STRESS COMPONENTS AT FAILURE FOR DIFFERENT TESTS'/
       +T5,'----------------------------------------------'//
       +T7,'SIGF-1',6X,'SIGF-2',6X,'SIGF-3'/)
        DO 12 L=1,NTEST
   12 WRITE(6,1002)STRF1(L),STRF2(L),STRF3(L)
 1002 FORMAT(1X,3E12.3)
        CALL HYDST(HVPH,SMIN,NPH)
C
        DO 100 IA=1,NTEST
        READ(5,200)(NAME(L),L=1,20)
  200 FORMAT(20A4)
        WRITE(6,300)(NAME(L),L=1,20)
  300 FORMAT(/////T5,20A4//)
        READ(5,*)NPT(IA),UMOD1(IA),UMOD2(IA),UMOD3(IA)
        IF(IA.GT.1) GO TO 13
        EMOD = UMOD1(IA)
        XNU = (EMOD/UMOD2(IA)+EMOD/UMOD3(IA))/2.
   13 NP=NPT(IA)
        DO 20 IB=1,NP
        READ(5,*)SX(IB),SY(IB),SZ(IB),EX(IB),EY(IB),EZ(IB)
        EX(IB)=EX(IB)/1.
        EY(IB)=EY(IB)/1.
```

```
   20 EZ(IB)=EZ(IB)/1.
      SMINP=(SX(1)+SY(1)+SZ(1))/3.
      DO 21 IX=1,NPH
      IF(SMINP.GT.SMIN(IX).AND.SMINP.LE.SMIN(IX+1))GO TO 22
   21 CONTINUE
   22 RM=(SMIN(IX+1)-SMINP)/(SMIN(IX+1)-SMIN(IX))
      EVPH(IA)=HVPH(IX+1)-RM*(HVPH(IX+1)-HVPH(IX))
C
      DO 25 IB=1,NP
      PJ1(IA,IB)=SX(IB)+SY(IB)+SZ(IB)
   25 PJ2D(IA,IB)=SQRT(0.5*(SX(IB)**2+SY(IB)**2+SZ(IB)**2-3.*(PJ1(IA,IB)
     +/3.)**2))
C
      IND=INDEX(IA)
      GO TO (40,50,60),IND
   40 CALL IPSTN(IA,SX,EX,EY,EZ,UMOD1,UMOD2,UMOD3,NP,DI1P,DI2DP,TI1P,
     +EVPH)
      GO TO 90
   50 CALL IPSTN(IA,SY,EX,EY,EZ,UMOD1,UMOD2,UMOD3,NP,DI1P,DI2DP,TI1P,
     +EVPH)
      GO TO 90
   60 CALL IPSTN(IA,SZ,EX,EY,EZ,UMOD1,UMOD2,UMOD3,NP,DI1P,DI2DP,TI1P,
     +EVPH)
   90 CONTINUE
      WRITE(6,1006)NPT(IA),UMOD1(IA),UMOD2(IA),UMOD3(IA)
 1006 FORMAT(
     +T5,'NO. OF DATA POINTS . . . . . . . . =',I3/
     +T5,'UNLOADING MODULUS IN X-DIR. . . . =',E12.5/
     +T5,'UNLOADING MODULUS IN Y-DIR. . . . =',E12.5/
     +T5,'UNLOADING MODULUS IN Z-DIR. . . . =',E12.5/)
C
      WRITE(6,1007)
 1007 FORMAT(//T5,'SIGMA-1',6X,'SIGMA-2',6X,'SIGMA-3',5X,
     +'EPSILON-1',4X,'EPSILON-2',4X,'EPSILON-3'/)
      NPN=NPT(IA)
      WRITE(6,1008)(SX(IX),SY(IX),SZ(IX),EX(IX),EY(IX),EZ(IX),IX=1,NPN)
 1008 FORMAT(T1,6(E10.3,3X))
      WRITE(6,1009)
 1009 FORMAT(//T10,'PJ1',10X,'PJ2D',8X,'DI1P',8X,'DI2DP',7X,'TI1P'/)
      NPM=NPN-1
      WRITE(6,1010) (PJ1(IA,IX),PJ2D(IA,IX),DI1P(IA,IX),DI2DP(IA,IX),
     +TI1P(IA,IX),IX=1,NPM)
 1010 FORMAT(T4,5E12.3      )
  100 CONTINUE
      WRITE(6,1011)EMOD,XNU
 1011 FORMAT(//
     +T5,'ELASTIC PARAMETERS'/
     +T5,'------------------'//
     +T5,'YOUNG S MODULUS    = ',E12.5/
     +T5,'POISSON S RATIO    = ',F5.2//)
      CALL CAPMOD(NTEST,PJ1,PJ2D,EVPH,NPT,TI1P,BB,PJ1M,XJ1F,SQJ2DF,
     +DI1P,DI2DP,XJMX)
      CALL HARDEN(NTEST,PJ1M,EVPH)
      STOP
      END
C
      SUBROUTINE IPSTN(IA,ST,EX,EY,EZ,UMOD1,UMOD2,UMOD3,NP,DI1P,DI2DP,
     +TI1P,EVPH)
C
      DIMENSION ST(1),EX(1),EY(1),EZ(1),UMOD1(1),UMOD2(1),UMOD3(1),
     +DI1P(20,20),DI2DP(20,20),TI1P(20,20),EVPH(1)
```

```
C
      DO 100 I=2,NP
      SD=ABS(ST(I)-ST(I-1))
      EXP=EX(I)-EX(I-1)
      TMX=SD/UMOD1(IA)
      IF(EXP.GT.0.)GO TO 15
      EXP=EXP+TMX
      GO TO 20
   15 EXP=EXP-TMX
   20 TMY=SD/UMOD2(IA)
      EYP=EY(I)-EY(I-1)
      IF(EYP.GT.0.) GO TO 25
      EYP=EYP+TMY
      GO TO 30
   25 EYP=EYP-TMY
   30 TMZ=SD/UMOD3(IA)
      EZP=EZ(I)-EZ(I-1)
      IF(EZP.GT.0.) GO TO 35
      EZP=EZP+TMZ
      GO TO 40
   35 EZP=EZP-TMZ
   40 DI1P(IA,I-1)=(EXP+EYP+EZP)/3.
      DI2DP(IA,I-1)=2.*SQRT(0.5*(EXP**2+EYP**2+EZP**2-3.*DI1P(IA,I-1)**2
     +))
      IF((I-1).NE.1) GO TO 45
      TI1P(IA,I)=DI1P(IA,I-1)*3.
      GO TO 50
   45 TI1P(IA,I)=TI1P(IA,I-1)+DI1P(IA,I-1)*3.
   50 CONTINUE
  100 CONTINUE
      TI1P(IA,1)=0.0
      DO 110 J=1,NP
  110 TI1P(IA,J)=TI1P(IA,J)+EVPH(IA)
      RETURN
      END
C
      SUBROUTINE FAILP(XJ1F,SQJ2DF,PJ1,PJ2D,DI1P,DI2DP,NTEST,NPT,TI1P,
     +BB,XJMX,KFLAG)
C
      DIMENSION XJ1F(1),SQJ2DF(1),A(20,20),B(20),W(20),AA(20,20),BB(20),
     +PJ1(20,20),PJ2D(20,20),DI1P(20,20),DI2DP(20,20),NPT(20),TI1P(20,20
     +)
      COMMON/FAIL/ND,XL,XLEN,YLEN,HGT
C
      READ(5,*)KFLAG
      IFLAG=0
      IF(KFLAG.NE.1) GO TO 13
      READ(5,*)ND,XL,XLEN,YLEN,HGT
   13 DO 30 I=1,NTEST
      XJ1=XJ1F(I)
      B(I)=SQJ2DF(I)
      DO 20 J=1,3
   20 A(I,J)=XJ1**(J-1)
   30 W(I)=1.0
      IF(IFLAG.NE.1) GO TO 40
      READ(5,*)(W(I),I=1,NTEST)
   40 CALL MULTP(A,W,B,NTEST,3,AA,BB,20,20)
      IF(BB(1).LT.0.)BB(1)=0.0
      XJA=XJMX*.75
      XJB=XJMX
      XDA=BB(1)+BB(2)*XJA+BB(3)*XJA**2
```

```
            XDB=BB(1)+BB(2)*XJB+BB(3)*XJB**2
            THETA=(XDB-XDA)/(XJB-XJA)
            ALPHA=XDA-THETA*XJA
            GAMA=ALPHA-BB(1)
            BETA=(BB(2)-THETA)/GAMA
        WRITE(6,1001)
  1001 FORMAT(
       +T5,'PARAMETERS FOR FIXED FAILURE ENVELOPE'/
       +T5,'------------------------------------'/)
        WRITE(6,1000)ALPHA,THETA,GAMA,BETA
  1000 FORMAT(
       +T5,'ALPHA . . . . . . =',E12.5/
       +T5,'THETA . . . . . . =',E12.5/
       +T5,'GAMA  . . . . . . =',E12.5/
       +T5,'BETA  . . . . . . =',E12.5)
        RETURN
        END
C
        SUBROUTINE MULTP(A,W,B,NTEST,M,AA,BB,NA,NB)
C
        DIMENSION A(NA,NB),W(NA),B(NA),AA(NB,NB),BB(NB),AT(20,20)
C
        DO 20 I=1,M
        DO 20 K=1,NTEST
   20   AT(I,K)=A(K,I)*W(K)
        DO 40 I=1,M
        SUM1=0.0
        DO 30 J=1,M
        SUM=0.0
        DO 25 K=1,NTEST
   25   SUM=SUM+AT(I,K)*A(K,J)
   30   AA(I,J)=SUM
        DO 35 K=1,NTEST
   35   SUM1=SUM1+AT(I,K)*B(K)
   40   BB(I)=SUM1
        CALL SOLVE(AA,BB,NB,M)
        RETURN
        END
C
        SUBROUTINE SOLVE(A,B,NN,N)
C                 **** SOLVES NXN LINEAR SIMULTANEOUS EQUS. USING
C                 GAUSSIAN ELIMINATION WITH PARTIAL PIVOTING.
        DIMENSION A(NN,NN),B(1)
        NM1=N-1
C                 **** CHANGE ROWS AND DO FORWARD ELIMINATION
        DO 30 I=1,NM1
            XLARGE=1.E-15
            JCOUNT=0
            DO 15 J=I,N
              IF(ABS(A(J,I)) .GT.XLARGE)GO TO 10
              GO TO 15
   10         XLARGE=ABS(A(J,I))
              JCOUNT=J
   15       CONTINUE
            IF(JCOUNT.EQ.0)GO TO 60
            DO 20 K=I,N
              TEMP=A(I,K)
              A(I,K)=A(JCOUNT,K)
              A(JCOUNT,K)=TEMP
   20       CONTINUE
            TEMP=B(I)
```

```
            B(I)=B(JCOUNT)
            B(JCOUNT)=TEMP
            IP1=I+1
            DO 30 K=IP1,N
            Q=-A(K,I)/A(I,I)
            B(K)=Q*B(I)+B(K)
            DO 30 J=IP1,N
            A(K,J)=Q*A(I,J)+A(K,J)
      30 CONTINUE
         IF(A(N,N).EQ.0.)GO TO 60
C                    **** BACK SUBSTITUTION
         B(N)=B(N)/A(N,N)
         NP1=N+1
         DO 50 K=1,NM1
         Q=0.0
         NMK=N-K
         DO 40 J=1,K
         Q=Q+A(NMK,(NP1-J))*B(NP1-J)
      40 CONTINUE
         B(NMK)=(B(NMK)-Q)/A(NMK,NMK)
      50 CONTINUE
         IERROR=1
         RETURN
      60 IERROR=2
         WRITE(6,300)
     300 FORMAT(/////' ***** MATRIX IS SINGULAR ********'/)
         RETURN
         END
C
         SUBROUTINE CAPMOD(NTEST,PJ1,PJ2D,EVPH,NPT,TI1P,BB,PJ1M,XJ1F,
        +SQJ2DF,DI1P,DI2DP,XJMX)
C
         DIMENSION PJ1(20,20),EVPH(1),SJ1(20),SJ2D(20),NPT(20),TI1P(20,2)),
        +A(20,20),W(20),B(20),AA(20,20),BM(20),BB(20),PJ1M(20),PJ2D(20,20)
         DIMENSION XJ1F(1),SQJ2DF(1),DI1P(20,20),DI2DP(20,20)
           COMMON/FAIL/ND,XL,XLEN,YLEN,HGT
           COMMON/SPS/TJ1(5),TJ2(5),EJ1(5),EJ2(5),KNT
C
         CALL FAILP(XJ1F,SQJ2DF,PJ1,PJ2D,DI1P,DI2DP,NTEST,NPT,TI1P,BB,
        +   XJMX,KFLAG)
         DO 10 I=1,NTEST
         PJ1M(I)=PJ1(I,1)
      10 CONTINUE
         NT1=NTEST-1
         DO 30 I=1,NT1
         IA=I+1
         DO 20 J=IA,NTEST
         SM=PJ1M(I)
         SV=PJ1M(J)
         SMIN=EVPH(I)
         EV=EVPH(J)
         IF(EV.GE.SMIN)GO TO 20
         PJ1M(I)=SV
         PJ1M(J)=SM
         EVPH(I)=EV
         EVPH(J)=SMIN
      20 CONTINUE
      30 CONTINUE
         READ(5,*)NCAP
         SDIN=(ABS(EVPH(NTEST))-ABS(EVPH(1)))/FLOAT(NCAP)
         EV=EVPH(1)
```

```
      SUM=0.0
      KNT=0
      DO 150 I=1,NCAP
      EV=EV+SDIN
      DO 35 J=2,NTEST
      IF(EV.GT.EVPH(J))GO TO 35
      GO TO 50
   35 CONTINUE
   50 CAPJ1=PJ1M(J)- (EVPH(J)-EV)/(EVPH(J)-EVPH(J-1))*(PJ1M(J)-PJ1M(J-1)
     +)
      SJ1(1)=CAPJ1
      SJ2D(1)=0.0
      KOUNT=1
      DO 65 K=1,NTEST
      NP=NPT(K)-1
      DO 55 L=1,NP
      IF(EV.GT.TI1P(K,L).AND.EV.LE.TI1P(K,L+1)) GO TO 60
      IF(EV.LT.TI1P(K,L).AND.EV.GE.TI1P(K,L+1)) GO TO 60
   55 CONTINUE
      GO TO 65
   60 KOUNT=KOUNT+1
      RT=ABS(TI1P(K,L+1)-EV)/(ABS(TI1P(K,L+1)-TI1P(K,L)))
      SJ1(KOUNT)=PJ1(K,L+1)-(PJ1(K,L+1)-PJ1(K,L))*RT
      SJ2D(KOUNT)=PJ2D(K,L+1)-(PJ2D(K,L+1)-PJ2D(K,L))*RT
   65 CONTINUE
      IF(KOUNT.LT.3) GO TO 150
      DO 85 K=1,KOUNT
      B(K)=SJ2D(K)
      XJ1=SJ1(K)
      DO 80 L=1,3
   80 A(K,L)=XJ1**(L-1)
   85 W(K)=1.0
      CALL MULTP(A,W,B,KOUNT,3,AA,BM,20,20)
      P2J1=(-(BB(2)-BM(2))+SQRT((BB(2)-BM(2))**2-4.*(BB(1)-BM(1))*(BB(3)
     +-BM(3))))/(2.*(BB(3)-BM(3)))
      P2J2=BB(1)+BB(2)*P2J1+BB(3)*P2J1**2
      P1J1=SJ1(1)
      P1J2=SJ2D(1)
      RX=(P1J1-P2J1)/P2J2
      IF(RX.GT.8.) GO TO 150
      SUM=SUM+RX
      KNT=KNT+1
        TJ1(KNT)=P1J1
        TJ2(KNT)=P1J2
        EJ1(KNT)=P2J1
        EJ2(KNT)=P2J2
  150 CONTINUE
      IF(KNT.EQ.0)GO TO 170
      R=SUM/FLOAT(KNT)
      WRITE(6,4010)
 4010 FORMAT(//
     +T5,'SHAPE FACTOR AND HARDENING PARAMETERS'/
     +T5,'------------------------------------')
      WRITE(6,2000)R
        IF(KFLAG.EQ.1)CALL PLOTER(XJ1F,SQJ2DF,PJ1,PJ2D,DI1P,DI2DP,
     =  NTEST,NPT,TI1P,BB)
      GO TO 180
  170 WRITE(6,3000)
 3000 FORMAT(//10X,'SHAPE FACTOR CANNOT BE DETERMINED DUE'
     1/10X,'TO INCONSISTENT DATA INPUT'//)
 2000 FORMAT(/T5,'SHAPE FACTOR . . . . . . =',E12.5)
```

```
  180 RETURN
      END
C
      SUBROUTINE HYDST(HVPH,SMIN,NPH)
C
      DIMENSION HVPH(20),SMIN(20),NAME(20)
C
      READ(5,300)(NAME(I),I=1,20)
  300 FORMAT(20A4)
      WRITE(6,400)(NAME(I),I=1,20)
  400 FORMAT(////T5,20A4)
      READ(5,*)NPH,UMH1,UMH2,UMH3
      WRITE(6,1001)NPH,UMH1,UMH2,UMH3
 1001 FORMAT(//
     +T5,'NO. OF POINTS TAKEN ON HC CURVE       = ',I3/
     +T5,'UNLOADING MODULUS FOR EPSILON-1        = ',E12.5/
     +T5,'UNLOADING MODULUS FOR EPSILON-2        = ',E12.5/
     +T5,'UNLOADING MODULUS FOR EPSILON-3        = ',E12.5///
     +T5,'MEAN PRESSURE AND STRAIN COMPONENTS'/
     +T5,'-----------------------------------'///
     +T2,'MEAN PRESSURE',5X,'EPSILON-1',8X,'EPSILON-2',8X,
     +'EPSILON-3'/)
C
      DO 100 I=1,NPH
      READ(5,*)SMIN(I),EXHC,EYHC,EZHC
      EXHC=EXHC/1.
      EYHC=EYHC/1.
      EZHC=EZHC/1.
      WRITE(6,1002)SMIN(I),EXHC,EYHC,EZHC
 1002 FORMAT(T2,E12.5,5X,E12.5,5X,E12.5,5X,E12.5)
      HVPH(I)=EXHC+EYHC+EZHC-SMIN(I)*(1./UMH1+1./UMH2+1./UMH3)
  100 CONTINUE
      RETURN
      END
C
      SUBROUTINE HARDEN(NTEST,PJ1M,EVPH)
C
      DIMENSION PJ1M(20),EVPH(20),WW(20),DD(20)
C
      NT2 = NTEST - 1
      DA=0.0
      WA=0.0
      KOUNT = 0
      DO 100 I = 1,NT2
      X1 = PJ1M(I)
      X2 = PJ1M(I+1)
      E1 = EVPH(I)
      E2 = EVPH(I+1)
      IF(X1.EQ.X2) GO TO 100
      KOUNT = KOUNT+1
      DX=X2-X1
      XM=E2/E1
      A=DX/X1
      D=-(ALOG((XM-1.)/A))/X1
      W1=E1/(1.-EXP(-X1*D))
      W2=E2/(1.-EXP(-X2*D))
      W=(W1+W2)/2.
      DA=DA+D
      WA=WA+W
  100 CONTINUE
      D=DA/FLOAT(KOUNT)
```

```
      W=WA/FLOAT(KOUNT)
      Z=0.0
      WRITE(6,3000)D,W,Z
 3000 FORMAT(
     +T5,'D . . . . . . . . . . . . =',E12.5/
     +T5,'W . . . . . . . . . . . . =',E12.5/
     +T5,'Z . . . . . . . . . . . . =',E12.5//)
      RETURN
      END
      SUBROUTINE PLOTER(XJ1F,SQJ2DF,PJ1,PJ2D,DI1P,DI2DP,NTEST,NPT,TI1P
     =,BB)
C
      DIMENSION XJ1F(1),SQJ2DF(1),A(20,20),B(20),W(20),AA(20,20),BB(20),
     +PJ1(20,20),PJ2D(20,20),DI1P(20,20),DI2DP(20,20),NPT(20),TI1P(20,20
     +)
      COMMON/FAIL/ND,XL,XLEN,YLEN,HGT
      COMMON/SPS/TJ1(5),TJ2(5),EJ1(5),EJ2(5),KNT
C
      SMAX1=XJ1F(1)
      DO 50 I=2,NTEST
   50 IF(ABS(XJ1F(I)).GT.SMAX1) SMAX1=ABS(XJ1F(I))
      SMAX2=0.0
      M=3
      DO 60 I=1,M
   60 SMAX2=SMAX2+BB(I)*SMAX1**(I-1)
      WRITE(6,2000)SMAX1,SMAX2,BB(1),BB(2),BB(3),BB(4)
 2000 FORMAT(///T5,'SMAX1 =',E12.5,'  SMAX2 =',E12.5//T5,4E14.5)
      MX=SMAX1/XLEN+1.
      NY=SMAX2/YLEN+1.
      XMX=MX*XLEN
      YNY=NY*YLEN
      DV1=XMX/XLEN
      DV2=YNY/YLEN
      WRITE(6,2002)NY,YNY,DV2
 2002 FORMAT(//T4,'NY =',I4,' YNY =',E12.5,' DV2 =',E12.5/)
      WRITE(6,2004)XLEN,YLEN,DV1,DV2
 2004 FORMAT(//T5,'XLEN =',E12.5,' YLEN =',E12.5,' DV1 =',E12.5,
     +' DV2=',E12.5//)
      JFLAG=0
      CALL INITIAL(0,99,0.4,0,0)
      GO TO 67
C
   65 CALL PLOT(10.,0.5,-3)
      GO TO 68
   67 CALL PLOT(.5,.5,-3)
      CALL FACTOR(1.0)
C  68 CALL AXIS(0.0,0.0,2HJ1,-2,XLEN,0.0,0.0,DV1)
C     CALL AXIS(0.0,0.0,5HSQJ2D,5,YLEN,90.,0.0,DV2)
   68     CALL PLOT(XLEN,0.0,2)
          CALL PLOT(0.0,0.0,3)
          CALL PLOT(0.0,YLEN,2)
          CALL PLOT(0.0,0.0,3)
      DO 70 I=1,NTEST
      X=XJ1F(I)/DV1
      Y=SQJ2DF(I)/DV2
      CALL PLOT(X,Y,3)
   70 CALL PLOT(X,Y,2)
C
      X=0.0
      Y=BB(1)/DV2
      CALL PLOT(X,Y,3)
```

```
      SM=SMAX1/FLOAT(ND)
      DO 80 I=1,ND
      X=X+SM
      XX=X
      XX=XX/DV1
      Y=0.0
      DO 85 J=1,M
   85 Y=Y+BB(J)*X**(J-1)
      YY=Y/DV2
   80 CALL PLOT(XX,YY,2)
C
      DO 90 I=1,NTEST
      X=PJ1(I,1)/DV1
      Y=0.0
      CALL PLOT(X,Y,3)
      X=XJ1F(I)/DV1
      Y=SQJ2DF(I)/DV2
   90 CALL PLOT(X,Y,2)
C
      IF(JFLAG.EQ.1) GO TO 101
C
      DO 100 I=1,NTEST
      NP=NPT(I)-1
      DO 95 J=1,NP
      X=PJ1(I,J)/DV1
      Y=PJ2D(I,J)/DV2
      CALL PLOT(X,Y,3)
      TH=ATAN(DI2DP(I,J)/ABS(DI1P(I,J)))
      X=X+XL*COS(TH)
      Y=Y+XL*SIN(TH)
   95 CALL PLOT(X,Y,2)
  100 CONTINUE
      CALL ENDPLT
      JFLAG=JFLAG+1
      IF(JFLAG.EQ.1) GO TO 65
  101 DO 200 I=1,KNT
      X1=TJ1(I)
      Y1=TJ2(I)
      X2=EJ1(I)
      Y2=EJ2(I)
      R=(X1-X2)/Y2
      X=X1/DV1
      Y=Y1/DV2
      CALL PLOT(X,Y,3)
      Y=0.0
      SM=Y2/FLOAT(ND)
      DO 110 J=1,ND
      Y=Y+SM
      X=SQRT((X1-X2)**2-R**2*Y**2)+X2
      XA=X/DV1
      YA=Y/DV2
      CALL PLOT(XA,YA,2)
  110 CONTINUE
  200 CONTINUE
      CALL ENDPLT
      RETURN
      END
```

INDEX

Indicial notation (*cont.*)
  second-order, 415
  third-order, 416
Induced anisotropy, 371
  joint invariants for, 371
Inelastic behavior, 204 (*see also* Plastic behavior)
Inelastic materials, 204 (*see also* Plastic materials)
Inherent anisotropy, 395
Initial anisotropy, 344
Initial yield surface, 346, 347
Interfaces, 62
Intermediate principal strain, 27
Intermediate principal stress, 46
Invariants:
  of deviatoric strain tensor, 22–25
  of deviatoric stress tensor, 39, 40
  joint, 371
  notation, 430
  of strain tensor, 19, 20
    first, 20
    second, 20
    third, 20
  of stress tensor, 36, 37, 207
    first, 37
    second, 37
    third, 37
  of a tensor, 426–36
    first, 429
    second, 429
    third, 429
  in yield criteria, 207
Inverse (stress-strain) relation, 89
Irreversibility condition, 229
Isotropic consolidation, 285
Isotropic hardening, 344–52, 364
Isotropic material, 136, 206, 344–46
  hardening, 346, 352
  initially, 344
  models, 282, 295, 381
Isotropic work-hardening model, 381

J

Joint invariants, 371
Joints, 62

K

Kinematic hardening, 345–49, 352, 364, 370, 371
  joint invariants in, 373
  nested surface models, 349, 352
  Prager's rule, 346, 348
  rules, 347
Kinematic hardening rules, 347, 370
Kinetic energy, 96

L

Laboratory:
  test devices, 58
  tests, 59–61
Laboratory test devices, 58
  cylindrical triaxial, 59
  direct shear, 60
  hollow cylinder, 60
  simple shear, 60
  truly triaxial, 62
  uniaxial, 58
Laboratory testing, 7, 58
  devices, 58
  importance of, 9, 58
  need for, 7, 58
Lame's constants, 88, 109, 137
Least-squares-fit technique, 115, 138, 150, 437
  for parameter evaluation, 115, 138, 437
  program, 437
Levy-Mises stress-strain relations, 223
Linear elastic, 83, 84, 336
  behavior, 83, 84
Linear hardening rule, 349
Load-deformation behavior, 3, 4, 53, 54
Loading, 189, 191, 192, 229, 232, 359, 396, 400
Lode parameter, 226

M

Major principal strain, 27
Major principal stress, 46